THOMAS L. HARMAN

CHARLES E. ALLEN

Guide to the National Electrical Code®

1987 Edition

Prentice-Hall, Inc., Englewood Cliffs, New Jersey 07632

Library of Congress Cataloging-in-Publication Data

HARMAN, THOMAS L. (date)
  Guide to the National Electrical Code.
    On t.p. the registered trademark symbol "R" within
a circle is superscript following "code" in the title.
    Bibliography: p.
    Includes Index.
    1. Electric engineering—Law and legislation—
United States. I. Allen, Charles E. II. National
Fire Protection Association. National Electrical
Code. III. Title.
KF5704.H37 1987   343.73'0786213   86-30643
ISBN 0-13-370404-1   347.303786213

Editorial/production supervision: Reynold Rieger
Interior design: Phyllis Hall
Cover design: 20/20 Services Inc.
Manufacturing buyer: S. Gordon Osbourne

Printed in the United States of America

10 9 8 7 6 5 4 3 2 1

ISBN 0-13-370404-1 025

Prentice-Hall International (UK) Limited, *London*
Prentice-Hall of Australia Pty. Limited, *Sydney*
Prentice-Hall Canada Inc., *Toronto*
Prentice-Hall Hispanoamericana, S.A., *Mexico*
Prentice-Hall of India Private Limited, *New Delhi*
Prentice-Hall of Japan, Inc., *Tokyo*
Prentice-Hall of Southeast Asia Pte. Ltd., *Singapore*
Editora Prentice-Hall do Brasil, Ltda., *Rio de Janeiro*

Contents

PART III—GENERAL ELECTRICAL THEORY

PART IV—FINAL EXAMINATIONS

Preface
to the First Edition

This book is in response to an increasing demand in the industrial and academic communities for a detailed guide to the *National Electrical Code* and the principles of electrical design based on the Code. In particular, the information presented should serve the aspiring Master Electrician as well as the practicing Master Electrician, and the electrical technology student. Each area of interest to the modern day electrician is discussed in detail with an ample number of example problems and their solutions. This Guide differs from other publications that treat the *National Electrical Code* in that the emphasis here is on the types of questions and problems that typically appear on Master Electrician's Examinations given throughout the country. Since these examinations reflect the knowledge expected of a Master Electrician in practice, their content has guided the authors in selecting material for this Guide.

The Master Electrician should be competent in three major areas, the first two of which are based on the *National Electrical Code.* The first two areas are (a) the design of electrical wiring systems, and (b) the construction and installation of electrical systems. The Guide covers these subjects in Parts I and Part II, respectively. The third major area, presented in Part III of the Guide, is basic electrical theory and practice. This material is, for the most part, outside the scope of the Code although the Master Electrician is expected to be familiar with the principles presented in Part III.

This guide can serve as a self-study text or it can form the basis for one- or two-semester course covering the rules of the *National Electrical Code* and related material. The Guide presents all rules and problem-solving techniques necessary to pass a Master Electrician's examination. The problems also treat practical situations arising in the design and construction of electrical installations. The only other reference text required is the *National Electrical Code* itself.

Part I of the Guide presents the rules and wiring design calculations required to determine the ratings of electrical services, feeders, and branch circuits for typical electrical installations. Beginning with a general discussion of these circuits, the chapters of Part I of the Guide present increasingly complex situations. The final two chapters in Part I, Chapters 4 and 5, present detailed calculations for the design of electrical systems in dwellings and in industrial or commercial occupancies, respectively. A quiz is given after each unit to summarize the knowledge in that unit and provide practice for the reader. A lengthy examination after each chapter covers the material presented in the chapter. Wherever a rule from the *National Electrical Code* affects a calculation, reference is made to the particular rule in the margin of the text.

Part II of this Guide covers the major sections of the *National Electrical Code* and provides summaries of important rules which govern the construction and installation of electrical equipment. A large number of tables, problems, and quizzes organize the material logically to aid the reader's understanding.

Part III of the Guide begins with a treatment of basic direct-current circuits. Subsequent sections present a review of the properties of conductors, basic alternating-current circuits, and equipment in ac circuits. The discussion and examples cover material useful for the solution of design problems presented in other parts of the Guide.

The final portion of the Guide, Part IV, contains two examinations covering the material presented in the first three parts. The final examinations included there are representative of examinations for the Master Electrician's license given by various city and state examination boards.

The Appendices contain information of general interest, such as a detailed list of useful electrical formulas. Most important, the Appendices contain the solutions to all quizzes, tests, and final examinations given in the Guide. The solutions to problems have been worked out in complete detail showing the method used and the appropriate references to the *National Electrical Code.*

Notes to the Student

Preparation for a Master Electrician's License is a long and difficult procedure involving practical experience and a thorough knowledge of the material presented in this Guide. A fundamental knowledge of elementary algebra and simple direct-current and alternating-current circuit theory is most helpful in fully understanding the approaches to problem solving taken in the Guide. If this material is not familiar to you, a self-study program or a course at a local community college covering these basic subjects would be helpful.

For students with the proper background knowledge, the Guide can be studied with the *National Electrical Code* to prepare for required city or state electrician's examinations. Each principle should be mastered before going on, although your study program may begin with Part I, Part II, or Part III of the Guide, depending on your previous knowledge or preference.

When you feel prepared, take the final examinations presented in Part IV of the Guide according to the directions and time limits specified for each test. The examinations may be scored by referring to the solutions given in the Appendices. Although a score of 70 percent is passing, a score of 80 percent or more on each examination indicates that you are well prepared for the real thing.

Notes to the Instructor

This Guide is unique because it presents the more difficult subject of electrical wiring design calculations in Part I and the general Code rules and basic electrical theory in Part II and Part III, respectively. This was done because many students who take courses covering the *National Electrical Code* are familiar with the general organization of the Code and its rules. It is necessary for those students to concentrate on problem solving rather than Code rules and basic theory which can be briefly presented and understood by the student. In addition, typical Master Electrician's examinations separate the questions concerned with wiring design from those dealing with installation and general electrical practice. To avoid confusing students, the instructor may divide this course similarly and begin with the most important subject, that of electrical design. The other Code rules and the basic theory support this activity.

If the students are not well prepared, the instructor might decide to begin with Part III, basic electrical theory, proceed to Part II, and finally present the design calculations of Part I. These three parts are not dependent upon each other; each contains quizzes and tests based on only the material presented in that part. The examinations in Part IV, of course, cover the entire range of topics presented in the Guide.

As an adjunct to the material in the Guide, it would be appropriate for the instructor to present rules which are covered by local ordinances. Such material could be added to the examinations given in the Guide.

Notes to the Practicing Electrician or Designer

This Guide has been prepared to explain in detail the use of the *National Electrical Code,* particularly as it applies to design of electrical wiring systems. With this intent, the Guide does not attempt to present design techniques that necessarily result in the most efficient or economical electrical system. For instance, no provisions are made for future expansion in the examples given in the Guide. The authors assume that experienced electricians or designers will use their own approaches to problem solving while using the Guide as a reference. In the same way, the selection of equipment such as circuit breakers and service equipment in the examples presented in the Guide is based on the minimum Code requirements. The rating of such equipment may be neither adequate nor convenient for a practical installation. The judgment of the designer must be relied upon to determine the design which best fits a particular installation.

The problems presented in Part I of the Guide deal with standard alternating-current circuits used as services, feeders, or branch circuits. Other special electrical systems such as two-phase alternating-current and direct-current installations are not discussed. The unique rules for these circuits concerning grounding, size of neutral, etc. will be found in the Code and should be used in addition to or in place of the rules presented in the Guide.

Finally, the design requirements for circuits supplying equipment such as X-ray machines are not covered in detail in the Guide. Reference is made when necessary to the appropriate section of the Code that covers such equipment.

Acknowledgments

The authors wish to acknowledge the significant contribution of Ken Richardson, Master Electrician, who has been a part of this project since its inception. We also wish to thank our illustrator, Vic Bowen; our art typographer, Lu Songer; our manuscript typists, Carolyn Block, Kathy Cannon, Nancy Harman, and Helen Hickey; and finally, our wives, Nancy and Dixie, for their patience and encouragement.

Thomas L. Harman
Charles E. Allen

Preface
to the 1987 Edition

This edition covers the 1987 *National Electrical Code.*® The text and questions and answers have been revised where appropriate to reflect the 1987 Code rule changes. The successful style and format of the previous editions have been retained.

The authors appreciate the various comments received on the previous edition. In particular, a number of students in the authors' *National Electric Code*® classes have made helpful suggestions that served to improve the presentation of the material. Comments concerning this edition may be forwarded to the publisher.

Thomas L. Harman
Charles E. Allen

1

Introduction

The purpose of this Guide is to present the broad range of knowledge related to the safe installation of electrical wiring and equipment in which an individual must demonstrate proficiency to become qualified as a Master Electrician. The Guide, therefore, also serves as a guide to the *National Electrical Code,* [1] hereafter also referred to as the Code, for anyone interested in the design and installation of electrical wiring systems.

To simplify and organize the material presented, the Guide has been divided into four separate and independent parts, as shown in Figure 1-1. Part I of the Guide presents discussions, sample calculations, and illustrations that will aid the reader to understand and become proficient in the application of the Code rules for electrical wiring design. Part II of the Guide summarizes Code rules covering the installation of electrical wiring and equipment. Part III presents study material covering electrical theory and practice consistent with the level reflected by the questions typically included in Master Electrician's examinations. Part IV contains tests and examinations that are representative of typical Master Electrician's examinations given by city and state licensing agencies.

[1] Boston, Mass.: National Fire Protection Association, 1983.

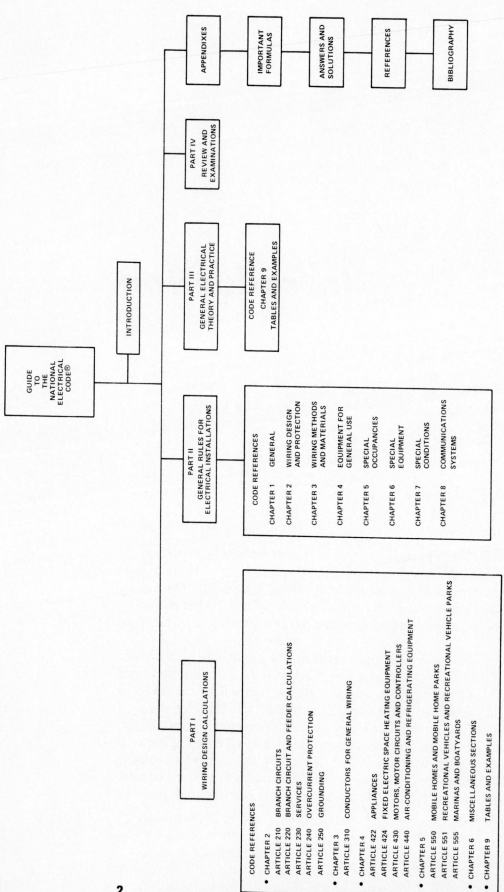

Figure 1-1. Organization of the Guide

2

1-1 THE MASTER ELECTRICIAN AND THE MASTER ELECTRICIAN'S EXAMINATION

Although the responsibilities of a Master Electrician vary somewhat from city to city or from state to state, the Master Electrician or his designated representative is normally the only person who may apply to the City Building Office or similar municipal regulatory or inspection agency for a permit to perform electrical contracting work. The Master Electrician is responsible for the supervision and control of the work for which the permit was obtained. The Master Electrician may own his or her own electrical contracting company, hiring other electricians to work for him or her, or the Master Electrician may be employed by an electrical contracting company and act as its representative. In many localities the Master Electrician is required to show not only technical competence but also financial responsibility, usually in the form of workmen's compensation insurance and a bond covering some percentage in dollars of the work to be performed.

The Electrical Board establishes requirements of eligibility of applicants for the Master Electrician's examination. The electrician is usually required to serve a certain number of years as a journeyman electrician. Eligible applicants are administered a test to determine their qualifications to design and safely install electrical wiring and equipment. Applicants who pass the examination and satisfy all requirements established by the Electrical Board are granted a Master Electrician's license.

1-1.1 The Examination

The Master Electrician's examination consists of questions and problems that the local examining agency prepares and administers. Most local examinations are given periodically at a time and place specified by the examining board. A usual passing grade is 70% correct for the total examination, although some local agencies administer the test in parts which are graded separately. The form of the examination should be discussed with the examining agency before the test is taken since it varies greatly from locality to locality.

Each local examining agency selects questions that it feels best represent the qualifications required of a Master Electrician in that area; therefore, it is not possible to define a "standard" examination and prepare for it. The topics to be covered on the Master Electrician's examination, however, can be divided into several broad areas from which questions appear on every examination regardless of the locality in which the test is given.

The areas of examination are illustrated in Figure 1-2. These areas include examination questions covering (a) the *National Electrical Code,* (b) general knowledge of electrical practice, (c) supporting theoretical knowledge, and (d) local ordinance rules. In the authors' experience, questions about the Code, including rules and design calculations, usually comprise from 70% to 80% of the examination. Some examining agencies include test questions on local ordinances and installation rules in a separate examination. By comparing Figure 1-2 with Figure 1-1, it is evident that this Guide covers all examination areas with the exception of local ordinances and rules. Information on the local material may be obtained from the proper local agency.

1-1.2 The Mechanics of Test Taking

The form of the Master Electrician's examination, the time allowed, and the reference material which the applicant may be allowed to take into the examination room vary with each locality. Typically, an applicant may be allowed from 6 to 8

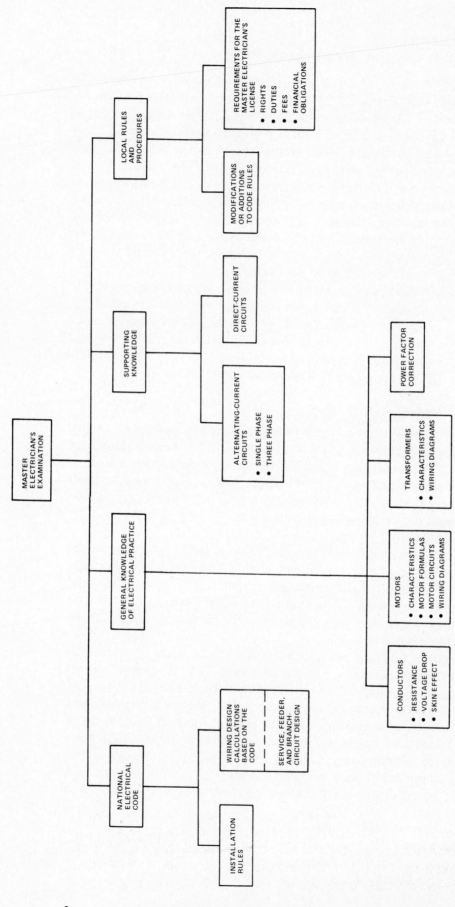

Figure 1-2. Major Examination Areas

4

hours to complete the examination. Most examinations include a closed-book portion for which no reference material, including the Code book itself, may be used. Another portion of the examination is usually open-book with the Code book serving as the primary reference. In some cases, textbooks, notes, and other aids are also allowed as reference material.

The materials an applicant should bring to an examination include:

a. The approved application form
b. A working hand-held calculator
c. A large tablet of paper
d. The latest edition of the *National Electrical Code*
e. Notes and other reference material (when permitted)

A well-prepared applicant should not have a great deal of difficulty passing the examination if the applicant *reads the questions carefully,* works in an efficient manner, and *checks all answers.*

1-1.3 Differences in Grading

Each member of the examining board prepares a certain number of questions on the examination and grades them. If there is any doubt about the meaning of a specific question, the applicant should ask for clarification from members of the board who are present during the examination. They are usually courteous and helpful in this regard.

The solutions to problems, especially when calculations based on Code rules are involved, may differ slightly according to local application or interpretation. The Code usually specifies a maximum or minimum size or rating of conductors and equipment, such as circuit breakers, to be used in an electrical installation. The specified ratings may not coincide with the standard ratings of equipment available from electrical manufacturers. If the applicant is unsure about the expected result to be given as the answer (calculated value or standard size), the applicant should ask a member of the examining board.

1-2 HOW TO USE THIS GUIDE

The first two parts of the Guide are designed to be studied in conjunction with the Code book. Each article or section of the Code referenced by the Guide should be studied thoroughly since the Guide summarizes the intent of the Code and does not usually repeat Code rules verbatim. Any article or section not covered or referenced by the Guide should nevertheless be read as part of the general study of the Code. The goal should be to have some knowledge of *every* paragraph in the Code even though certain areas might receive greater emphasis by the Guide.

Many provisions of the Code that are applicable to the manufacturing of electrical equipment, for instance, are not as important to the practicing electrician as are the rules which apply to the installation of electrical equipment. The rules that apply to the manufacturing of electrical equipment are not covered by this Guide in great detail.

An ample number of quizzes, tests, and final examinations has been provided to help the reader measure his or her progress. The answers and method of solution for each question are contained in Appendix B. All problems in the three parts of the Guide should be worked and the results checked carefully. The Code rules tabulated in Part II, General Rules for Installations, must be memorized (there is no other way) by repeated study and test taking. The study course outlined by this Guide may be supplemented by reading the publications listed in the References and Bibliography if further study is necessary.

1-3 THE NATIONAL ELECTRICAL CODE

According to the interesting and informative booklet, *The National Electrical Code and Free Enterprise,*[2] the first "National Electrical Code" was begun in 1896 and published in 1897. The current sponsor of the Code, the National Fire Protection Association, began preparing the Code in 1911. The Code has been revised every few years since then; the latest editions were published in 1978 and 1981. Each edition of the Code provides a timetable showing the approximate date for the next edition. A few months before the new Code is published, a preprint of the proposed amendments to the last Code is distributed so that local agencies and other interested parties can participate in the rule making or in preparing for the new release of the Code. The latest edition of the Code applied in the reader's locality should be used to accompany this Guide.

1-3.1 Purpose, Scope, and Enforcement

The purpose, scope, and enforcement of the Code is described in the introductory Article of the Code. Several very important items are mentioned there that bear on the use and interpretation of the Code.

The Code is written for the purpose of safeguarding persons and property from hazards arising from the use of electricity. As is stated in the Code, the Code is *not* a design specification and it is not an instruction manual. Provisions for future expansion and special considerations must be provided for in an electrical wiring system by the designer.

The Code provides rules covering most electrical installations but excludes those in automobiles and properties not normally accessible to the general public or where other rules are applicable, such as in aircraft. The exact coverage is clearly specified in the Code. The provisions of the Code are advisory as far as the National Fire Protection Association is concerned but they form the basis of local and state ordinances and statutes regulating electrical installations. The interpretation and administration of the Code provisions are usually handled by the electrical inspection agency of the governmental body having local jurisdiction.

1-3.2 Organization

The text of the Code includes the introduction (Article 90) and nine chapters. The rules governing the design and construction of a specific electrical system may be taken from more than one chapter and, in fact, this is usually the case. To simplify the task of the designer and installer of an electrical system, the chapters of the Code are organized into three major groups.

The first group consists of Chapters 1, 2, 3, 4, and 9. The first four chapters present the rules for the design and installation of electrical wiring systems for most of the situations encountered by the Master Electrician. Chapter 9 contains tables which specify the properties of conductors and rules for the use of conduit to enclose

[2]Merwin M. Brandon, *The National Electrical Code and Free Enterprise* (Boston, Mass.: National Fire Protection Association, 1971).

the conductors. The examples of Chapter 9 demonstrate the use of the rules for design given in the first four chapters.

The second group consists of Chapters 5, 6, and 7. These chapters are concerned with special occupancies, equipment, and conditions. Rules in these chapters may modify or amend those in the first four chapters.

Chapter 8 is the third group and is independent of the other chapters. This chapter covers communications systems, such as telephone and telegraph systems, as well as radio and television receiving equipment.

1-3.3 Classification of Code Rules

The Code rules may be divided into three general categories as follows:

a. Wiring *design* rules used to determine sizes or rating of circuit conductors and equipment
b. Rules that specify *installation* requirements for various conditions and occupancies
c. Rules for *manufacturing* electrical equipment

Figure 1-3 illustrates the application of the three categories of rules. The figure depicts a situation in which a motor is installed in a location in which combustible

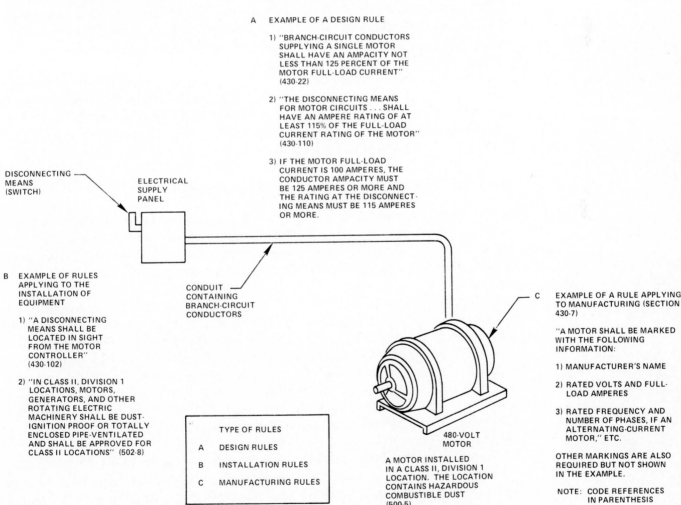

A   EXAMPLE OF A DESIGN RULE

1) "BRANCH-CIRCUIT CONDUCTORS SUPPLYING A SINGLE MOTOR SHALL HAVE AN AMPACITY NOT LESS THAN 125 PERCENT OF THE MOTOR FULL-LOAD CURRENT" (430-22)

2) "THE DISCONNECTING MEANS FOR MOTOR CIRCUITS . . . SHALL HAVE AN AMPERE RATING OF AT LEAST 115% OF THE FULL-LOAD CURRENT RATING OF THE MOTOR" (430-110)

3) IF THE MOTOR FULL-LOAD CURRENT IS 100 AMPERES, THE CONDUCTOR AMPACITY MUST BE 125 AMPERES OR MORE AND THE RATING AT THE DISCONNECTING MEANS MUST BE 115 AMPERES OR MORE.

DISCONNECTING MEANS (SWITCH)

ELECTRICAL SUPPLY PANEL

B   EXAMPLE OF RULES APPLYING TO THE INSTALLATION OF EQUIPMENT

1) "A DISCONNECTING MEANS SHALL BE LOCATED IN SIGHT FROM THE MOTOR CONTROLLER" (430-102)

2) "IN CLASS II, DIVISION 1 LOCATIONS, MOTORS, GENERATORS, AND OTHER ROTATING ELECTRIC MACHINERY SHALL BE DUST-IGNITION PROOF OR TOTALLY ENCLOSED PIPE-VENTILATED AND SHALL BE APPROVED FOR CLASS II LOCATIONS" (502-8)

CONDUIT CONTAINING BRANCH-CIRCUIT CONDUCTORS

TYPE OF RULES	
A	DESIGN RULES
B	INSTALLATION RULES
C	MANUFACTURING RULES

480-VOLT MOTOR

A MOTOR INSTALLED IN A CLASS II, DIVISION 1 LOCATION. THE LOCATION CONTAINS HAZARDOUS COMBUSTIBLE DUST (500-5)

C   EXAMPLE OF A RULE APPLYING TO MANUFACTURING (SECTION 430-7)

"A MOTOR SHALL BE MARKED WITH THE FOLLOWING INFORMATION:

1) MANUFACTURER'S NAME

2) RATED VOLTS AND FULL-LOAD AMPERES

3) RATED FREQUENCY AND NUMBER OF PHASES, IF AN ALTERNATING-CURRENT MOTOR," ETC.

OTHER MARKINGS ARE ALSO REQUIRED BUT NOT SHOWN IN THE EXAMPLE.

NOTE: CODE REFERENCES IN PARENTHESIS

Figure 1-3. Various Categories of Code Rules Which Apply to a Motor and Its Circuits

dust may be present. This type of hazardous location is defined by the Code as a Class II, Division 1 location and certain mandatory provisions are specified for such locations.

One example of a *design* rule in the Code specifies the conductor ampacity[3] (current-carrying capacity of electric conductors expressed in amperes) which must be at least 125% of the full-load current of the motor. The rating of the disconnecting means that disconnects the conductors of a circuit from their source of supply must be at least 115% of the full-load current rating of the motor according to the Code. Thus, if a motor with a full-load current of 100 amperes were used, the conductor ampacity must be at least 125 amperes. The disconnecting means must have an ampere rating of 115 amperes or more. Other design rules not shown in Figure 1-3 apply to the size or rating of other circuit elements such as the fuses or circuit breaker protecting the circuit.

The rules for *installation* of the motor given in the figure pertain to the location of the disconnecting means within sight of the motor and to the type of motor which may be installed in the Class II, Division 1 location. The rating of the motor (100 amperes) has no effect on these rules for the 480-volt motor used in the example. The design rules and the installation rules, therefore, may be treated separately.

The manufacturer must mark the motor on its nameplate according to the applicable Code rules for the marking of equipment. The information on the nameplate aids both the electrical designer and the installer. The designer specifies the characteristics which the motor must have for a particular installation; the installer must be sure that the correct motor is supplied.

The purpose of separating the rules in this manner is to aid the reader in learning the Code. The design of complete electrical systems must take into account all the rules that apply. The Master Electrician should be thoroughly familiar with the design rules and installation rules presented in the Code. These rules are covered in Parts I and II, respectively, of the Guide.

1-4 STATE AND LOCAL CODES AND ORDINANCES

The *National Electrical Code* is not enforced as a nationwide set of rules governing electrical construction practices. Instead, the adoption and enforcement of the Code are under the jurisdiction of local governmental organizations. Many state and local governments adopt the Code and also adopt a supplementary code in the form of state statutes or local ordinances.

The local ordinances are normally separated into several sections covering legal requirements and technical provisions. The legal section deals with organizational, financial, and other such aspects of electrical construction work. For instance, the responsibilities of the local electrical board would be defined, and the requirements and fees for a Master Electrician's license would be stated. The second portion is of a technical nature and states that the *National Electrical Code* is adopted as the standard code to govern electrical construction. Any local rules that change or supplement the provisions of the *National Electrical Code* would be stated and explained in the technical section.

Local rules should be applied by the Master Electrician whenever they differ from those in the Code. Since the interpretation of the *National Electrical Code* is left to local agencies having jurisdiction, any rule that may be subject to differences in interpretation should be discussed with members of the local electrical board.

[3]These terms are defined in Code Article 100.

TEST CHAPTER 1

1. Define a Master Electrician according to the local electrical Code enforced in your area.
2. State the qualifications for a Master Electrician in your area.
3. If you are preparing for a Master Electrician's examination, do the following by contacting the appropriate local agency:
   (a) Determine the major areas the Master Electrician's examination will cover.
   (b) Find out the details of the examination and determine the materials an applicant may bring to the examination room.
4. (a) The *National Electrical Code* is not intended as a design specification. True or false?
   (b) The Code is written to safeguard persons and property from hazards arising from the use of electricity. True or false?
5. What agency enforces the Code in your area?
6. Name the members of the electrical board in your area.
7. Obtain a copy of the local electrical Code ordinance to answer the following questions:
   (a) State as many differences as you can between the rules of the *National Electrical Code* and the local Code.
   (b) How may the Master Electrician's license be obtained?
   (c) How and for what reasons may the Master Electrician's license be revoked?
   (d) How are local permits for electrical construction obtained?
   (c) What type of electrical construction work requires a permit?

Wiring Design Calculations

I

Part I is concerned primarily
with calculations that lead to an electrical design
in conformance with the National Electrical Code requirements
with respect to ratings or capacities of electrical equipment and conductors.
Special emphasis is given
to the calculations for service, feeder, and branch-circuit loads.
All rules and definitions directly related to calculations
are presented with the calculations.
Numerous examples are given
to demonstrate the use of the Code rules in performing the required calculations.
The complete examples progress from relatively simple problems
involving the one-family dwelling
to the more complex problems encountered in a large industrial installation.
Quizzes and problems follow each section
and a summary examination is given after each chapter.

2

Services, Feeders, and Branch Circuits

This chapter presents an introduction to the wiring design rules and related calculations used to determine electrical capacities and ratings of conductors and equipment required for the wiring systems in dwelling, commercial, and industrial occupancies. The material presented is grouped according to the three major elements of a premises wiring system: (a) services, (b) feeders, and (c) branch circuits. The utility company's power distribution system will also be described briefly in order to discuss the entire electrical system.

The Code rules concerned with the electrical capacity of interior or premises wiring systems govern the design of these systems. The minimum or maximum size or rating of each element in the system as appropriate is determined by the applicable Code rule. In most cases, the Guide provides tables that summarize these design rules and gives references to the Code rules from which the tables were taken. These tables, and other Code tables not repeated here, allow the selection of the proper conductors and other equipment which make up the system.

The Code definitions for electrical terms should be studied carefully since the **ARTICLE 100** Code rules are very specific for each element of the wiring system.

The sample design cases which are presented are summarized by figures and tables. These usually include a simplified electrical diagram of the example, the calculations, and the Code references. Examples in this chapter do not cover motor loads or continuous loads, which are treated in Chapter 3.

2-1 THE ELECTRICAL SYSTEM

The electrical system to supply an occupancy consists of the distribution system and the premises wiring system according to the Code. The premises wiring system extends from the connection to the distribution system to the equipment to be supplied. The Code considers the premises wiring system to begin at the *service* for a

230-2* building. Each building (with a few exceptions) has *one* service that supplies the interior electrical equipment referred to as the *electrical load* or simply as the *load*. The service may supply *feeder* circuits if they are required for the electrical system. The circuits which extend from the service or feeder and supply the load are *branch circuits*. The Code has special rules that apply to each portion of the premises wiring system.

Figure 2-1 illustrates a typical electrical system. The various elements of the premises wiring system are defined according to the Code definitions. These definitions serve as a basis for discussing wiring systems in the Guide.

The distribution system begins at the generating station or distribution substation operated by the utility company. The distribution system supplies electrical power to each customer through transmission lines normally operating at relatively high voltages, a typical example being 138 000 volts. Transformers at or near the customer's location step down the supply voltage to the value required by the customer's premises wiring system. This reduced voltage is referred to as the *service voltage* and may typically range from 13 800 volts (for special industrial services) to the common 240-volt or 120-volt, single-phase service. The customer may install additional transformers if several different voltages are required or if both single-phase and three-phase loads are to be supplied. That portion of an electrical system

90-2 installation under the exclusive control of the electrical utility company is not covered by the Code; however, the transformers supplied by the customer, as well as all conductors connected to or derived from the supply system, must meet Code requirements.

The premises wiring system extends from the service connection to the main

240-3 supply to the outlets for the equipment. The Code requires that every conductor be protected against excessive current that could damage the conductor. The protection,

240-21 in the form of an overcurrent protective device such as a fuse or circuit breaker, must, in most cases, be located where the conductors are electrically connected to their source of supply. Thus, each feeder and branch circuit in the premises wiring system must be protected. The service enclosure and the feeder panel shown in Figure 2-1 enclose the overcurrent protective devices for these circuits.

ARTICLE 100 The Code defines the service as the conductors and equipment which deliver energy from the supply system to the wiring system of the premises served. The service conductors extend from the main supply or a transformer to the service equipment. The Code considers the service conductors to consist of the service drop

*References in the margins are to the specific applicable rules and tables in the *National Electrical Code*.

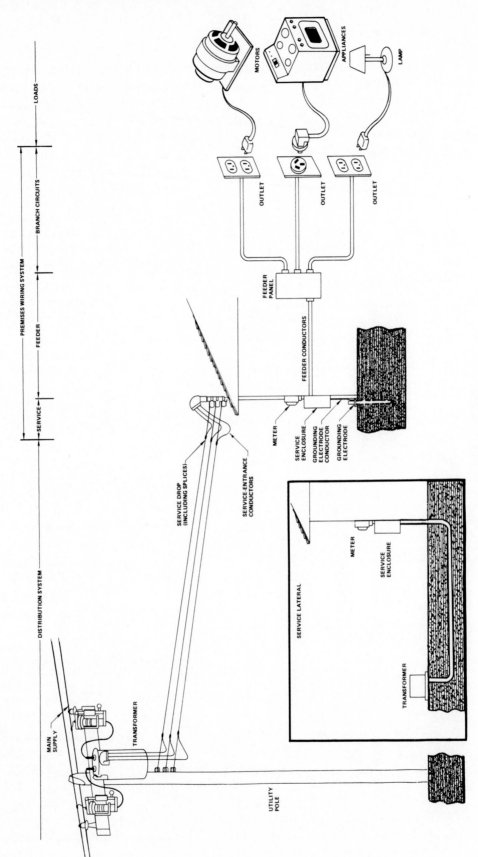

Figure 2-1. Typical Electrical System

(or service lateral for underground conductors) and the service-entrance conductors. The distinction is important since the Code provides separate rules for each portion of the service circuit.

The service-entrance conductors connect the service drop or service lateral to **230-90** the service equipment. The service equipment consists of the main overcurrent **230-70** protection for the entire premises wiring system and the means to disconnect all conductors in a building from the service-entrance conductors. These devices, as well as overcurrent devices protecting any feeder circuits, are usually enclosed in the service enclosure.

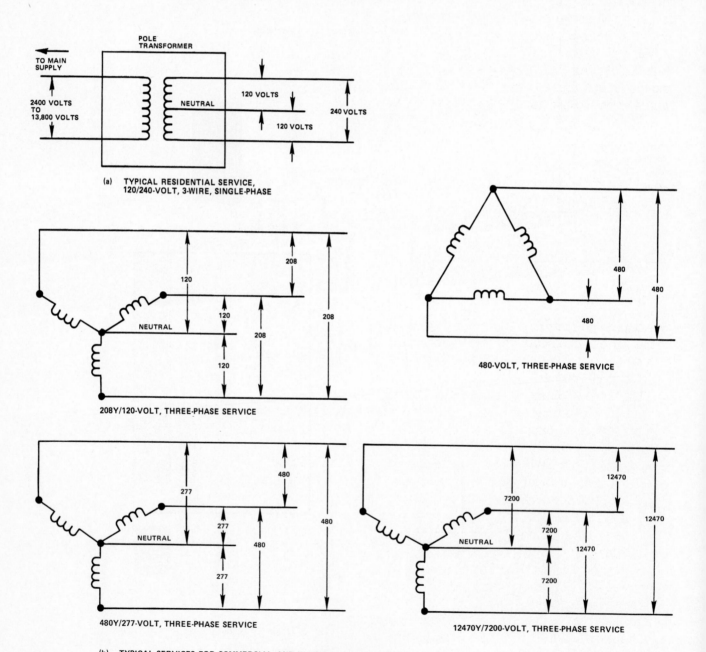

Figure 2-2. Typical Service Configurations

The typical services are shown schematically in Figure 2-2. The most common residential service is 120/240 volts, single-phase. Commercial and industrial establishments require three-phase services in many cases. The common 208Y/120-volt service supplies both 208-volt, three-phase loads such as motors and 120-volt, single-phase loads. Motor loads and 277-volt fluorescent lighting loads can be supplied by a 480Y/277-volt service. A high-voltage 12470Y/7200-volt service would be used for special industrial applications.

Feeder circuits are primarily used in industrial and commercial occupancies. As shown in Figure 2-1, the feeder conductors extend from the service equipment enclosure to a feeder panel, or similar unit, that contains overcurrent protective devices for a group of branch circuits. The feeder conductors themselves are protected by circuit breakers or fuses located in the service equipment enclosure.

Branch circuits extend from the final overcurrent device protecting the circuit to the outlets for lights, appliances, and other equipment. In one-family dwellings branch circuits usually originate at the overcurrent protective device in the service equipment enclosure since feeder circuits are seldom used in such occupancies.

QUIZ
(Closed-Book)

1. Draw a diagram showing the difference between the service drop, service lateral, and service-entrance conductors.
2. What are the standard services provided by the utility company in your area?
3. Describe the premises wiring system for a one-family dwelling.

2-2 SERVICES

This section presents rules and related calculations for determining sizes and ratings of the various conductors and equipment required by the Code to be included in each service installation. The elements of the service installation to be discussed are depicted in Figure 2-3.

The service-entrance conductors are normally enclosed in a service raceway such **230-43** as conduit. The conductors terminate at the service disconnecting means, which is usually a switch or circuit breaker. The service or main overcurrent protection is a **230-90** set of fuses or a circuit breaker that protects the service-entrance conductors. Each **240-21 EXCEPTION 4** metallic part of the service must be bonded together by an equipment bonding jumper.

If one conductor of the circuit, such as the neutral conductor, is grounded, a **250-23** grounding electrode conductor that connects the grounded conductor to a grounding electrode is required. Each feeder circuit may require an equipment grounding conductor to ground the noncurrent-carrying metal parts of equipment. If this is the **250-53** case, the main bonding jumper is used to electrically connect the equipment grounding conductor, the service-equipment enclosure, and the grounded conductor of the system. In the figure the terminal block bonded to the service enclosure serves as the main bonding jumper. As shown in Figure 2-3, the water piping system of the **250-80** building must be bonded to the grounding electrode conductor by a bonding jumper.

Each conductor, the disconnecting means, and the overcurrent protective device must satisfy Code rules which specify the size or rating as appropriate for the element. These rules form the basis for the electrical design of services.

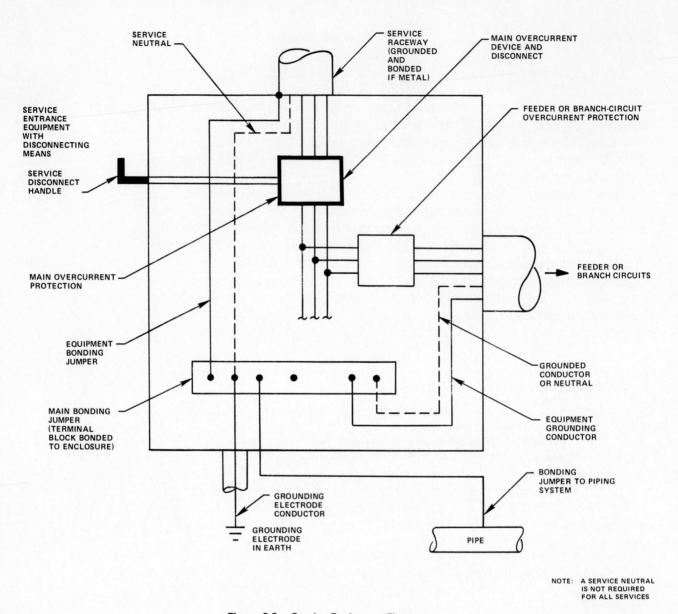

SERVICE NEUTRAL

SERVICE RACEWAY (GROUNDED AND BONDED IF METAL)

MAIN OVERCURRENT DEVICE AND DISCONNECT

SERVICE ENTRANCE EQUIPMENT WITH DISCONNECTING MEANS

SERVICE DISCONNECT HANDLE

FEEDER OR BRANCH-CIRCUIT OVERCURRENT PROTECTION

MAIN OVERCURRENT PROTECTION

FEEDER OR BRANCH CIRCUITS

EQUIPMENT BONDING JUMPER

GROUNDED CONDUCTOR OR NEUTRAL

MAIN BONDING JUMPER (TERMINAL BLOCK BONDED TO ENCLOSURE)

EQUIPMENT GROUNDING CONDUCTOR

BONDING JUMPER TO PIPING SYSTEM

GROUNDING ELECTRODE CONDUCTOR

GROUNDING ELECTRODE IN EARTH

PIPE

NOTE: A SERVICE NEUTRAL IS NOT REQUIRED FOR ALL SERVICES

Figure 2-3. Service Equipment Elements

2-2.1 General Design Rules for Services

Table 2-1 defines for each element of the service those characteristics which are important in electrical design. The size of the conductors and equipment is determined by the Code rules and Code tables referenced in Table 2-1. For example, the size of the service-entrance conductors is given either as current-carrying capacity in amperes or American wire gauge (AWG) size and is determined by the load to be served. The electrical load must be computed in amperes to determine the proper size of the conductors according to the Code rules that apply. The size of the service-entrance conductors then determines the size or ratings of the main overcurrent protective device, the grounding electrode conductor, and the main bonding jumper. The service raceway trade size depends on the number of conductors in the service as well as the size of the conductors. Finally, the rating of the service disconnecting means depends on the load to be served in amperes. The calculations of this subsection determine the electrical load in amperes when the load in watts is given, as is normally the case. Each element of the service is then chosen based on the Code references shown in Table 2-1.

Table 2-1. Summary of Requirements for Service Equipment and Conductors

Service Element	Size Determined By	Rating Units	Code Ref. Section	Code Ref. Table
Service-Entrance Conductors	Load to be served (Article 220)	Amperes or AWG size	230-42	310-16 thru 310-31
Service Raceway	Number and size of conductors (Chapter 9)	Trade size in inches	Chapter 9	Tables 3, 4, and 5 of Chapter 9
Service Disconnect	Load to be served (Article 220)	Amperes	230-79	—
Main Overcurrent Protection	Size of service-entrance Conductors*	Amperes	230-90 240-6	—
Grounding Electrode Conductor	Size of service-entrance conductors	AWG size	250-94	250-94
Main Bonding Jumper	Size of service-entrance conductors	AWG size	250-79(c)	250-94

*An exception is allowed for services supplying motor circuits.

Minimum Service Ratings. The Code specifies a minimum size or minimum ampacity for service conductors and the service disconnecting means as shown in Table 2-2. The smallest service rating is 60 amperes unless only one or two branch circuits are served. The minimum sizes or current ratings do not apply to the grounded conductor of a service. The grounded conductor may not be smaller than the grounding electrode conductor which, for an alternating-current system, is based on the size of the service-entrance conductors. **ARTICLE 230** **250-23(b)** **TABLE 250-94**

In a one-family dwelling with six or more two-wire branch circuits, or a load of 10 000 watts or more, the minimum service rating is 100 amperes. This is one of many cases in which the Code specifies rules that depend on the type of occupancy. **230-42**

If the service voltage is greater than 600 volts, the special rules that apply to higher voltage circuits must be used. The minimum service conductor sizes are given in the Code. **ARTICLE 230 PART H** **230-202**

Table 2-2. Minimum Sizes or Ampacities for Service Equipment and Conductors

Conductor or Equipment	Service With Single Branch Circuit	Service With Not More Than Two 2-Wire Branch Circuits	Any Other Service
Service-Drop Conductor	No. 12 Hard Drawn Copper (230-23)	—	No. 8 Copper No. 6 Alum. (230-23)
Service Lateral	No. 12 Copper No. 10 Alum. (230-31)	—	No. 8 Copper No. 6 Alum. (230-31)
Service-Entrance Conductors	No. 12 Copper No. 10 Alum. (230-42)	No. 8 Copper No. 6 Alum. (230-42)	60 amperes (230-42)
Service Disconnecting Means (230-79)	15 amperes	30 amperes	60 amperes

Note: Special rules apply to services operating at over 600 volts and services in one-family dwellings.

230-70   *Service Disconnecting Means and Overcurrent Protection.* The main or service disconnecting means disconnects all conductors in a building from the service-entrance conductors. The service disconnecting means is not intended to protect equipment, as the overcurrent protective device does, and it is not intended to be opened to interrupt short-circuit currents; therefore, a minimum size is specified no smaller than that listed in Table 2-2. Normally, the service disconnecting means is rated in amperes and has the same rating as the service-entrance conductors. If motors are supplied, special rules must be applied.

230-91
240-21   The service overcurrent protective device must be an integral part of the service disconnecting means or be located immediately adjacent to it. Although the overcurrent protection is at the load end of the service-entrance conductors, an exception to the rule that requires conductors to be protected at the supply end is made for service-entrance conductors. For nonmotor loads, the overcurrent protection device **240-3** must not have a higher ampacity rating than the conductors except that the next higher standard-rated overcurrent device (up to 800 amperes) may be used if the **240-6** ampacity of the conductor does not correspond to a standard ampere rating.

ARTICLE 430 PART H

    Service Grounding and Bonding. The grounding electrode conductor connects the grounding electrode to the grounded conductor (or neutral) for grounded systems. **250-94** When the grounding electrode is not a *made electrode,*[1] the minimum size of the grounding electrode conductor is determined by the size of the largest service-entrance **250-94 EXCEPTION 1** conductor. The connection to a made electrode is not required to be larger than a No. 6 copper wire or equivalent. Other rules apply where the connection is to a concrete-encased electrode or a ground ring.

250-79
250-80   Several equipment bonding jumpers may be required at the service to bond or electrically connect service conduit to the service enclosure or to bond the grounding electrode to the water piping system in a building. The size of these jumpers is determined by the size of the service conductors. Equipment bonding on the *load* side of the service overcurrent devices requires a jumper size based on the setting of the overcurrent devices, not the conductor size.

    Design Technique and Rules. To determine the rating or size of the service elements, the load current is first determined and then the size of the ungrounded and grounded service-entrance conductors is chosen based on the load current. The other elements, including the service raceway, disconnecting means, overcurrent protection, and grounding and bonding conductors, are then selected.

    The ampacity of the conductors depends on the type of conductor material and the type of insulation. For a given AWG size, copper conductors will carry safely more current than aluminum conductors because copper has a lower resistance. The insulation, which can be damaged by excessive heat, determines the ampacity of a **TABLES 310-16** given conductor. If the conductor is in a raceway or cable and not free in the air, its **THRU 310-31** ampacity is reduced. The Code provides tables which relate AWG size to the ampacity based on the type of material and the type of insulation. Individual tables are provided for conductors in a raceway or cable and for conductors in free air. All of these factors must be considered in order to select the table that fits the situation.[2]

---

[1]A made electrode is constructed of metal rods or plates. The metal structure of a building or the water piping system are other types of grounding electrodes.

[2]For calculation purposes, all conductors are assumed to be in a raceway or cable unless otherwise stated in the Guide.

2-2.2 Sample Service Calculation

An example of the calculations required for designing a simple service installation is given in Figure 2-4. The applicable rules for determining the size or rating of the various elements of the service are contained in the Code references cited in that figure. In following the example given in the Guide, the Code book should also be used and each Code section should be read at the time it is referenced by the Guide.

The service is chosen to be a 120/240-volt, single-phase service. A load of 40 000 volt-amperes is to be served for which the size or rating of all the elements of the service must be selected. Type THW copper conductors in conduit are selected for the service-entrance conductors.

First, the load in volt-amperes (or watts) is converted to amperes.[3] The load current is

$$\frac{\text{power in volt-amperes}}{\text{voltage}} = \frac{40\ 000\ \text{VA}}{240\text{V}} = 167\ \text{A}$$

for this single-phase service. The service-entrance conductors must have an ampacity of at least 167 amperes. Type THW copper conductors of AWG size No. 2/0 (175 amperes) are required according to the Code table of ampacities for these conductors. The service is a three-wire circuit consisting of the two ungrounded conductors and a neutral conductor.
TABLE 310-16

The service raceway must conform to the Code rules for percent of conduit fill. Since there are three conductors in the conduit, their cross-sectional area, including insulation, must not exceed 40% of the total cross-sectional area of the conduit. This is specified in the Code to prevent damage to the conductors when they are pulled in during installation of the wiring system. Since all of the conductors are the same size, the Code table "Maximum Number of Conductors in Trade Sizes of Conduit or Tubing" may be used to find the required conduit size. According to that table, three No. 2/0 THW conductors are allowed in a 1½-inch trade size conduit.
CHAPTER 9 TABLE 1

CHAPTER 9 TABLE 3A

The service disconnecting means and the overcurrent device must be sized according to the rating of the service-entrance conductors. The disconnecting means must have a rating *not less* than the size of the load, while the overcurrent device cannot be rated higher than the ampacities of the conductors in the example given. A 175-ampere disconnecting means would satisfy the first requirement. The overcurrent protective device may be rated at 175 amperes also since this corresponds to the conductor ampacity.
230-79

230-90

240-3

The Code table "Grounding Electrode Conductor for AC Systems" determines the size of the grounded electrode conductor and the main bonding jumper. The grounded electrode conductor and the main bonding jumper are sized as No. 4 copper because of the No. 2/0 service-entrance conductors.
TABLE 250-94

2-2.3 Three-Phase Services

In principle, the calculations for three-phase services are the same as for single-phase services when the loads between phase conductors or between phase conductors and neutral are balanced. The load in amperes on each phase conductor is given by the formula:

---

[3]For loads with a power factor of less than 1, the volt-ampere rating should be used for load calculations. For purely resistive loads (i.e., power factor equal to 1), watts are equivalent to volt amperes. Refer to Section 3-2.5 and to Chapter 12 of the Guide for a discussion.

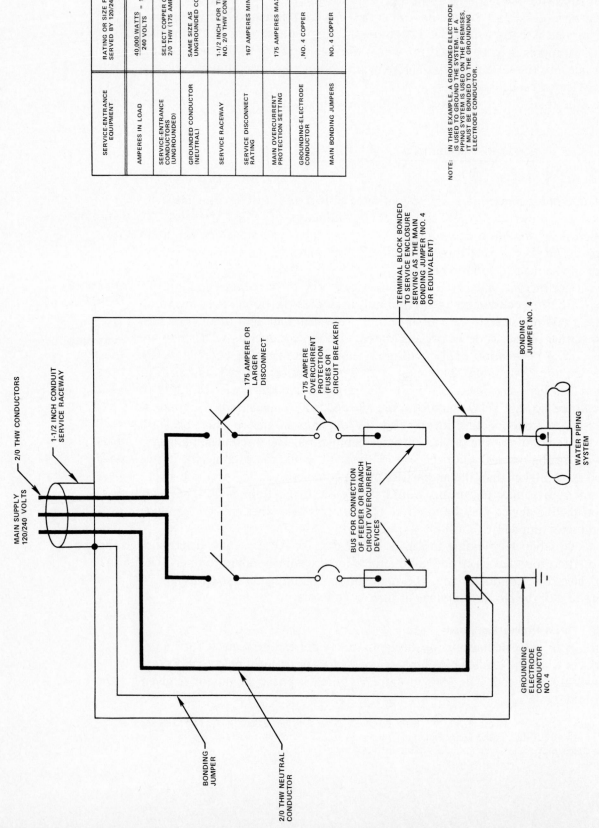

SERVICE-ENTRANCE EQUIPMENT	RATING OR SIZE FOR 40-KW LOAD SERVED BY 120/240-VOLT SERVICE	CODE REFERENCE
AMPERES IN LOAD	$\dfrac{40,000 \text{ WATTS}}{240 \text{ VOLTS}} = 167$ AMPERES	—
SERVICE-ENTRANCE CONDUCTORS (UNGROUNDED)	SELECT COPPER CONDUCTORS: 2/0 THW (175 AMPERES)	TABLE 310-16
GROUNDED CONDUCTOR (NEUTRAL)	SAME SIZE AS UNGROUNDED CONDUCTORS	—
SERVICE RACEWAY	1-1/2 INCH FOR THREE NO. 2/0 THW CONDUCTORS	TABLE 3A CHAPTER 9
SERVICE DISCONNECT RATING	167 AMPERES MINIMUM	SAME AS LOAD 230-79
MAIN OVERCURRENT PROTECTION SETTING	175 AMPERES MAXIMUM	240-6
GROUNDING-ELECTRODE CONDUCTOR	NO. 4 COPPER	TABLE 250-94
MAIN BONDING JUMPERS	NO. 4 COPPER	TABLE 250-94

NOTE: IN THIS EXAMPLE, A GROUNDED ELECTRODE IS USED TO GROUND THE SYSTEM. IF A PIPING SYSTEM IS USED ON THE PREMISES, IT MUST BE BONDED TO THE GROUNDING ELECTRODE CONDUCTOR.

Figure 2-4. Sample Service Calculation

$$\text{load per phase (amperes)} = \frac{\text{load (volt-amperes or watts)}}{\sqrt{3} \times \text{voltage between phases (volts)}}$$

For example, a 40-kilovolt-ampere load evenly distributed on a 208Y/120-volt, three-phase service (13.3 kilovolt-amperes per phase) places a line load in amperes on each phase conductor of:

$$\frac{40\ 000\ \text{VA}}{1.732 \times 208\ \text{V}} = 111.0\ \text{A}$$

This is a four-wire service and requires either three fuses (one in each ungrounded   **230-90**
conductor) or a three-pole circuit breaker for the main overcurrent protection.

QUIZ
(Closed-Book)

1. What are the minimum size requirements for service-drop conductors?
2. What is the minimum rating for service-entrance conductors supplying a one-family dwelling with four branch circuits?
3. What is the purpose of the main overcurrent device?
4. What is the purpose of the main disconnecting means?
5. Describe each item and its rating in the service to a home having a 10-kilovolt-ampere load.
6. Draw a diagram showing the grounding electrode conductor, bonding jumper, and the grounded conductor for a standard service.
7. The size of the grounding electrode conductor depends on which portion of the service?
8. What is the maximum size required for a grounding electrode conductor connected to a made electrode?
9. Under what conditions does a dwelling require a 100-ampere, three-wire service?
10. Which service conductor is intentionally grounded?

QUIZ
(Open-Book)

1. What is the ampacity of the conductors and the size of the service raceway for THW copper conductors if the service is 120/240 volts and supplies a load of 46 kVA?
2. What is the conduit size in problem 1 above if THHN conductors are used? (*Hint:* These are 90 °C conductors as specified in the appropriate Code table.)
3. Compare the ampacity of the conductors and the size of the service conduit for THW copper conductors if the service is 120/240 volts in the following cases:
   (a) 10-kilovolt-ampere load on three branch circuits
   (b) 20-kilovolt-ampere load
   (c) 20-kilovolt-ampere load in a one-family dwelling
   (d) 100-kilovolt-ampere load
4. Three No. 2/0 THW copper conductors in a 120/240-volt service will serve what load in volt-amperes?
5. Select the size of the standard fuse that would be used in the following cases:
   (a) 900-ampere load
   (b) 130-ampere load
6. If the service conductors are No. 2 THW copper and the main circuit breaker is set at 115 amperes, what size is required for the following:
   (a) The jumper bonding the service conduit to the neutral terminal block
   (b) The grounding electrode conductor
   (c) The equipment bonding jumper to the water piping system
7. Determine the required size of type THW copper conductors in conduit to supply a 40-kilovolt-ampere load if the voltage is 480 volts, three-phase.

2-3 FEEDERS

This section presents the basic techniques involved in the design of feeder circuits and introduces the concept of feeder demand factors.

Feeder circuits are primarily used in apartment buildings and commercial and industrial occupancies. A typical feeder circuit system is illustrated in Figure 2-5 in which the service supplies separate feeders to each of four apartment units. The feeder circuits originate at the service enclosure and supply panelboards containing overcurrent protective devices for branch circuits. Each feeder circuit conductor is protected by a circuit breaker or set of fuses located within the service enclosure.

2-3.1 General Feeder Design Rules

To determine the size or rating of each element of a feeder circuit, the load is calculated; then the size or rating of conductors, raceways, and the overcurrent protective device is determined based on the results of the load calculations. The calculations for feeders are similar to those for services. Also, the main bonding jumper and the grounding electrode conductor are used only for services.

If several feeders originate at a service, the conductor sizes and the rating of the overcurrent protective device for each feeder circuit should first be calculated separately before the load for the entire service is calculated. In the apartment shown in Figure 2-5, the load for each apartment would be calculated to determine the feeder rating. The service capacity is determined by the total load from all the apartments.

215-2 Feeder conductors must have an ampacity not lower than that required to supply the computed load. The minimum feeder conductor sizes specified by the Code are summarized in Table 2-3. As was the case with services, minimum sizes for feeders are specified by the Code under certain conditions.

The feeder conductor ampacity shall not be lower than that of the service-entrance conductors if the feeder conductors carry the total load supplied by service-entrance conductors of 55 amperes or less. If the feeder supplies a number of branch circuits, the smallest allowable ampacity is 30 amperes, whether it is a two-wire feeder (such as a 240-volt feeder) or a three-wire feeder (such as a 120/240-volt feeder).

225-6 The Code also requires that overhead feeders meet the minimum size requirements shown in Table 2-3.

Table 2-3. Minimum Sizes for Feeder Conductors

Feeder Type	Two-Wire Feeder Supplying Two or More 2-Wire Branch Circuits	Three-Wire Feeder Supplying 1) More Than Two 2-Wire Branch Circuits, or 2) Two or More 3-Wire Branch Circuits	600 Volts or Less		Over 600 Volts	
			Spans Up To 50 Feet	Spans Longer Than 50 Feet	Open Individual Conductors	In Cable
All Feeders[1]	30 amperes (215-2)	30 amperes (215-2)	—	—	—	—
Overhead Feeders (225-6)			No. 10 Cu No. 8 Al	No. 8 Cu No. 6 Al	No. 6 Cu No. 4 Al	No. 8 Cu No. 6 Al

Notes:
1) The feeder conductor ampacity shall not be lower than that of the service-entrance conductors where the feeder conductors carry the total load supplied by service-entrance conductors with an ampacity of 55 amperes or less. (215-2)

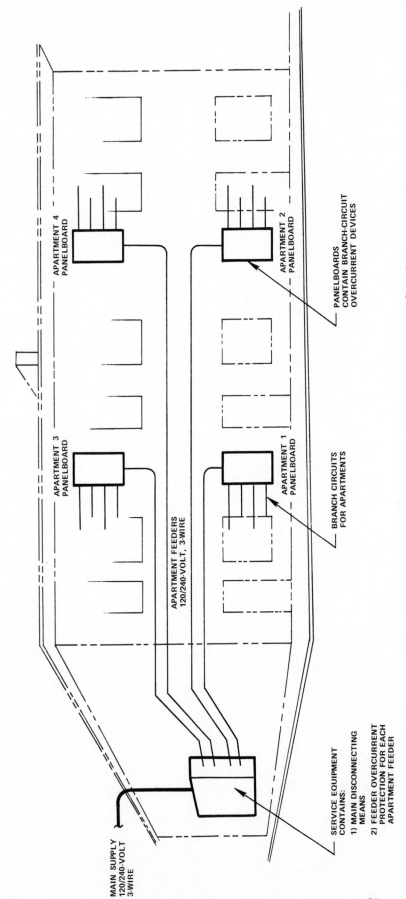

Figure 2-5. Typical Feeder Arrangement for an Apartment Complex

MAIN SUPPLY
120/240-VOLT
3-WIRE

SERVICE EQUIPMENT
CONTAINS:
1) MAIN DISCONNECTING
   MEANS
2) FEEDER OVERCURRENT
   PROTECTION FOR EACH
   APARTMENT FEEDER

APARTMENT FEEDERS
120/240-VOLT, 3-WIRE

APARTMENT 3
PANELBOARD

APARTMENT 4
PANELBOARD

APARTMENT 1
PANELBOARD

APARTMENT 2
PANELBOARD

BRANCH CIRCUITS
FOR APARTMENTS

PANELBOARDS
CONTAIN BRANCH-CIRCUIT
OVERCURRENT DEVICES

25

Feeder Calculations. The feeder load in amperes is calculated from the total load in watts to determine the minimum rating of the feeder circuit. Since one service may supply several feeders, the load for each feeder (if the loads are **220-10(a)** different) must be determined. The feeder load is the sum of the loads on the branch circuits when no demand factors are applied. The feeder conductor ampacity, the raceway size, the rating of the overcurrent protective device, and the size of the equipment grounding conductor are calculated to complete the feeder circuit design. An example will serve to illustrate the calculation techniques involved.

Assume the feeder circuits of Figure 2-5 are 120/240-volt circuits that supply a 10-kilovolt-ampere computed load in each apartment. The feeder capacity for each feeder must be at least

$$\frac{10\ 000\ \text{VA}}{240\ \text{V}} = 41.7\ \text{A}$$

220-10(b)
TABLE 310-16

TABLE 3A
CHAPTER 9

240-3
240-6
TABLE 250-95

250-91

The minimum standard rating of the feeder circuit is 45 amperes. If type THW copper conductors in conduit supply the apartments, size No. 8 is required according to the Code table of ampacities. The conduit to enclose three No. 8 THW conductors must be at least ¾ inch, as specified in the Code table "Maximum Number of Conductors in Trade Sizes of Conduit or Tubing." A standard 45-ampere or 50-ampere overcurrent device could be used to protect the circuit.[4] The Code table "Size of Equipment Grounding Conductors for Grounding Raceway and Equipment" specifies a No. 10 copper conductor for the equipment grounding conductor since the overcurrent device is rated at less than 60 amperes. In this installation the conduit could serve as the equipment grounding conductor, in which case an additional conductor would not be required.

The load given in this example is referred to as the *demand load*. For a feeder, this is not necessarily the same value as the load that is directly connected to the feeder through the branch circuits when a feeder demand factor is allowed.

Feeder Demand Factors. In most cases, the feeder ampacity must be sufficient **220-10(a)** to supply power to the total connected load. Under special conditions, however, the Code allows the feeder circuit ratings to be reduced (or derated) according to specified demand factors. The maximum demand load for the feeder is

maximum demand = connected load × demand factor

The Code specifies the demand factors for each situation in which they may be used. The Code rules are very precise and demand factors must not be applied unless they are specifically allowed.

As an example of the use of a demand factor which is allowed *only* in dwellings such as apartment buildings, Figure 2-6 shows ten 5-kilowatt electric clothes dryers connected to a 240-volt feeder. The connected load is 50 kilowatts (10 × 5 kilowatts) and would require a feeder ampacity of 50 000 watts/240 volts = 208 amperes. **TABLE 220-18** Since all the dryers would not normally be operating simultaneously at their full-rated load, the Code table "Demand Factors for Household Electric Clothes Dryers" allows a demand factor of 50% (.50) to be applied. The demand load is then

demand load = 10 units × 5000 W × .50 = 25 000 W

---

[4]In practice, a 60-ampere device may be more readily available and would be used with No. 6 conductors. The Guide uses the smallest standard ratings given in the Code.

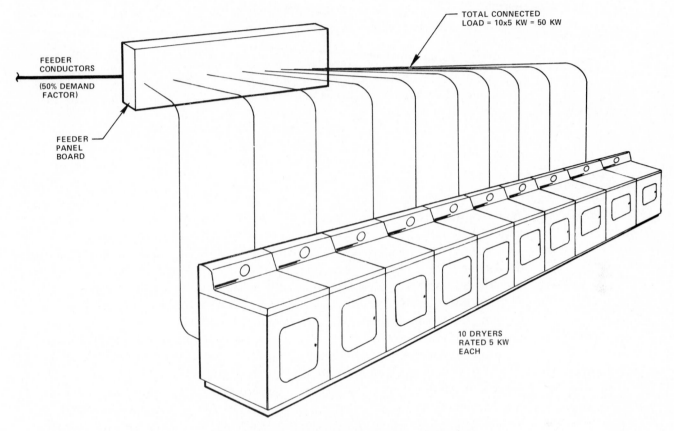

MAXIMUM DEMAND = CONNECTED LOAD x DEMAND FACTOR
= 50,000 WATTS x 0.5
= 25,000 WATTS

TOTAL CONNECTED
LOAD = 10x5 KW = 50 KW

FEEDER
CONDUCTORS

(50% DEMAND
FACTOR)

FEEDER
PANEL
BOARD

10 DRYERS
RATED 5 KW
EACH

Figure 2-6. Example of Demand Factors Application

The required feeder ampacity must be only 25 000 watts (volt-amperes)/240 volts = 104 amperes in this case.

The Code allows a demand factor to be applied to the feeder neutral load in certain cases. Normally, the feeder neutral load is considered to be the maximum connected load between the neutral and any one ungrounded conductor. If the load is balanced between the ungrounded conductors and the neutral, the neutral load is then the same as the load on any one ungrounded conductor. If the feeder is a three-wire, single-phase feeder or a four-wire, three-phase feeder, the neutral load may be reduced if the load exceeds 200 amperes and the load does not consist of electric discharge lighting. The Code specifies a demand factor of 70% (.70) for the load that exceeds 200 amperes. Thus, if the neutral load is computed to be 450 amperes, the neutral demand load is

220-22

$$
\begin{aligned}
200 \text{ amperes (first) at } 100\% &= 200 \text{ A} \\
250 \text{ amperes (excess) at } 70\% &= \underline{175} \text{ A} \\
&= 375 \text{ A}
\end{aligned}
$$

This demand factor applies to *any* feeder neutral that meets the requirements stated above, regardless of the type of occupancy served.

It is important to note that reductions based on demand factors are not always possible, i.e., the application of demand factors depends on the type of load being served by the feeder and on the type of occupancy involved. When a feeder supplies only motor circuits, for example, the application of demand factors is generally not permitted; in fact, the feeder circuit ampacity will be required to be greater than the sum of the full-load current ratings of all the motors connected to the feeder.

430-24

2-3.2 Sample Feeder Calculation Using Demand Factors

Figure 2-7 summarizes a feeder calculation based on a 110-kilowatt connected load and a 35% feeder demand factor. Type THW copper conductors are used for the 120/240-volt feeder circuit that is enclosed in conduit.

The demand load in amperes is

$$\frac{110\ 000\ VA}{240\ V} \times .35 = 160\ A$$

TABLE 310-16   This requires size No. 2/0 THW copper conductors since they will be enclosed in a raceway (conduit). The conductors are capable of carrying 175 amperes.

240-3   The overcurrent device is selected to be 175 amperes, which protects the conductors and corresponds to a standard rating.

250-91   The equipment grounding conductor is required unless the raceway is metal and of a type allowed to be used for grounding. It electrically connects the noncurrent-carrying metal portions of equipment to the grounded conductor and the grounding

TABLE 250-95   electrode conductor at the service. A No. 6 copper conductor is selected based on a 175-ampere overcurrent protective device.

CHAPTER 9
TABLE 1   The conduit must enclose three No. 2/0 conductors and the No. 6 equipment grounding conductor. The total area of the conductors must not exceed 40% of the cross-sectional area of the conduit since more than two conductors are enclosed.

CHAPTER 9
TABLE 5   To determine the area of the conductors, the Code table "Dimensions of Rubber-Covered and Thermoplastic-Covered Conductors" must be used since several different size conductors are enclosed. Column 5 of that table gives the area of type THW conductors as follows:

Three No. 2/0 THW = 3 × .2781 square in. = .8343
One No. 6 THW   = .0819

Total   .9162 square in.

CHAPTER 9 TABLE 4   The conduit must enclose conductors with an area of .9162 square inches and be 40% or less full. The Code table "Dimensions and Percent Area of Conduit and of Tubing" relates the trade size of conduit to the 40% area in square inches. The 40% area of the conduit must be at least .9162 square inches to meet the Code requirement. A 2-inch conduit with a 40% fill area of 1.34 square inches is the smallest permissible size in this case.

QUIZ
(Closed-Book)

1. A 120/240-volt feeder supplies two 120/240-volt branch circuits. What is the minimum size feeder conductor allowed?
2. What size feeder conductors are required if the feeder carries the total load of a service whose conductors are size No. 8 (THW copper)?

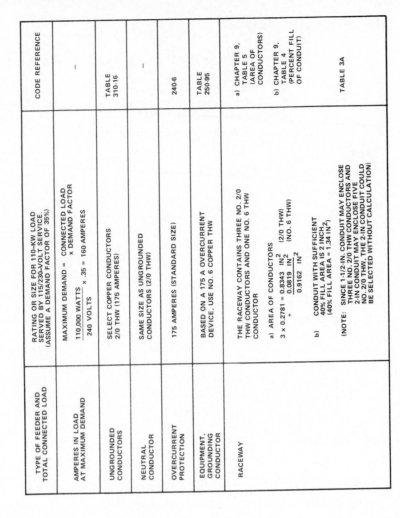

TYPE OF FEEDER AND TOTAL CONNECTED LOAD	RATING OR SIZE FOR 110-KW LOAD SERVED BY 115/230-VOLT SERVICE. (ASSUME A DEMAND FACTOR OF 35%)	CODE REFERENCE
AMPERES IN LOAD AT MAXIMUM DEMAND	MAXIMUM DEMAND = CONNECTED LOAD × DEMAND FACTOR $$\frac{110,000 \text{ WATTS}}{240 \text{ VOLTS}} \times .35 = 160 \text{ AMPERES}$$	—
UNGROUNDED CONDUCTORS	SELECT COPPER CONDUCTORS 2/0 THW (175 AMPERES)	TABLE 310-16
NEUTRAL CONDUCTOR	SAME SIZE AS UNGROUNDED CONDUCTORS (2/0 THW)	—
OVERCURRENT PROTECTION	175 AMPERES (STANDARD SIZE)	240-6
EQUIPMENT, GROUNDING CONDUCTOR	BASED ON A 175 A OVERCURRENT DEVICE, USE NO. 6 COPPER THW	TABLE 250-95
RACEWAY	THE RACEWAY CONTAINS THREE NO. 2/0 THW CONDUCTORS AND ONE NO. 6 THW CONDUCTOR a) AREA OF CONDUCTORS $$3 \times 0.2781 = 0.8343 \text{ IN}^2 \quad (2/0 \text{ THW})$$ $$+ \quad 0.0819 \text{ IN}^2 \quad (\text{NO. 6 THW})$$ $$\overline{\quad\quad 0.9162 \text{ IN}^2}$$ b) CONDUIT WITH SUFFICIENT 40% FILL AREA IS 2 INCH (40% FILL AREA = 1.34 IN²) (NOTE: SINCE 1-1/2-IN. CONDUIT MAY ENCLOSE THREE NO. 2/0 THW CONDUCTORS AND 2-IN CONDUIT MAY ENCLOSE FIVE NO. 2/0 THW, THE 2-IN CONDUIT COULD BE SELECTED WITHOUT CALCULATION)	a) CHAPTER 9, TABLE 5 (AREA OF CONDUCTORS) b) CHAPTER 9, TABLE 4 (PERCENT FILL OF CONDUIT) TABLE 3A

Figure 2-7. Sample Feeder Calculation

29

3. A 480-volt feeder circuit is run as an overhead span 100 feet in length. What is the minimum size of the feeder conductors?
4. Define feeder according to the Code.
5. Describe the important differences between feeders and services.
6. What is the maximum demand load if a feeder supplies a connected load of 100 kilowatts with a demand factor of 50%?

QUIZ
(Open-Book)

1. A 120/240-volt feeder supplies a 30-kilovolt-ampere load. Find the following:
   (a) The line load in amperes
   (b) The size of THW copper conductors to carry the load
   (c) The total cross-sectional area of the wires
   (d) The required conduit size if the conduit is used for equipment grounding
2. If the 30-kVA load in problem 1 is supplied by a 480Y/277-volt feeder, what are the answers to the questions of problem 1? Compare the two results.
3. A 120/240-volt feeder supplies a load of 100 kVA. Find the following:
   (a) The line load in amperes
   (b) The required ampacity of the neutral after derating
   (c) The size of the ungrounded conductors and the neutral conductor if THW copper conductors are used
4. A four-wire, three-phase feeder is required to supply a 1000-ampere load. What is the required ampacity of the neutral conductor?

2-4 BRANCH CIRCUITS

The branch circuit is that portion of the premises wiring system that extends from the last overcurrent device protecting the branch-circuit conductors to the outlets supplying the utilization equipment connected to the branch circuit. Normally, the overcurrent device is a set of fuses or a circuit breaker located in a panelboard supplied by a feeder or service. In a one-family dwelling the branch-circuit over-current devices are located in the service equipment enclosure. Typical branch circuits supplied from a feeder panelboard are illustrated in Figure 2-8. Figure 2-8(a) shows a wiring system that might be found in a modern home. The large appliances have individual branch circuits and the lights and receptacles are supplied by a number of 15- or 20-ampere circuits. Figure 2-8(b) and (c) show circuits typical of industrial installations. The lighting circuits and the receptacle circuits are supplied by the 120/240-volt panel and the motor by a separate 480-volt panel.

ARTICLE 100 DEFINITIONS   The Code distinguishes branch circuits according to their use, and, in some cases, different rules apply to appliance branch circuits, general-purpose branch circuits, or individual branch circuits. The term *utilization equipment* is taken to be the same as the *load to be served.* The Code implies that the load is supplied at an *outlet,* which is simply a connection to supply the conductors. A *receptacle outlet* is used for the common cord and plug attachment to a branch circuit.

2-4.1 General Branch-Circuit Design Rules

Unlike services that supply the total electrical requirements for a building, or feeders that supply portions of a building, the branch circuit provides electrical energy to specific equipment or loads such as room lighting, electric ranges, or motors. With few exceptions, the branch-circuit rating must be as great or greater than the full-load current rating of the equipment attached to the branch circuit.

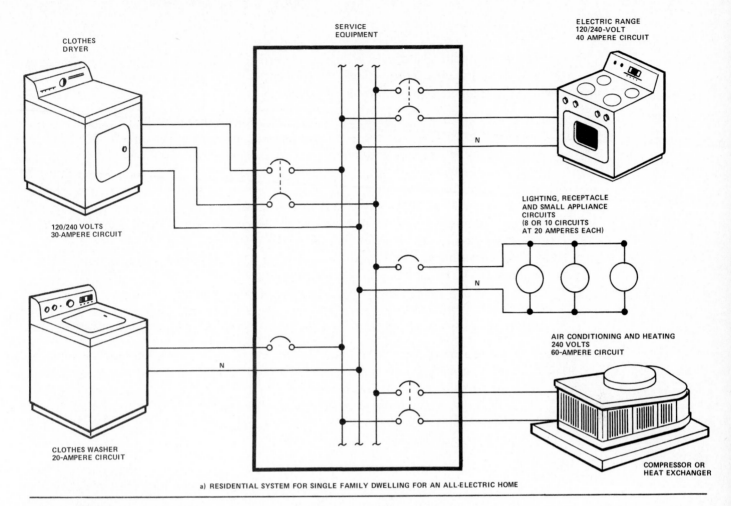

CLOTHES
DRYER

SERVICE
EQUIPMENT

ELECTRIC RANGE
120/240-VOLT
40 AMPERE CIRCUIT

N

120/240 VOLTS
30-AMPERE CIRCUIT

LIGHTING, RECEPTACLE
AND SMALL APPLIANCE
CIRCUITS
(8 OR 10 CIRCUITS
AT 20 AMPERES EACH)

N

AIR CONDITIONING AND HEATING
240 VOLTS
60-AMPERE CIRCUIT

N

CLOTHES WASHER
20-AMPERE CIRCUIT

COMPRESSOR OR
HEAT EXCHANGER

a) RESIDENTIAL SYSTEM FOR SINGLE FAMILY DWELLING FOR AN ALL-ELECTRIC HOME

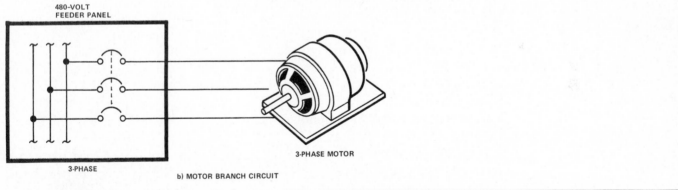

480-VOLT
FEEDER PANEL

3-PHASE MOTOR

3-PHASE

b) MOTOR BRANCH CIRCUIT

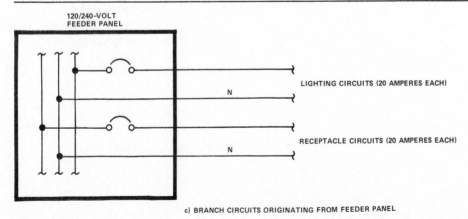

120/240-VOLT
FEEDER PANEL

LIGHTING CIRCUITS (20 AMPERES EACH)

N

RECEPTACLE CIRCUITS (20 AMPERES EACH)

N

c) BRANCH CIRCUITS ORIGINATING FROM FEEDER PANEL

Figure 2-8. Typical Branch Circuits

Demand factors, such as are applied to feeders or services, are generally not applicable to branch circuits. (Exceptions to this rule include household electric ranges and equipment that operates intermittently.)

220-19
ARTICLE 630

210-3 ***Classification of Branch Circuits.*** A branch circuit is classified according to the rating or setting of the overcurrent device protecting the circuit. For example, if a 15-ampere circuit breaker protects a circuit of No. 10 THW conductors that have a 35-ampere rating, the branch circuit is still rated at 15 amperes. The use of the larger conductor in this case may have been necessary in order to reduce an excessive voltage drop from the feeder or service to the equipment being supplied. The 15-ampere circuit breaker may have been required to protect a small appliance; a larger circuit breaker would not have been permitted.

210-3
210-23

An *individual* branch circuit may have any rating and it may supply any load for which it is rated.

210-3

If two or more outlets are supplied by a branch circuit, the rating of the branch circuit is restricted to those ratings listed in the Code, which are 15, 20, 30, 40, and 50 amperes. (An exception is provided for certain industrial installations.)

210-19
430-22

Ampacity of Branch-Circuit Conductors. Except for a branch circuit supplying a motor, branch-circuit conductors must have an ampacity not less than the rating of the branch circuit as determined by the overcurrent protective device and not less than the maximum load to be served. The Code specifies the minimum size for branch-circuit conductors as No. 14. The frequently used No. 14 THW copper conductors are suitable for such a circuit.

TABLE 310-16

210-22
210-23

Maximum and Permissible Loads. The load on a branch circuit cannot exceed the branch-circuit rating. As stated previously, an individual branch circuit may supply any load for which it is rated. When a load is supplied by branch circuits with two or more outlets, the load must be divided so that there is a sufficient number of branch circuits of the proper rating. Too many branch circuits, such as a separate circuit for each lighting fixture, would be impractical and uneconomical.

210-21
210-23

For certain loads, the Code defines a maximum rating for the branch circuit that will be used to supply the load. For example, standard lampholders may not be connected to a branch circuit having a rating in excess of 20 amperes. The total lighting load of standard lampholders must therefore be divided into a number of 15- or 20-ampere branch circuits.

The branch circuit is limited in the power it may supply to a load by both the specified voltage and current rating. For example, a 120-volt, 15-ampere branch circuit may supply a maximum load of

$$120 \text{ V} \times 15 \text{ A} = 1800 \text{ VA}$$

In the case of a three-phase branch circuit, the power that may be supplied to a load is

$$\sqrt{3} \times \text{line voltage} \times \text{circuit rating (amperes)}$$

A three-phase, 480-volt, 15-ampere circuit may supply a maximum load of

$$1.732 \times 480 \text{ V} \times 15 \text{A} = 12\,470 \text{ VA}$$

Multiwire Branch Circuits. A typical branch circuit for single-phase loads consists of two circuit conductors and an equipment grounding conductor. One circuit conductor is usually grounded as is the case with the commonly used 120-volt, two-wire circuit. If a circuit has two or more ungrounded conductors with a potential difference between them, and a grounded conductor with equal potential difference between it and each ungrounded conductor, it is referred to as a *multiwire circuit.* A typical multiwire circuit used for single-phase loads is the 120/240-volt, three-wire circuit.

A single three-wire circuit has a capacity equivalent to two two-wire circuits. Three-wire branch circuits, therefore, are frequently used to supply lighting and receptacle loads because of the cost factors involved, especially in commercial and industrial occupancies where large loads and long distribution distances are present.

A 120/240-volt, three-wire circuit with a 15-ampere rating can supply a total load of

$$240 \text{ V} \times 15 \text{ A} = 3600 \text{ W}$$

or twice the capacity of 1800 watts provided by a 120-volt, two-wire circuit with the same ampere rating.

Multiwire circuits may be used to supply only line to neutral loads except in the cases in which only a single load is supplied and in which all ungrounded conductors of the circuit are opened simultaneously by the branch-circuit protective device.   **210-4**

2-4.2 Branch-Circuit Calculation Techniques

The standard calculations for individual branch circuits include determining the conductor ampacity, the rating of the overcurrent device, the size of the equipment grounding conductor, and, finally, the conduit size (if used). The techniques involved are the same as those shown previously for services and feeder circuits.

When the load is given in kilovolt-amperes or watts, the load in amperes is easily calculated to determine the minimum required circuit rating. A 10-kilovolt-ampere load connected to a 120/240-volt branch circuit, for example, represents a load in amperes of

$$\frac{10\ 000 \text{ VA}}{240 \text{ V}} = 41.7 \text{ A}$$

Once the load in amperes has been determined, each element of the branch circuit can be selected. The 120/240-volt circuit with a load of 41.7 amperes could   **TABLE 310-16**
use No. 8 THW copper conductors and an overcurrent protective device rated at 50 amperes or less. A No. 10 copper equipment grounding conductor would be re-   **TABLE 250-95**
quired if other means of grounding were not used.

In the example, an individual branch circuit with a minimum rating of 45 amperes could supply the 10-kilovolt-ampere load. An individual branch circuit is normally used for such large single loads.

QUIZ
(Closed-Book)   1. Define branch circuit according to the Code.
                    2. What is the minimum size of branch-circuit conductors for general installation?

3. What are the standard ratings for branch circuits having two or more outlets?
4. Can a 60-ampere branch circuit be used in a single-family dwelling?
5. A 10-kilovolt-ampere load is to be supplied by a 240-volt circuit. What standard rating branch circuit must be used?

(Open-Book)

1. Determine the branch-circuit elements to supply a 20-kilovolt-ampere, 120/240-volt load. Use type THW copper conductors.
2. Determine the branch-circuit elements to supply a 20-kilovolt-ampere, 480Y/277-volt, three-phase load. Use type THW copper conductors.

TEST CHAPTER 2

I. True or False *(Closed-Book)*

			T	F

1. The service drop is part of the service-entrance conductors. [ ] [ ]
2. The minimum rating of service-entrance conductors is 60 amperes. [ ] [ ]
3. The smallest neutral conductor for a 120/240-volt service is No. 8 copper if the service supplies more than one branch circuit. [ ] [ ]
4. A service disconnecting means may have a rating as small as 15 amperes in certain cases. [ ] [ ]
5. The grounding conductor is intentionally grounded. [ ] [ ]
6. The cross-sectional area of three or more THW conductors in conduit must not exceed 40% of the cross-sectional area of the conduit. [ ] [ ]
7. The Code gives the maximum ratings for service disconnecting means. [ ] [ ]
8. Standard lighting fixtures and duplex receptacles may be served by a 30-ampere branch circuit. [ ] [ ]
9. A 30-ampere branch circuit requires a 30-ampere overcurrent device to protect it. [ ] [ ]
10. If a one-family dwelling has a demand load of 10 kilovolt-amperes, a 150-ampere service is required. [ ] [ ]

II. Multiple Choice *(Closed-Book)*

1. The rating of the service for a one-family residence shall not be less than 100 amperes when the initial load is at least:
    (a) 8 kilovolt-amperes
    (b) 10 kilovolt-amperes
    (c) 12 kilovolt-amperes
    (d) 15 kilovolt-amperes

    1. _____

2. When it is allowed, the demand factor applied to the service neutral for a load in excess of 200 amperes is:
    (a) 50%
    (b) 60%
    (c) 70%
    (d) 90%

    2. _____

3. If the phase-to-neutral voltage in a three-phase system is 2400 volts, the phase-to-phase voltage is approximately:
    (a) 4800 volts
    (b) 4160 volts
    (c) 2400 volts
    (d) 12 470 volts

    3. _____

4. The equipment bonding jumper on the supply side of the service is sized by the rating of:
    (a) The overcurrent protective device
    (b) The service-entrance conductors
    (c) The service drop
    (d) The load to be served

    4. _____

5. A 400-ampere load supplied by a 120/240-volt feeder requires a feeder neutral with an ampacity of:
    (a) 400 amperes
    (b) 340 amperes
    (c) 280 amperes
    (d) 360 amperes

    5. _____

6. Which of the following is not required for a 480Y/277-volt feeder circuit and panelboard?
   (a) Grounded conductor
   (b) Neutral conductor
   (c) Equipment grounding conductor
   (d) Grounding electrode conductor

7. Which of the following is not always required in a typical electrical installation?
   (a) The service
   (b) Feeders
   (c) Overcurrent protection for conductors
   (d) Main disconnecting means

6. _____

8. Which of the following is not a standard classification for a branch circuit supplying several loads?
   (a) 20 amperes
   (b) 25 amperes
   (c) 30 amperes
   (d) 50 amperes

7. _____

9. The power used by a load supplied by a 20-ampere, 120-volt branch circuit cannot exceed:
   (a) 2400 watts
   (b) 2000 watts
   (c) 5000 watts
   (d) 4600 watts

8. _____

9. _____

III. Problems *(Open-Book)*

1. A 120/240-volt service supplies a total load of 92 kilovolt-amperes. Determine the size or rating of the following:
   (a) Service-entrance conductors of type THHN copper
   (b) Neutral conductor
   (c) Main overcurrent protection
   (d) Grounding electrode conductor

2. The maximum demand load for 208Y/120-volt, three-phase feeder is 20 kilovolt-amperes. What size THW aluminum conductors are required and how large is the conduit to enclose them?

3. Calculate the load in amperes and the type THW copper conductor size required for 100-kVA loads supplied by the following services with the voltage of the load as shown:
   (a) 120/240-volt, single-phase   (230-volt load)
   (b) 208Y/120-volt, three-phase   (208-volt load)
   (c) 480Y/277-volt, three-phase   (480-volt load)
   (d) 480-volt, three-phase   (480-volt load)

4. Two 120/240-volt feeders connected to one service supply a total connected load of 20 kilovolt-amperes each. If the feeder demand factor is 50% and the service demand factor is 35%, find the following:
   (a) The maximum demand load and the load in amperes for each feeder
   (b) The maximum demand load and current for the service
   (c) The conductor size for THW copper conductors for each feeder and the service

5. What size conduit is required to enclose six No. 10 TW conductors, three No. 14 THW conductors, and two No. 12 THHN conductors?

3

General Design
Calculations

This chapter discusses the design aspects of branch circuits and feeder circuits that depend on the type of load to be served. In the previous chapter the electrical system from the distribution station to the service, feeders, and branch circuits was described and a number of simple problems served to introduce the use of Code rules and tables which apply to electrical wiring design. The circuits were designed based on the amount of load to be served and there was little emphasis on the type of load or the type of occupancy. Section 3-1 discusses the effects the type of load to be served has on branch-circuit calculations. Section 3-2 provides a similar discussion for feeder circuit calculations and also covers special feeder design problems involving parallel conductors, transformer circuits, and unbalanced loads on three-phase feeders.

The Code provides certain rules for each type of load such as lighting units, receptacles, motors, appliances, heating equipment, and air-conditioning units. To complicate matters further, the rules for a particular load may differ according to the type of occupancy. This chapter introduces the design techniques that are required to satisfy these various Code requirements, although the differences based on occupancy are not fully developed until Chapter 4 of the Guide.

Except for lighting and receptacle loads and for electric-range loads, the design of branch circuits is independent of the type of occupancy involved. These particular circuits in dwellings are governed by rules that do not apply to commercial or industrial buildings. The lighting and receptacle circuits discussed in this chapter would generally apply to any occupancy since lighting circuits and receptacle circuits are covered separately. On the other hand, the examples for electric ranges given in this chapter apply only to ranges in dwellings unless specifically stated otherwise. This subject is introduced in this chapter in order to reduce the number of new concepts to be presented in Chapter 4, which deals exclusively with dwelling occupancies.

The calculations for feeder and service circuits follow the same pattern. As with branch circuits, feeders supplying motors, most appliances, heating equipment, and air-conditioning equipment may be designed without regard to occupancy. Feeders supplying electric clothes dryers or electric ranges in dwellings are subject to rules that allow the application of demand factors to the feeder load to reduce the demand on the feeder.

3-1 BRANCH-CIRCUIT LOAD CALCULATIONS

The Code presents a large number of rules that pertain to the design of branch circuits. Figure 3-1 classifies the rules into those that apply to the type of circuit, the type of load, and finally the type of occupancy. The specific Code references are presented for each entry in the figure. As we shall see, individual branch circuits, general-purpose branch circuits, outside branch circuits, and other types of circuits each have specific rules that must be followed. The rules for appliances, electric ranges, heating equipment, motors, and other types of loads also are separated. Finally, the type of occupancy must be considered in order to determine the proper design rules that apply.

Despite these classifications and differences, a typical branch circuit contains at least the elements shown in Figure 3-2. The Code rules that apply to design indicate the method of determining the size or rating of the various elements.

The disconnecting means is not always required to be a separate switch as shown. In fact, a branch circuit supplying lighting or receptacle loads does not require a disconnecting means since such a rule for this is not mentioned in the Code. Appliance circuits require a means to disconnect the appliance from the branch circuit. The Code specifies the types of disconnecting means that can be used, but it does not specify the rating of the device. Circuits serving motor-driven appliances or motors require a disconnecting means with a rating that satisfies Code requirements.

422-20*
ARTICLE 422, PART D
ARTICLE 430

The location of the disconnecting means in the circuit must be such that it disconnects the *equipment* to be supplied from all ungrounded conductors of the circuit. Although it is shown in Figure 3-2 as disconnecting the entire branch circuit from the feeder, this is not a Code requirement.

An overcurrent device protects the circuit conductors and equipment from the extremely high current resulting from a short circuit between ungrounded conductors or a ground fault. For circuits that do not contain motors, the overcurrent protection also prevents excessive heating of the conductors caused by overloads by *tripping* or *blowing* and thus disconnecting the branch circuit from the feeder or service. Motor circuits require additional protection for the circuit conductors and the motor itself to prevent overloads.

ARTICLE 430

*References in the margins are to the specific applicable rules and tables in the *National Electrical Code.*

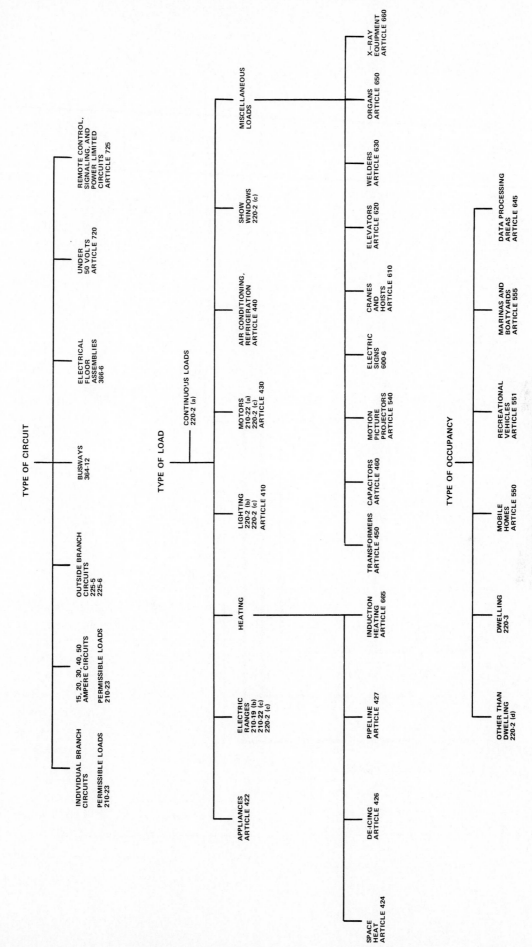

Figure 3-1. Branch-Circuit Classification

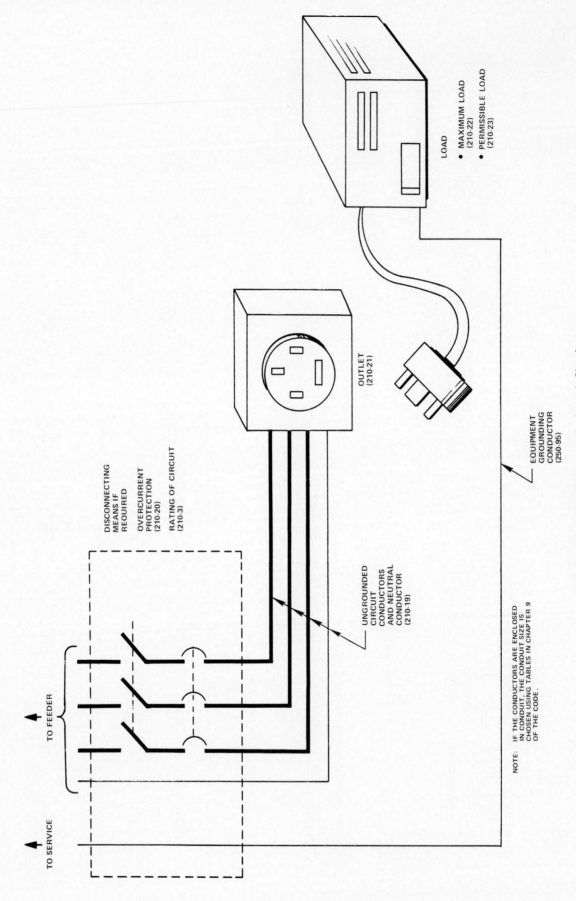

LOAD
- MAXIMUM LOAD (210-22)
- PERMISSIBLE LOAD (210-23)

OUTLET (210-21)

EQUIPMENT GROUNDING CONDUCTOR (250-95)

DISCONNECTING MEANS IF REQUIRED

OVERCURRENT PROTECTION (210-20)

RATING OF CIRCUIT (210-3)

UNGROUNDED CIRCUIT CONDUCTORS AND NEUTRAL CONDUCTOR (210-19)

TO FEEDER

TO SERVICE

NOTE: IF THE CONDUCTORS ARE ENCLOSED IN CONDUIT, THE CONDUIT SIZE IS CHOSEN USING TABLES IN CHAPTER 9 OF THE CODE.

Figure 3-2. Basic Elements of a Branch Circuit

40

The circuit conductor size is determined by the load current of the equipment served. Various rules explained in this chapter must be applied in order to determine the correct size.

The neutral conductor, if present, is treated differently from the ungrounded conductors when the circuit supplies an electric range in a dwelling. In other branch circuits no special rules apply to the size of the neutral conductor.   **210-19(b)** **220-19**

Figure 3-2 shows an equipment grounding conductor used to connect the non-current-carrying metal parts of equipment to the grounding terminal at the service. A separate conductor is not necessarily required if another suitable form of grounding conductor is provided; for example, the conduit that encloses the circuit conductors.   **250-91**

The Code also provides rules that govern the rating and the type of load that may be supplied by a branch circuit. The maximum load and the permissible load are specified in certain cases. These rules, which affect the design of the branch circuit, will be discussed in greater detail in this chapter.   **210-22, 210-23**

The branch-circuit examples in this chapter cover lighting and receptacle circuits, motor circuits, appliance circuits, circuits for heating equipment, and branch circuits for air-conditioning equipment. The separate discussions of these topics consider only a single type of load being supplied by the branch circuit. When several types of loads, such as lights and appliances, are supplied by one branch circuit, the branch circuit is said to supply a *mixed load*. The special rules for these loads are also treated in this chapter.

If a load operates continuously for more than 3 hours, it is considered a *continuous load*. The continuous load must not exceed 80% of the branch-circuit rating. The intent of this provision is to reduce the heating effect of a continuous load on the branch-circuit overcurrent device if it is not listed for continuous operation at 100% of its rating.   **ARTICLE 100** **220-3** **210-22(c)**

3-1.1 Lighting and Receptacle Branch Circuits

General-purpose branch circuits are provided in all occupancies to supply lighting outlets for illumination and receptacle outlets for small appliances and office equipment. When lighting circuits are separate from circuits that supply receptacles the Code provides rules for the design of each type of branch circuit.[1] This is usually the case in commercial and industrial occupancies.

Applicable Rules. The lighting load to be used in the branch-circuit calculations for determining the required number of circuits must be the larger of the values obtained by using one of the following: **220-3**

   a. The actual load
   b. A minimum load in volt-amperes or watts per square foot as specified in the Code

The Code provides a table that specifies the minimum unit load in volt-amperes per square foot of floor area based on outside dimensions for the occupancies listed. If the actual lighting load *is known* and if it exceeds the minimum determined by the volt-amperes per square foot basis, the actual load must be used because the Code specifies that branch-circuit conductors shall have an ampacity not less than the maximum load to be served. A bank building, for example, with 2000 square feet of floor space (outside dimensions of 40 feet by 50 feet) would have a minimum   **TABLE 220-3(b)** **TABLE 220-3(b)** **210-19** **TABLE 220-3(b)**

---

[1]Receptacle outlets of 20 amperes or less in a one-family or multifamily dwelling or in guest rooms of hotels and motels (except for special circuits) are considered as part of the lighting load.

lighting load of 7 kilovolt-amperes based on the 3.5 volt-ampere per square foot unit load specified by the Code. If the actual connected load were one hundred 150-watt lamps, or 15 kilowatts, the actual load, being the larger value, would be used to compute the required number of branch circuits.

210-23 *Branch Circuits for Lighting.* The Code permits only 15- or 20-ampere branch circuits to supply lighting units with standard lampholders. Branch circuits of greater than 20-ampere rating are permitted to supply fixed lighting units with heavy-duty lampholders in other than dwelling occupancies. In other words, branch circuits of greater than 20-ampere rating are *not* permitted to supply lighting units in dwellings.

In certain problems it is required to determine the number of branch circuits that are necessary to supply a given load. The number of branch circuits as determined by the load is

$$\text{number of circuits} = \frac{\text{total load in volt-amperes (or watts)}}{\text{capacity of each circuit in volt-amperes (or watts)}}$$

A 15-ampere, 120-volt circuit has a capacity of 15 amperes $\times$ 120 volts = 1800 volt-amperes. If the circuit is rated at 20 amperes, the capacity is 20 amperes $\times$ 120 volts = 2400 volt-amperes. By comparison, a 480-volt, three-phase circuit rated at 20 amperes has a capacity of $\sqrt{3} \times$ 480 volts $\times$ 20 amperes = 16 627 volt-amperes, a considerable increase.

Figure 3–3 presents the results of a calculation to determine the number of 120-volt, 20-ampere branch circuits to supply a 60 000 volt-ampere lighting load. The 20-ampere circuits have a capacity of 2400 volt-amperes. Thus, the number of circuits is

$$\frac{60\ 000\ \text{VA}}{2400\ \text{VA}} = 25 \text{ circuits}$$

If the number of lamps per circuit is known to be four hundred 150-watt lamps, two methods may be used to determine the result. When the watts per lamp is known and when the capacity of the circuit has been determined, the number of lamps per circuit is

$$\frac{\text{capacity of each circuit in watts}}{\text{watts per lamp}} = \frac{2400\ \text{W}}{150\ \text{W}}$$
$$= 16 \text{ lamps per circuit}$$

With each circuit supplying 16 lamps, the total number of circuits required would be

$$\frac{400\ \text{lamps}}{16\ \text{lamps/circuit}} = 25 \text{ circuits}$$

This problem may be checked by noting that each circuit ampacity of 20 amperes must not be exceeded. The current drawn by each lamp at 120 volts is

$$I = \frac{150\ \text{W}}{120\ \text{V}} = 1.25 \text{ A}$$

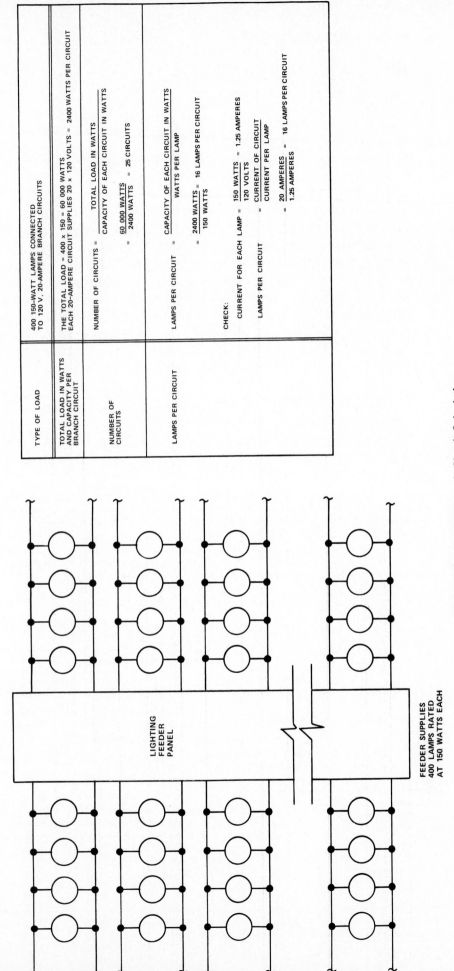

TYPE OF LOAD	400 150-WATT LAMPS CONNECTED TO 120 V, 20-AMPERE BRANCH CIRCUITS
TOTAL LOAD IN WATTS AND CAPACITY PER BRANCH CIRCUIT	THE TOTAL LOAD = 400 x 150 = 60 000 WATTS EACH 20-AMPERE CIRCUIT SUPPLIES 20 x 120 VOLTS = 2400 WATTS PER CIRCUIT
NUMBER OF CIRCUITS	NUMBER OF CIRCUITS = $\dfrac{\text{TOTAL LOAD IN WATTS}}{\text{CAPACITY OF EACH CIRCUIT IN WATTS}}$ = $\dfrac{60\ 000\ \text{WATTS}}{2400\ \text{WATTS}}$ = 25 CIRCUITS
LAMPS PER CIRCUIT	LAMPS PER CIRCUIT = $\dfrac{\text{CAPACITY OF EACH CIRCUIT IN WATTS}}{\text{WATTS PER LAMP}}$ = $\dfrac{2400\ \text{WATTS}}{150\ \text{WATTS}}$ = 16 LAMPS PER CIRCUIT CHECK: CURRENT FOR EACH LAMP = $\dfrac{150\ \text{WATTS}}{120\ \text{VOLTS}}$ = 1.25 AMPERES LAMPS PER CIRCUIT = $\dfrac{\text{CURRENT OF CIRCUIT}}{\text{CURRENT PER LAMP}}$ = $\dfrac{20\ \text{AMPERES}}{1.25\ \text{AMPERES}}$ = 16 LAMPS PER CIRCUIT

LIGHTING FEEDER PANEL

FEEDER SUPPLIES 400 LAMPS RATED AT 150 WATTS EACH

Figure 3-3.   Sample Branch-Circuit Calculation

The 20-ampere circuit, then, can supply

$$\frac{20 \text{ A}}{1.25 \text{ A/lamp}} = 16 \text{ lamps}$$

or 16 lamps as before. The confidence in the result is high since the answer has been determined several independent ways.

Table 3-1 summarizes the Code rules used in the branch-circuit calculations for lighting circuits. In the case of a show window, the code requires at least one receptacle outlet to be installed directly above a show window for each 12 linear feet of show-window length and specifies that the unit load for such outlets shall be considered as not less than 180 volt-amperes. The Code also requires, however, that a load of not less than 200 volt-amperes be included for each foot of length of the show window and permits this value to be used instead of the specified unit load per receptacle. The load used for branch-circuit calculations, then, becomes the largest of the unit load per receptacle, 200 volt-amperes per linear foot, or the actual load. The 200 volt-amperes per linear foot value is considered to be a continuous load value by the Code and, therefore, does not have to be increased by 25% if the load is continuous.

210-62

220-3(c)

220-12

220-3(c) EXCEPTION 3

CHAPTER 9
EXAMPLE 3

Branch Circuits for Receptacles. The major Code rules used in branch-circuit calculations for receptacle circuits are summarized in Table 3-2.

General-purpose receptacles are specified as a 180-volt-ampere load for each single or multiple receptacle. If a multioutlet assembly is used, the minimum load

220-3(c)

Table 3.1 Summary of Code Rules for Computing Branch-Circuit Lighting Loads

Type of Load	Method of Calculating Load Value	Branch-Circuit Rating	Number of Circuits Required	Code Ref.
General Illumination*	Larger of: • volt-amperes per square foot area of listed occupancies, or • actual load if known** (increase by 25% if continuous)	15- or 20 ampere circuit	15-ampere circuit: $$\frac{\text{total load (volt-amperes)}}{15 \text{ amperes} \times 120 \text{ volts}}$$ 20-ampere circuit: $$\frac{\text{total load (volt-amperes)}}{20 \text{ amperes} \times 120 \text{ volts}}$$ (If 120/240 volt, three-wire circuits are used, only half as many are needed.)	220-3(b)
Heavy-Duty Lampholders in Fixed Lighting Units	Larger of: • 600 volt-amperes per unit, or • actual load if known (increase by 25% if continuous)	30-, 40-, or 50-ampere circuit	$$\frac{\text{total load (volt-amperes)}}{\text{circuit rating (amperes)} \times \text{circuit voltage (volts)}}$$ (In circuits supplied by three-phase feeders, the circuit voltage is the phase-to-neutral voltage since lighting is a single-phase load.)	220-3(c)
Show Window Illumination	Larger of: • 180 volt-amperes per outlet for each outlet above show window (increase by 25% if continuous), or • 200 volt-amperes per linear foot, or • actual load if known (increase by 25% if continuous)	Rating depends on type of lampholder	— Note: If the receptacles above the show window are not for illumination, they are treated as part of the building receptacle load. The lighting load is then 200 volt-amperes per linear foot, minimum.	210-62 220-3(c) Exception 3

*If lighting load is inductive, the load value is based on the nameplate amperes of unit not the total lamp wattage.

**For recessed fixtures, use the maximum rating. Track lighting is covered in Section 410-102 of the Code.

Table 3-2. Summary of Code Rules for Computing Branch-Circuit Receptacle Loads

Type of Load	Method of Calculating Load Value	Branch-Circuit Rating	Number of Circuits Required	Code Ref.
General Receptacles	Larger of: • 180-volt-amperes per unit, or • actual load if known (increased by 25% if continuous)	15- or 20-ampere circuit	15-ampere circuit: $$\frac{\text{number of units} \times 180 \text{ volt-amperes}}{15 \text{ amperes} \times 120 \text{ volts}}$$ 20-ampere circuits: $$\frac{\text{number of units} \times 180 \text{ volt-amperes}}{20 \text{ amperes} \times 120 \text{ volts}}$$	220-3(c)(5)
Specific Outlets	Actual load	Any rating	–	220-3(c)(1)
Multioutlet Assemblies	a) 180 volt-amperes per each 5 feet for general loads b) 180 volt-amperes per each foot for appliance loads	15- or 20 ampere circuit	–	220-3(c) Exception 1

is 180 volt-amperes ($1\frac{1}{2}$ amperes at 120 volts) per each 5 feet of length. This factor must be increased to 180 volt-amperes per foot of length if several appliances are likely to be used simultaneously. The decision in this case is obviously a matter of judgement, although a multioutlet assembly in a machine shop or similar location would probably require the increased load capacity.

Continuous Loads. When the load is continuous, the load value used to select the branch-circuit rating must be increased by 25%. This assures that the total load will not exceed 80% of the branch-circuit rating. This Guide assumes industrial and commercial lighting loads are continuous. Since the lighting is connected to a multioutlet circuit, the rating of the circuit and the ampacity of the circuit conductors will be the same in most designs. **220-3(a)** **210-24**

Design Example. A design example is presented in Table 3-3 for an occupancy containing the following lighting and receptacle loads:

a. Floor area of 100 feet by 200 feet with 3.5 volt-amperes/square foot load
b. Four hundred 150-watt incandescent lamps for general illumination
c. 50-foot long show window
d. Special lighting load of one hundred 277-volt fluorescent lighting fixtures; each unit draws 2 amperes
e. Two hundred 120-volt duplex receptacles (not continuous)
f. 100 feet of multioutlet assembly along baseboards for appliance loads

In the example, 20-ampere rated branch circuits are used to supply the various loads, all of which are assumed to be continuous except the duplex receptacle circuits. The total number of circuits is to be determined.

The value used for the continuous general lighting load becomes either

$$100 \text{ ft} \times 200 \text{ ft} \times 3.5 \text{ VA/ft}^2 \times 1.25 = 87\ 500 \text{ VA}$$
(based on the given volt-amperes/sq ft unit load)

or

$$400 \text{ lamps} \times 150 \text{ W/lamp} \times 1.25 = 75\ 000 \text{ W}$$
(actual load)

The Code requires that the larger value of 87 500 volt-amperes be used in determining the required number of branch circuits. This requires a distribution panel to hold 37 two-wire, 120-volt circuits rated at 20 amperes each. Actual circuits need only be installed to serve the connected load. **220-3(b)** **220-4(d)**

220-3(c)   Since no other information is given for the show window, a load of 200 volt-amperes per linear foot must be used. The load for the 50-foot window is

$$200 \text{ VA/ft} \times 50 \text{ ft} = 10 \text{ kVA}$$

Five 20-ampere, 120-volt circuits will supply the load.

The special 277-volt lighting load is assumed to be for other than general illumination, perhaps for lighting a factory assembly area. The load is

$$100 \text{ units} \times 2 \text{ A/unit} \times 277 \text{ V} \times 1.25 = 69\ 250 \text{ VA}$$

Thirteen 277-volt circuits are required.

The calculations for the receptacles and multioutlet assembly are based on the rules referenced in the table. When the result for the number of circuits results in a fraction, the number is increased to the next larger integer. Sixteen 20-ampere circuits are required to supply the 36-kilovolt-ampere receptacle load. Ten additional 20-ampere circuits are required to supply the 21 563-volt-ampere load of the multioutlet assembly.

The total number of circuits is thirteen 277-volt circuits and 67 two-wire, 120-volt circuits (or 35 three-wire, 120/240-volt circuits).

Table 3-3. Design Example for Lighting and Receptacle Circuits

Type of Load	Load Value	Number of 20-ampere Branch-Circuits Required	Code Reference
General Lighting Load: Building 100 × 200 ft (3.5 VA/ft²) containing 400 bulbs (150 watts each)	Larger of: 100 ft × 200 ft × 3.5 volt-amperes/ft² × 1.25 = 87 500 volt-amperes, or 400 × 150 watts × 1.25 = 75 000 watts (neglect)	$\dfrac{87\ 500 \text{ volt-amperes}}{20 \text{ amperes} \times 120 \text{ volts}} = 36.46$ or thirty-seven 2-wire circuits or nineteen 3-wire circuits	220-3(b)
50-foot Show Window	50 feet × 200 volt-amperes/foot = 10 000 volt-amperes	$\dfrac{10\ 000 \text{ volt-amperes}}{20 \text{ amperes} \times 120 \text{ volts}} = 4.17$ or five 2-wire circuits, or three 3-wire circuits	220-3(c) Exception 3
277-Volt Lighting Load (100 Units rated at 2 amperes each)	100 units × 2 amperes/unit × 277 volts × 1.25 = 69 250 volt-amperes	$\dfrac{69\ 250 \text{ volt-amperes}}{20 \text{ amperes} \times 277 \text{ volts}} = 12.5,$ or thirteen 277-volt circuits	
Duplex Receptacles (200 units)	200 units × 180 volt-amperes/unit = 36 000 volt-amperes	$\dfrac{36\ 000 \text{ volt-amperes}}{20 \text{ amperes} \times 120 \text{ volts}} = 15$ fifteen 2-wire circuits, or eight 3-wire circuits	220-3(c)
Multioutlet Assembly (100 feet)	100 feet × 180 volt-amperes × 1.25 = 22 500 volt-amperes	$\dfrac{22\ 500 \text{ volt-amperes}}{20 \text{ amperes} \times 120 \text{ volts}} = 9.38$ or ten 2-wire circuits, or five 3-wire circuits	220-3(c) Exception 1

Total number of circuits is thirteen 277-volt circuits and either sixty-seven 2-wire, 120 volt circuits, or thirty-five 3-wire, 120/240 volt circuits.

QUIZ

(Closed-Book)
1. The branch-circuit loads specified by the Code for lighting and receptacles are considered:
   (a) Minimum loads
   (b) Maximum loads
   (c) The load to be served
2. The branch-circuit conductor ampacity must never be less than the maximum load to be served. True or false?
3. Lighting fixtures using standard bases must be supplied by 15- or 20-ampere circuits. True or false?
4. What is the total load for 10 feet of multioutlet assembly?
5. What is the computed branch-circuit load for a show window 20 feet in length?
6. Do you think that most industrial lighting and receptacle loads are continuous?
7. Are 30-ampere circuits permitted to supply lighting units in industrial occupancies?

(Open-Book)
1. How many two-wire, 120-volt, 15-ampere branch circuits are required to supply a show window that is 30 feet long?
2. An office building has an area of 5000 square feet. What is the computed lighting load if the unit load is 3.5 volt-amperes per square foot?
3. How many 120-volt, 20-ampere branch circuits are required to supply the computed lighting load for the office building in problem 2?
4. A small building has the following connected load:
   (a) 10-kilovolt-ampere load for incandescent lighting
   (b) 8-kilovolt-ampere fluorescent lighting load to be supplied by 30-ampere circuits
   (c) 10 feet of show window
   (d) 100 duplex receptacles (continuous)
   If the loads are continuous, find the number of branch circuits required. Use 20-ampere circuits except for fluorescent lighting load. The voltage of all the units is 120 volts.

3-1.2 Motor Branch Circuits

The Code devotes an entire article to the rules governing motor installations where branch circuits supplying motors are treated extensively. The design requirements for motor branch circuits are independent of the type of occupancy, which simplifies the design of motor circuits. Only continuous duty motors are considered here, but, other applications, such as intermittent duty, periodic duty, etc., are considered in the Code.

ARTICLE 430

430-22(a)

Applicable Rules. A basic motor branch circuit shown in Figure 3-4 contains the following elements:

ARTICLE 430

a. A motor *disconnecting means* to disconnect the motor and its controller from the branch circuit
b. *Overcurrent protection* designed to protect the branch-circuit conductors, motor controller, and motor against overcurrent caused by *short circuits and ground faults*
c. Branch-circuit *conductors*
d. A motor *controller* to start and stop the motor
e. Motor *overload protection* to protect the motor, motor controller, and branch-circuit conductors against excessive heating from motor overloads or failure to start. This is *running* protection.
f. The *motor* which represents the electrical load for the branch circuit

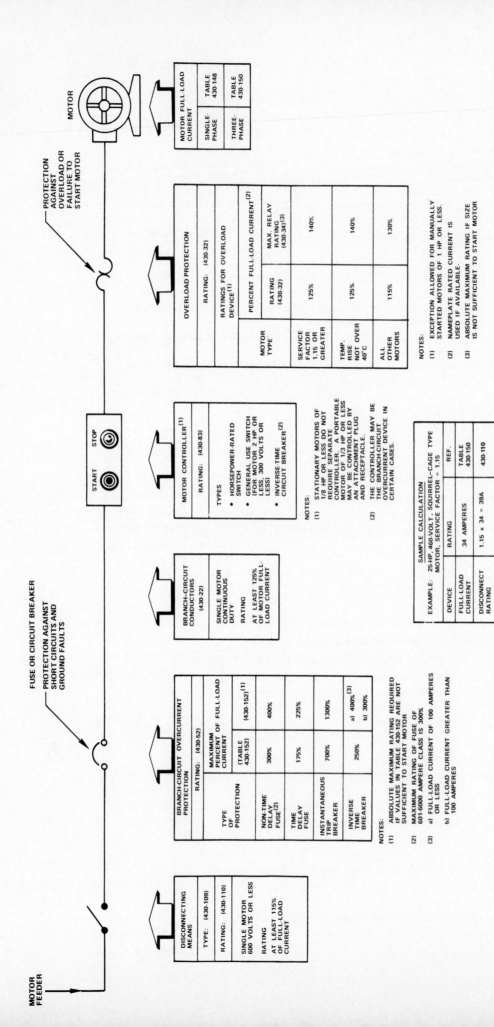

Figure 3-4. Rules for Single-Motor Branch Circuit

The rules shown in Figure 3-4 apply to standard motor circuits for either single-phase or three-phase motors. Motor circuits differ from other branch circuits because they always require a disconnecting means, a motor controller, and an overload protective device for the motor, in addition to a branch-circuit protective device. Also, the branch-circuit protective device may have a rating much larger than that of the circuit conductors in order to allow the motor to start.

The example of Figure 3-4 depicts the branch-circuit elements that would be required for a 460-volt, 25-horsepower, squirrel-cage motor with a service factor of 1.15. The service factor of 1.15 indicates that the motor may deliver 15% more horsepower than its 25-horsepower rating and not be damaged. Thus, the motor may draw more than its full-load current when delivering more than 25 horsepower. The table in Figure 3-4 shows the method used to determine the size or ratings of the various elements of the motor branch circuit with references to the applicable Code rules used.

The first step in the design of a motor branch circuit is to determine the full-load current drawn by the motor. The appropriate table in the Code should be used. According to the Code table, the 25-horsepower, 460-volt, three-phase motor draws 34 amperes. This current and the horsepower rating are used as the basis for determining the size or ratings of the other branch-circuit elements. The full-load nameplate current, if known, is used only in determining the rating of a separate overload protective device. **430-6** **TABLE 430-150** **430-6(a)**

The disconnecting means is required to be a motor circuit switch rated in horsepower or a circuit breaker for a 25-horsepower motor of the type in the example. The disconnecting means must have an ampere rating of at least 115% of the motor's full-load current. The horsepower rating must be 25 horsepower or more. The switch must disconnect both the motor and the controller from the branch-circuit conductors. In the example the disconnecting means disconnects the entire branch circuit from the motor feeder and therefore meets the requirement. Exceptions to the rule requiring a motor circuit switch or a circuit breaker for the disconnecting means are allowed for certain types of motors. **430-109** **430-110(a)** **430-103** **430-109**

Branch-circuit protection is provided by a circuit breaker or a set of fuses. The rating or setting of the branch-circuit protective device is chosen to be some percentage greater than the motor's full-load current in order to carry the inrush current while the motor is starting. The actual rating depends on the characteristics of the protective device itself as well as on the characteristics of the motor. The branch-circuit protective device provides protection against short circuits or ground faults.

Maximum ratings for branch-circuit protective devices based on full-load current are given for nontime-delay fuses, time-delay fuses, instantaneous trip circuit breakers, and inverse-time circuit breakers. A Code table lists the maximum rating or setting of the protective device for various types of motors. **430-52** **TABLE 430-152**

If a motor has a locked-rotor code letter marked on its nameplate indicating the inrush current when the motor attempts to start or the current drawn when the motor stalls, this code letter must be used to select the rating of the branch-circuit protective device. Since no code letter was given for the 25-horsepower, squirrel-cage motor in the example of Figure 3-4, it must be assumed that it had none; consequently, a nontime-delay fuse selected for the motor circuit may have a maximum rating of 300% of the full-load current. (According to the Code table, if the nameplate of the motor were marked with a code letter B, a nontime-delay fuse could have a maximum rating of only 250% of the full-load current.) **TABLE 430-152**

430-22   The branch-circuit *conductors* supplying a single motor must have an ampacity of at least 125% of the full-load current of the motor when the motor is operated as a continuous duty motor.

430-83   The motor *controller* for a 25-horsepower motor would normally be a controller with a rating of 25 horsepower. Certain exceptions are allowed depending on the type of controller.

430-32   The motor overload protective device is selected on the basis of:

a. Motor horsepower and starting method
b. Nameplate rated current or full-load current if nameplate rating is not known
c. Service factor or temperature rise

If this protection is provided by the manufacturer, it is not considered in the design of the branch circuit.

The Code classifies continuous duty motors for the purposes of determining the rating or type of overload protective device as follows:

a. More than 1 horsepower
b. One horsepower or less, manually started
c. One horsepower or less, automatically started

430-32   The 25-horsepower motor with a service factor of 1.15 or greater would require an overload protective device rated or set at not more than 125% of the motor nameplate full-load current rating. Since the nameplate rating is not given, the full-load current given in the Code tables should be used. The setting for the motor, then, must be no more than 1.25×34 amperes = 42.5 amperes. If an overload relay
430-34   were selected in accordance with the applicable Code rules but if it were insufficient in rating to allow the motor to start, the next standard size could be used, provided it did not exceed 140% of the full-load current.

Design Examples. Table 3-4 presents design examples for various types of single-phase motors. A clock motor; ¼-horsepower portable fan; ⅓-horsepower attic fan; 2-horsepower, 115-volt motor; and two 230-volt motors are considered. The characteristics and type of branch-circuit protective device for each motor are defined in the table. Each element of the branch circuits was chosen based on the applicable Code rules. The resulting branch circuits are depicted in Figure 3-5.[2]

430-109   The clock motor can be controlled and protected by the branch-circuit protective
430-81   device. A motor as small as a clock motor would not enter into branch-circuit design calculations because it would probably be connected by taps to a 15- or 20-ampere general-purpose branch circuit.

A portable fan connected by cord and plug would also not affect branch-circuit
430-32(b)   calculations since its use would not likely be foreseen. The fan would be plugged
430-81(c)   into an outlet of a 15- or 20-ampere general-purpose circuit and would be protected
430-109   by the branch-circuit protective device. The attachment plug and receptacle would
EXCEPTION 5   serve both as the controller and the disconnecting means.
430-32(b)(2)
The ⅓-horsepower attic fan protected by an inverse-time breaker qualifies as a motor of 1 horsepower or less which is manually started; but since it is permanently
TABLE 430-148   installed, it requires overload protection. The fan, with a full-load current of 7.2 amperes, could be connected to a 15- or 20-ampere branch circuit and would probably be controlled by a snap switch on the top floor of the building. An inverse-time

[2]The voltage range of the supply circuit can be 110 to 120 volts for 115-volt motors and 220 to 240 volts for 230-volt motors.

Table 3-4. Branch-Circuit Design Requirements for Various Single-Phase Motors

Motor Type	Type of Branch-Circuit Device	Full-Load Current (Table 430-148)	Minimum Rating of Disconnect (430-109)	Maximum Rating of Branch-Circuit Protective Device (Table 430-152)	Minimum Ampacity of Conductors (430-22)	Minimum Size of THW Copper Conductors (Table 310-16)	Controller Size (430-83)	Maximum Rating of Overload Device (430-32)
Small Electric Clock	Connect to 15- or 20-ampere branch circuit	—	—	—	—	—	Branch-circuit protective device	—
¼-hp Portable Fan	Connected by cord and plug to branch circuit	5.8 amperes	Attachment plug and receptacle	—	—	—	Attachment plug and receptacle	—
1/3-hp 115-Volt Attic Fan (Manual Start)	Inverse Time Circuit Breaker	7.2 amperes	• ac switch rated: 1.25 × 7.2 = 9.0 amperes • Circuit Breaker: 1.15 × 7.2 = 8.3 amperes	2.5 × 7.2 = 18 amperes (Attach to 15- or 20-ampere branch circuit)	1.25 × 7.2 = 9 amperes	No. 14	ac switch rated: 1.25 × 7.2 = 9 amperes	1.15 × 7.2 = 8.3 amperes
2-hp, 115-Volt Motor (Nameplate Current = 20 amperes)	Nontime-delay Fuse	24 amperes	• 2-hp switch • ac switch rated: 1.25 × 24 = 30 amperes	3 × 24 = 72 amperes	1.25 × 24 = 30 amperes	No. 10	• 2-hp switch • ac switch rated: 1.25 × 24 = 30 amperes	1.15 × 20 = 23 amperes
2-hp, 230-Volt Motor	Time Delay Fuse	12 amperes	• 2-hp switch or circuit breaker • ac switch rated: 1.25 × 12 = 15 amperes	1.75 × 12 = 21 amperes	1.25 × 12 = 15 amperes	No. 14	• 2-hp switch	1.15 × 12 = 13.8 amperes
10-hp, 230-Volt Motor	Inverse Time Circuit Breaker	50 amperes	• 10-hp switch • Circuit breaker rated: 1.15 × 50 = 58 amperes	2.5 × 50 = 125 amperes	1.25 × 50 = 62.5 amperes	No. 6	10-hp switch	1.15 × 50 = 58 amperes

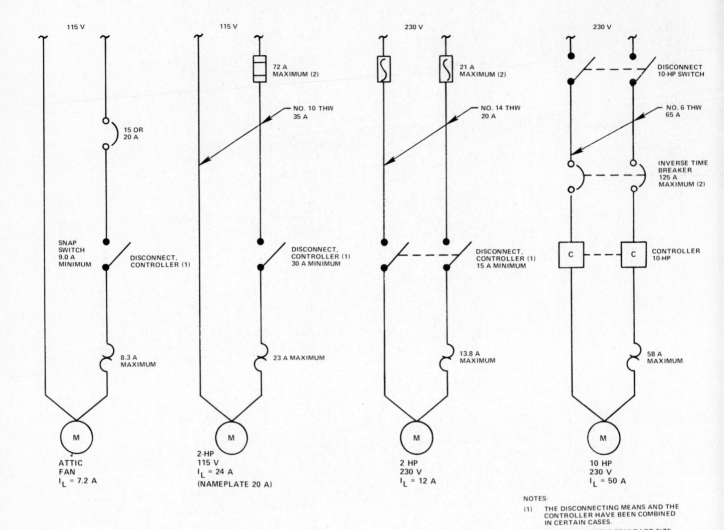

Figure 3-5. Branch Circuits for Single-Phase Motors

TABLE 430-148 circuit breaker rated at 250% of the full-load current is allowed to protect the branch circuit. This rating may be needed for motors that start slowly under heavy load, but a fan or similar motor would not require such an oversized circuit breaker. The Code only specifies the maximum rating allowable, not the most practical value for a specific application. The overload protection could be provided by a device set at not more than 1.15 × 7.2 amperes = 8.3 amperes.

TABLE 430-148 A 2-horsepower, 115-volt motor has a full-load current of 24 amperes. When the nameplate current is specified as 20 amperes, the overload protective device must 430-32 be set at no more than 1.15 × 20 amperes = 23 amperes. A nontime-delay fuse TABLE 430-152 serving as the branch-circuit protective device may be as large as 3 × 24 amperes = 430-83 72 amperes. The controller can be a 2-horsepower motor controller or an ac snap 430-109 switch rated at least 1.25 × 24 amperes = 30 amperes. The disconnecting means has the same rating.

A 2-horsepower, 230-volt motor whose branch circuit is protected by a time-TABLE 430-148 delay fuse has requirements similar to those of the 2-horsepower, 115-volt motor. TABLE 430-152 The full-load current of 12 amperes allows the fuse to be rated as large as 1.75 × 430-109, 12 amperes = 21 amperes. The disconnecting means and the controller can be 430-83 EXCEPTIONS motor circuit switches rated in terms of horsepower or ac general-use snap switches rated at 1.25 × 12 amperes = 15 amperes or more.

The final example is a 10-horsepower, 230-volt motor protected by an inverse-time circuit breaker. The full-load current of the motor is 50 amperes; therefore, the circuit breaker can be set at no more than 2.5 × 50 amperes = 125 amperes. An overload protective device cannot be set at greater than 1.15 × 50 amperes = 58 amperes. The controller and the disconnecting means should be rated at 10 horsepower or more. If the disconnecting means is a circuit breaker, it must be rated at least 1.15 × 50 amperes = 58 amperes. The controller would probably be a standard NEMA size controller for a 10-horsepower motor.[3]

TABLE 430-148
TABLE 430-152

430-110

The conductor ampacity in each example must be at least 125% of the full-load current of the motor. The size of the THW copper conductors is taken from the appropriate Code table. When the size of the branch-circuit overcurrent protective device is not a standard size, the next larger standard size may be used.

430-22
TABLE 310-16
430-52

Three-Phase Motors. Table 3-5 presents design examples for various three-phase motors. The motors included as examples represent different types of motor characteristics as well as horsepower ratings. The branch circuits required to supply the motors are shown in Figure 3-6.

Three-phase motors must meet similar Code rule requirements as single-phase motors. With three-phase motors, the Code letter, service factor, temperature rise, and motor type (synchronous, squirrel-cage, or wound-rotor) must be observed carefully in order to use the Code tables and rules correctly.

The first example is a 5-horsepower, 230-volt, three-phase motor with code letter A and a full-load current of 15.2 amperes. The branch circuit may be protected by an inverse-time circuit breaker set at only 1.5 × 15.2 amperes = 22.8 amperes because of the low locked-rotor current (Code letter A) of this motor. A 5-horsepower motor controller and a 5-horsepower motor circuit switch would be used to control and disconnect the motor, respectively. The maximum setting of the overload device is 1.15 × 15.2 amperes = 17.5 amperes.

TABLE 430-150

TABLE 430-152,
TABLE 430-7(b)

A 25-horsepower, 460-volt, three-phase motor has a full-load current of 34 amperes. A service factor of 1.2 allows the overload protective device to be set as high as 1.25 × 34 amperes = 42.5 amperes. A nontime-delay fuse protecting the motor circuit could be rated as high as 3 × 34 amperes = 102 amperes. A 25-horsepower motor starter with a disconnecting means could be used with this motor.

TABLE 430-150,
430-32
TABLE 430-152

A 30-horsepower, 460-volt, three-phase wound-rotor motor with a temperature rise of 40°C has characteristics that affect both the rating of the branch-circuit protective device and the overload device. Since the motor has a wound rotor, a nontime-delay fuse used as the branch-circuit protective device cannot be set higher than 150% of the full-load current, or 60 amperes. The 40°C temperature rise allows the overload device to be set as high as 1.25 × 40 amperes = 50 amperes. A 30-horsepower motor starter with a disconnecting means would probably be used to control this motor.

TABLE 430-150
TABLE 430-152
430-32

As a final example, a 50-horsepower, 440-volt, three-phase squirrel-cage motor with a full-load current of 65 amperes is presented. The nontime-delay fuses protecting the branch circuit may be rated as high as 195 amperes. The overload device cannot be set higher than 1.15 × 65 amperes = 75 amperes. A 50-horsepower motor starter would be used.

TABLE 430-150
TABLE 430-152
430-32

---

[3]NEMA: National Electric Manufacturers Association. The organization publishes standards for devices such as motor controllers.

Table 3-5. Branch-Circuit Design Requirements for Various Three-Phase Motors

Motor Type	Type of Branch-Circuit Protective Device	Full-Load Current (Table 430-150)	Minimum Rating of Disconnect (430-109)	Maximum Rating of Protective Device (Table 430-152)	Minimum Ampacity of Conductors (430-22)	Minimum Size of THW Copper Conductors (Table 310-16)	Controller Size (430-83)	Maximum Rating of Overload Device (430-32)
5-hp, 230-Volt, Three-phase Motor, Code Letter A	Inverse time breaker	15.2 amperes	• 5-hp switch • Circuit breaker rated: 1.15×15.2 = 17.5 amperes	1.5×15.2 = 22.8 amperes	1.25×15.2 = 19 amperes	No. 12	5 hp	1.15×15.2 = 17.5 amperes
25-hp, 460-Volt, Squirrel Cage, Service Factor 1.2	Nontime-delay Fuse	34 amperes	• 25-hp switch • Circuit breaker rated; 1.15×34 = 39 amperes	3×34 = 102 amperes	1.25×34 = 42.5 amperes	No. 8	25 hp	1.25×34 = 42.5 amperes
30-hp, 460-Volt, Wound-rotor, 40° C Rise	Nontime-delay Fuse	40 amperes	• 30-hp switch • Circuit breaker rated: 1.15×40 = 46 amperes	1.5×40 = 60 amperes	1.25×40 = 50 amperes	No. 8	30 hp	1.25×40 = 50 amperes
50-hp, 440-Volt, Squirrel Cage	Nontime-delay Fuse	65 amperes	• 50-hp switch • Circuit breaker rated: 1.15×65 = 75 amperes	3×65 = 195 amperes	1.25×65 = 81.25 amperes	No. 4	50 hp	1.15×65 = 75 amperes

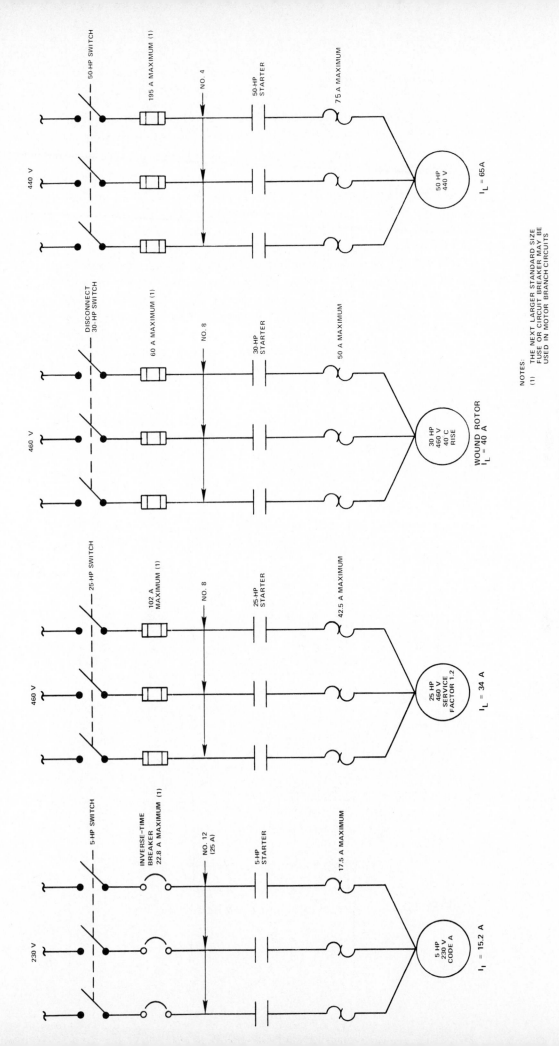

Figure 3-6. Branch Circuits for Three-Phase Motors Used as Examples

NOTES:
(1) THE NEXT LARGER STANDARD SIZE FUSE OR CIRCUIT BREAKER MAY BE USED IN MOTOR BRANCH CIRCUITS

430-22
430-52
The branch-circuit conductors are selected based on a required ampacity of 125% of the full-load current of the motor. The next larger standard branch-circuit protective device rating may be used if the calculated rating does not correspond to a standard size.

Two Motors Supplied by a Single Branch Circuit. The example shown in Figure 3-7 illustrates two 2-horsepower, 460-volt, three-phase motors supplied by one branch circuit. An inverse-time circuit breaker is used to protect the circuit.

TABLE 430-150
The Code table lists the full-load current for each motor as 3.4 amperes. If both motors are provided with individual overload protection, they may both be connected
430-53(b) to a 15-ampere circuit, which is the smallest circuit rating recognized by the Code. The branch-circuit protective device must protect the smaller motor (either one in
TABLE 430-152 this case). Each motor is allowed to have an inverse-time circuit breaker rated at 2.5 × 3.4 amperes = 8.5 amperes; the next higher standard size is 15 amperes. The branch-circuit protective device rated 15 amperes may protect *both* motors.

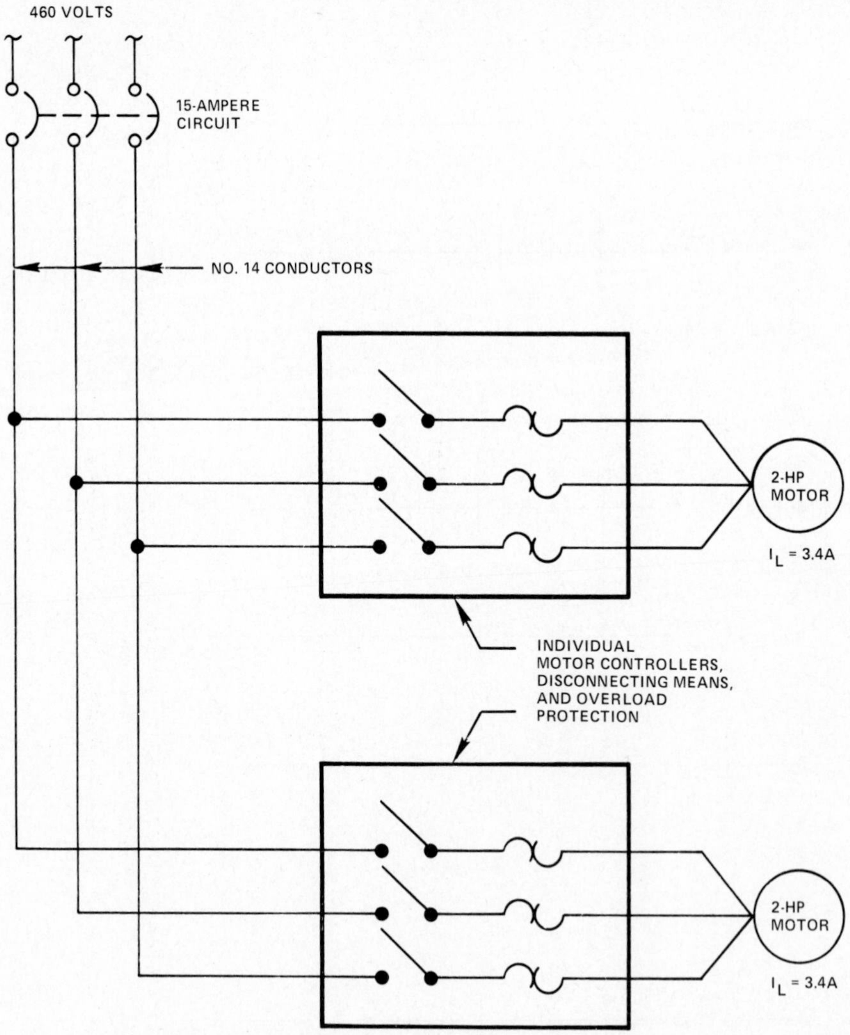

Figure 3-7. Two Motors Supplied by a Single Branch Circuit

The branch-circuit conductors must be sized to carry a load of at least the sum   **430-24**
of the full-load currents plus 25% of the highest rated motor current, or

$$
\begin{array}{l}
\quad 3.4 \ \text{A} \\
+\ 3.4 \ \text{A} \\
+\ \underline{\ .85} \ \text{A} \quad (.25 \times 3.4) \\
\quad 7.65 \ \text{A}
\end{array}
$$

The 15-ampere circuit with No. 14 conductors satisfies this requirement.

QUIZ
(Closed-Book)

1. Define:
   (a) Circuit breaker
   (b) Disconnecting means
   (c) Motor-circuit switch
2. Draw a diagram of a motor branch circuit and define each element of the circuit.
3. The motor full-load current from Code tables is *not* used to determine the rating of which element of a motor branch circuit if a rating is given on the nameplate?
   (a) Ampacity of conductors
   (b) Branch-circuit protective device
   (c) Motor overload protection
4. In general, the overload protective device cannot be rated or set at more than what percentage of the motor nameplate full-load current?
   (a) 115%
   (b) 130%
   (c) 150%
5. A 10-horsepower motor requires what type of disconnecting means?
6. A 3-horsepower motor may be controlled by a general-use switch. True or false?
7. Any automatically started motor requires overload protection. True or false?

(Open-Book)

1. A 20-horsepower, 230-volt, three-phase motor requires what rating for the following branch-circuit elements?
   (a) Branch-circuit conductors
   (b) Overload relay
2. If the motor in problem 1 had a marked temperature rise of 40°C, what is the maximum setting of the overload relay?
3. What size TW copper branch-circuit conductors are required for a 3-horsepower 230-volt, three-phase motor?
4. Design the branch circuit for a 25-horsepower, 480-volt, three-phase, wound-rotor motor with a temperature rise of 40°C. Use THW aluminum conductors in conduit and a nontime-delay fuse to protect the branch circuit.
5. Design the branch circuit for a 10-horsepower, 230-volt, single-phase motor, Code letter F. Use TW copper conductors and an inverse-time breaker to protect the branch circuit.

3-1.3 Individual Appliance Branch Circuits

The discussion that follows deals with individual branch circuits that supply single household or commercial appliances except those used for heating or air conditioning. Specific rules for electric ranges and other cooking equipment are covered separately.

Applicable Rules. For purposes of branch-circuit design, appliances can be classified as follows:

a. Motor-operated appliances
b. Nonmotor appliances
c. Appliances containing motors and other loads such as heaters and lights

ARTICLE 430
If a single appliance is motor operated, such as a garbage disposal or a trash compactor, the rules for motors must be applied as well as the rules for appliances. This implies that the branch circuit has a disconnecting means, a motor controller, and an overload protective device in addition to the branch-circuit overcurrent protective device. Although every appliance must have a disconnecting means, the disconnecting means for some motor operated appliances must satisfy the more stringent Code rules for motor-circuit disconnecting means.

ARTICLE 422
210-22
250-42
The Code distinguishes between cord- and plug-connected appliances and appliances which are permanently connected. Also, appliances that are fastened in place (fixed) require application of rules not considered for portable appliances. Small portable appliances are not considered in branch-circuit design since they will be plugged into outlets of general-purpose branch circuits.

422-5
210-22
Basically, an appliance branch circuit that serves a single appliance must be rated to serve the load. The rules for determining the load of the single appliance and selecting the branch-circuit protective device, however, vary according to the type of appliance served. The major Code rules for individual appliance branch circuits are summarized in Table 3-6. The table demonstrates the method of calculating the load and determining the rating of the protective device and provides examples for each case. The branch circuits for three of the cases are illustrated in Figure 3-8.

Table 3-6. Summary of Branch-Circuit Design Rules for Various Types of Appliances

Type of Appliance	Branch-circuit Load	Rating of Protective Device	Examples	Code Ref.
Motor only	1.25 × Full load current	Rules for motor circuits apply	½-hp disposal = 1.25 × 9.8A = 12.25A connect to 15A or 20A branch circuit	210-22(a) Article 430
Single non-motor-operated appliance	Ampere rating of unit (125% of ampere rating if continous)	If unit is rated 13.3 amperes or more, the protective device rating shall not exceed 150% of appliance rating	10-kVA hot plate at 240V • = 41.7 52.1A if continuous • Maximum breaker size is 1.5 × 41.7 = 62.5 Use 60A standard size	210-22 220-3 422-27(e)
Appliance with motor larger than 1/8 hp with other loads (fixed appliance)	125% of largest motor load plus the sum of other loads.	Rules for motor circuits apply unless appliance has protective device rating marked on nameplate	240-volt machine with 5A motor and 10A lighting load 1.25 × 5A + 10A = 16.25A connect to 20A circuit	210-22(a) Article 430 422-6
Water-heater	Continuous if capacity is 120 gallons or less	Sized to protect conductors	4.5-kVA heater, 240V = 18.75 × 1.25 =23.4A connect to 25 or 30A branch circuit	422-14

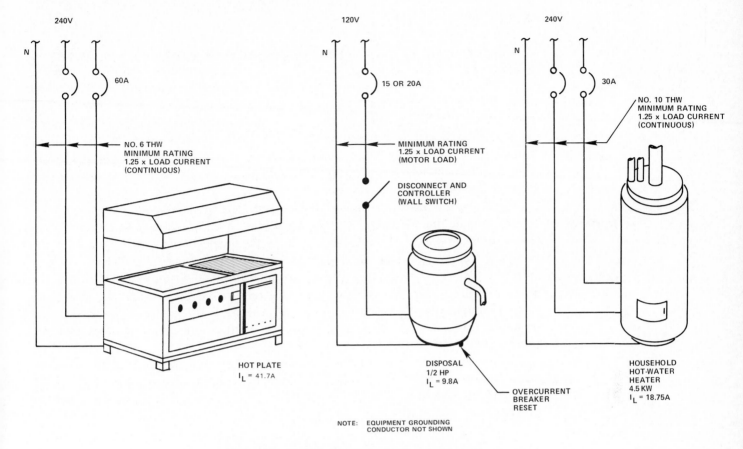

240V

N

60A

NO. 6 THW
MINIMUM RATING
1.25 x LOAD CURRENT
(CONTINUOUS)

HOT PLATE
I_L = 41.7A

120V

N

15 OR 20A

MINIMUM RATING
1.25 x LOAD CURRENT
(MOTOR LOAD)

DISCONNECT AND
CONTROLLER
(WALL SWITCH)

DISPOSAL
1/2 HP
I_L = 9.8A

OVERCURRENT
BREAKER
RESET

240V

N

30A

NO. 10 THW
MINIMUM RATING
1.25 x LOAD CURRENT
(CONTINUOUS)

HOUSEHOLD
HOT-WATER
HEATER
4.5 KW
I_L = 18.75A

NOTE: EQUIPMENT GROUNDING
CONDUCTOR NOT SHOWN

Figure 3-8. Branch Circuits for Various Appliances

For motor operated appliances, the branch-circuit conductors must have an ampacity of at least 125% of the full-load current of the motor. The branch-circuit protective device is selected based on the motor's full-load current. The ½-horse-power disposal unit of Figure 3-8 draws 9.8 amperes at 120 volts. A 15- or 20-ampere circuit would supply the unit. **210-22(a)** **TABLE 430-152** **TABLE 430-148**

A nonmotor operated appliance must be supplied by a branch circuit with a rating at least as large as the ampere rating of the appliance. If the appliance is a continuous load, its rating cannot exceed 80% of the branch-circuit rating. A 10-kilowatt, 240-volt hot plate, for example, requires a current of 10 000 watts ÷ 240 volts = 41.7 amperes. If the hot plate is a continuous load, the branch-circuit must have a rating of at least 1.25 × 41.7 amperes = 52.1 amperes. This requires a 60-ampere circuit. The conductors need to be sized for only 41.7 amperes. However, No. 6 THW conductors must be used with a 60-ampere overcurrent protective device. **210-22** **240-3**

If a fixed appliance contains a motor of ⅛-horsepower or larger in combination with other loads, the branch-circuit load is 125% of the motor load plus the sum of the other loads. A machine with a 5-ampere motor and a 10-ampere lighting load, for example, requires a branch circuit rated at least 1.25 × 5 amperes + 10 amperes = 16.25 amperes. **210-22(a)**

The Code requires that fixed storage water heaters having a capacity of 120 gallons or less be considered a continuous load. Individual branch circuits serving such water heaters must therefore have an ampacity rating of 125% of the water heater load. A 4.5-kilowatt, 240-volt water heater has a branch-circuit demand load of 1.25 × 18.75 amperes = 23.4 amperes. **422-14(b)**

Electric Ranges in Dwellings. Branch-circuit computations for household electric ranges and other such cooking equipment are more involved than for other appliances and are therefore discussed separately and in more detail. The Code allows the branch-circuit load for a single household electric range to be computed from a table of demand factors that reduces the demand load. The Code also allows the neutral conductor to be sized on the basis of 70% of the ampacity of the branch-circuit rating, but it may be no smaller than No. 10. Again, these rules apply only to household electric ranges and other household cooking appliances and not to such equipment in commercial use.

TABLE 220-19
NOTE 4

210-19(b)
EXCEPTION 1

TABLE 220-19

TABLE 310-16
210-19(b)

250-60

TABLE 220-19
NOTE 1

Several examples are given in Table 3-7 and illustrated in Figure 3-9.

Any range rated between 8¾ kilowatts and 12 kilowatts may be considered to require a demand load of only 8 kilowatts for the purposes of branch-circuit design. The load for the standard 120/240-volt circuit is 8000 watts ÷ 240 volts = 33.3 amperes. This load requires a branch-circuit rating of 40 amperes, which is the smallest rating allowed by the Code. The required neutral ampacity would be .7 × 40 amperes = 28 amperes. A No. 10 TW or THW neutral conductor could be used. In the case of an electric range, the neutral conductor may also serve as the equipment grounding conductor.

For large size ranges, the maximum demand of 8 kilowatts for a single range must be increased by 5% for each additional kilowatt the range rating exceeds 12 kilowatts. A 15-kilowatt range, as an example, constitutes a maximum demand load of:

$$
\begin{aligned}
8000 \text{ W (first)} &= 8000 \text{ W} \\
(.05 \times 8000 \text{ W}) \times (15-12) & \\
= 400 \text{ W} \times 3 \text{ (excess)} &= \underline{1200 \text{ W}} \\
& \quad\ \, 9200 \text{ W}
\end{aligned}
$$

Table 3-7. Branch-Circuit Calculations for Electric Ranges and Cooking Equipment in Dwellings

Type of Appliance	Connected Load	Demand Load	Branch-Circuit Rating	TW Copper Conductor Size
8-3/4 kW to 12 kW Range	$\frac{12,000W}{240V} = 50A$	$\frac{8000W}{240V} = 33.3A$	40A standard	Two No. 8 One No. 10
15 kW Range	$\frac{15,000W}{240V} = 62.5A$	Load = 8000W + 400W × [15-12] = 9200W $\frac{9200W}{240V} = 38.3A$	40A standard	Two No. 8 One No. 10
8 kW Range	$\frac{8000W}{240V} = 33.3A$	Load = 8 kW × .8 = 6400W $\frac{6400W}{240V} = 26.7A$	30A standard	Three No. 10
Oven 6 kW	$\frac{6000W}{240V} = 25A$	$\frac{6000W}{240V} = 25A$	30A standard	Three No. 10
10 kW Cooking Unit and Two Wall-Mounted Ovens (4 kW)	$\frac{10kW+4kW+4kW}{240V} =$ 75A	Load = 8000W + 400W × [18-12] = 10 400W = 10 400W $\frac{10\,400W}{240V} = 43.3A$	50A standard	Two No. 6 One No. 8

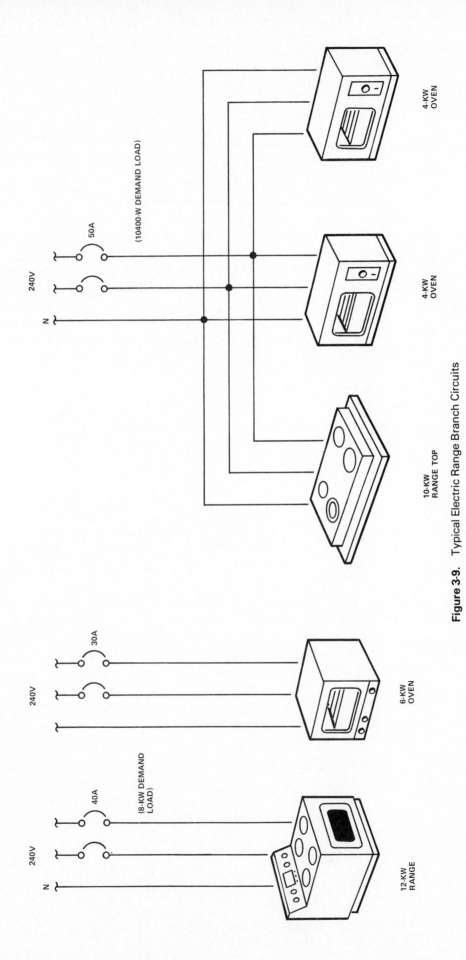

240V

50A

(10400-W DEMAND LOAD)

N

4-KW OVEN

4-KW OVEN

10-KW RANGE TOP

240V

30A

6-KW OVEN

240V

40A

(8-KW DEMAND LOAD)

N

12-KW RANGE

Figure 3-9. Typical Electric Range Branch Circuits

61

The branch-circuit ampacity must be at least

$$\frac{9200\ \text{W}}{240\ \text{V}} = 38.3\text{A (40 A minimum)}$$

210-19(b) No. 8 TW copper conductors with an ampacity of 40 amperes can be used for the ungrounded conductors. The neutral conductor ampacity must be at least

$$.7 \times 40\ \text{A} = 28\ \text{A}$$

TABLE 220-19 NOTE 4 since the Code allows the neutral to be reduced to 70% of the branch-circuit rating for a household range. In this case, a No. 10 TW copper conductor would be used for the neutral.

A somewhat different situation arises when an individual branch circuit supplies a counter-mounted cooking unit and one or two wall-mounted ovens. In this case, the nameplate ratings of the units are added together and the total load is treated as one range. Consider as an example three such units served by an individual branch circuit with the following ratings:

$$\text{Counter-mounted cooking unit} = 10\ \text{kW}$$
$$\text{Two wall-mounted ovens} = 4\ \text{kW each}$$

The total connected load is considered as 18 kilowatts (10 + 4 + 4). The maximum demand load on the branch circuit will then be

$$
\begin{aligned}
8000\ \text{W (first)} &= 8000\text{W} \\
.05 \times 8000\ \text{Watts} \times (18\text{-}12)\ \text{(excess)} &= 2400\text{W} \\
\hline
&\quad 10\ 400\ \text{watts}
\end{aligned}
$$

The load in amperes is 10 400 watts ÷ 240 volts = 43.3 amperes. The circuit **TABLE 220-19 COLUMN C** may be supplied by two No. 6 TW copper ungrounded conductors and a No. 8 neutral conductor.

An 8-kilowatt electric range is subject to an 80% demand factor and represents a load of 6400 watts. The load of 26.7 amperes may be supplied by No. 10 TW conductors.

A single oven or single cooking unit, such as the 6-kilowatt oven shown, must be considered at the full nameplate rating. The oven load is 25 amperes, which requires a 30-ampere branch circuit.

Commercial Electric Ranges. An electric range used for commercial cooking requires a branch circuit sized by the full-rated current of the range. No reduction in the ampacity of the ungrounded conductors or the neutral conductor is allowed.

A 20-kilowatt, three-phase electric range supplied by a 208Y/120-volt circuit draws a current of

$$1.25 \times \frac{20\ 000\ \text{W}}{\sqrt{3} \times 208\ \text{V}} = 69.3\ \text{A}$$

if the range is a continuous load. A 70- or 75-ampere circuit is required using No. 4 THW copper conductors for the branch circuit.

QUIZ

(Closed-Book)   **1.** Define:
  (a) Appliance
  (b) Cooking unit
  (c) Appliance branch circuit

2. What household appliances are considered continuous?

3. The demand factors in the Code range table apply to a single counter-mounted cooking unit. True or false?

4. For a single household electric range rated above 12 kilowatts, the maximum demand load for the branch circuit is 8 kilowatts plus for each kilowatt above 12 kilowatts:
  (a) 5% of excess above 12 kilowatts
  (b) 400 watts
  (c) 5% of 12 kilowatts

5. Electric ranges in a restaurant are permitted to have a branch-circuit demand factor applied. True or false?

6. If an appliance has a motor load of 10 amperes and a heating load of 20 amperes, what is the required ampacity of the branch-circuit conductors?

(Open-Book)   **1.** Determine the branch-circuit load for household ranges with the following ratings:
  (a) 10 kilowatts
  (b) 12 kilowatts
  (c) 14 kilowatts
  (d) 5 kilowatts

2. Design the branch circuit for a 1-horsepower, 230-volt, single-phase food chopper.

3. Determine the branch circuit required to supply a 6-kilowatt clothes dryer. Use THW aluminum conductors.

4. What size THW copper branch-circuit conductors are required to supply a 9-kilowatt counter-mounted cooking unit and an 8-kilowatt oven on the same circuit? The appliances require a 120/240-volt circuit and are located in a home.

5. A 10-kilowatt load is supplied by a 480Y/277-volt circuit using type MI copper conductors. What is the required branch-circuit rating and conductor size? (MI is mineral-insulated metal-sheathed cable with a temperature rating of 85°C.)

3-1.4 Branch Circuits for Space-Heating Equipment

This subsection covers branch circuits that supply fixed electric equipment used for space heating.

Applicable Rules. Individual branch circuits are permitted to supply fixed electric space-heating equipment of any rating. If several heating units, such as baseboard heaters, are supplied by one branch circuit, the branch circuit must be rated at 15, 20, or 30 amperes.   **424-3(a)**

The branch circuits supplying fixed electric space heaters must be rated at 125% of the total load. The load may include both resistive heating elements and motors for distributing the heated air.   **424-3(b)**

Design Examples for Space-Heating Equipment. Examples of the application of branch-circuit requirements rules for fixed electric space-heating equipment are shown in Figure 3-10.

A 4500-watt, 240-volt heater requires a branch circuit rated at least

$$\frac{4500 \text{ W}}{240 \text{ V}} \times 1.25 = 23.4 \text{ A}$$

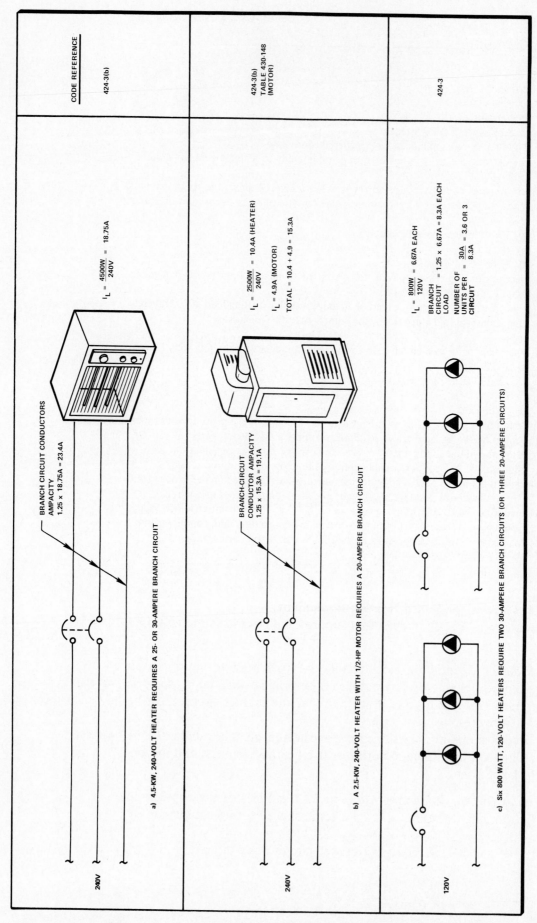

CODE REFERENCE

424-3(b)

$I_L = \dfrac{4500W}{240V} = 18.75A$

BRANCH CIRCUIT CONDUCTORS
AMPACITY
1.25 x 18.75A = 23.4A

a) 4.5-KW, 240-VOLT HEATER REQUIRES A 25- OR 30-AMPERE BRANCH CIRCUIT

240V

424-3(b)
TABLE 430-148
(MOTOR)

$I_L = \dfrac{2500W}{240V} = 10.4A$ (HEATER)

$I_L = 4.9A$ (MOTOR)

TOTAL = 10.4 + 4.9 = 15.3A

BRANCH-CIRCUIT
CONDUCTOR AMPACITY
1.25 x 15.3A = 19.1A

b) A 2.5-KW, 240-VOLT HEATER WITH 1/2-HP MOTOR REQUIRES A 20-AMPERE BRANCH CIRCUIT

240V

424-3

$I_L = \dfrac{800W}{120V} = 6.67A$ EACH

BRANCH
CIRCUIT = 1.25 x 6.67A = 8.3A EACH
LOAD

NUMBER OF
UNITS PER = $\dfrac{30A}{8.3A}$ = 3.6 OR 3
CIRCUIT

c) Six 800 WATT, 120-VOLT HEATERS REQUIRE TWO 30-AMPERE BRANCH CIRCUITS (OR THREE 20-AMPERE CIRCUITS)

120V

Figure 3-10. Branch Circuits for Space-Heating Equipment

This heating unit would require a 25- or 30-ampere branch circuit.

A heater with a resistive element of 2500 watts and a ½-horsepower motor requires a branch circuit rated at 125% of the total load. The heater draws

$$\frac{2500 \text{ W}}{240 \text{ V}} = 10.4 \text{ A}$$

According to the applicable Code table, the motor has a full-load current of 4.9 **TABLE 430-148** amperes; therefore, the branch-circuit rating must be at least

$$(10.4 + 4.9) \text{ A} \times 1.25 = 19.1 \text{ A}$$

which can be supplied by a 20-ampere circuit.

A group of six 800-watt, 120-volt heaters requires several branch circuits. The branch-circuit load per heater unit is

$$\frac{800 \text{ W}}{120 \text{ V}} \times 1.25 = 8.3 \text{ A}$$

Since the maximum allowable branch-circuit rating is 30 amperes, only three heating **424-3** units could be served by one 30-ampere circuit. The number of units per circuit is given by

$$\frac{\text{circuit rating (amperes)}}{\text{unit load (amperes)}} = \frac{30 \text{ A}}{8.3 \text{ A}} = 3.6$$

or three units per circuit. The number of circuits required is

$$\frac{\text{number of units}}{\text{units per circuit}} = \frac{6}{3} = 2 \text{ circuits}$$

Thus, two 30-ampere circuits are required. The load could also be supplied by three 20-ampere circuits, in which case two units would be connected to each circuit.

QUIZ
(Closed-Book)

1. A space-heating unit is rated at 20 amperes. What is the required ampacity of the branch-circuit conductors?
2. What are the permissible branch-circuit ratings if the circuit is to supply two or more fixed electric space heaters?
3. The branch-circuit conductors for a resistance heater and motor unit must be rated at what percentage of the load?
   (a) 125% of the total load
   (b) 125% of the motor load plus 100% of the heating load
   (c) 100% of the total load

(Open-Book)

1. A 10-kilowatt, 240-volt space heater requires what size branch circuit?
2. Design the branch circuit for a 5-kilowatt heater with a 3-horsepower blower motor. The circuit operates at 240 volts and the conductors are type THW copper.
3. Determine the various branch circuits that could supply four 1.2-kilowatt, 240-volt, fixed space heaters.

3-1.5 Branch Circuits for Air-Conditioning Equipment

Electric motor-driven air-conditioning and refrigerating equipment is subject to Code rules that differ from those that apply to conventional motors. The discussion that follows is separated into two parts: the first part covers room air conditioners and refrigerating appliances; the second part covers the branch-circuit design requirements for equipment containing hermetic refrigerant motor-compressors.

440-60

210-23(a)

440-62

Room Air Conditioners and Refrigerating Appliances. Room air conditioners and household refrigerators are usually connected by cord and attachment plug to receptacle outlets of general-purpose branch circuits. The rating of any such appliances must not exceed 80% of the branch-circuit rating if connected to a 15-, 20-, or 30-ampere general-purpose branch circuit. The rating of cord and plug connected room air conditioners must not exceed 50% of the branch-circuit rating if lighting units and other appliances are also supplied.

440-13

440-63

430-81(c)

Figure 3-11 illustrates the application of the rules for air-conditioning and refrigerating units that are also considered appliances. The disconnecting means is allowed to be the attachment plug and receptacle. In some cases, the attachment plug and receptacle may also serve as the controller, or the controller may be a switch that is an integral part of the unit. The required overload protective device may be supplied as an integral part of the appliance and need not be included in the branch-circuit calculations.

Hermetic Refrigerant Motor-Compressors. The basic branch-circuit elements for air-conditioning and refrigerating equipment are identical to those of motor circuits. A disconnecting means, a controller, an overload protection device, and a branch-circuit protective device are required in each branch circuit. The Code modifies certain rules for motor circuits since the motor-compressor combination is hermetically sealed within the same enclosure with the refrigerant. The motor, therefore, exhibits characteristics different from those of a conventional motor that is cooled by air.

440-4

440-6

The branch-circuit design for hermetic motor-compressors must be based on the nameplate rated-load current and locked-rotor current values. If the branch-circuit selection current is also given on the nameplate, that value must be used in determining the size or rating of the disconnecting means, the branch-circuit conductors, the controller, and the branch-circuit protective device.[4]

Table 3-8 and Figure 3-12 present the branch-circuit design for a motor-compressor rated at 25 amperes. The 240-volt, single-phase motor-compressor used as an example has the following nameplate data:

$$\text{Rated-load current} = 25 \text{ amperes}$$
$$\text{Locked-rotor current} = 100 \text{ amperes}$$

440-12

If a circuit breaker is used as the disconnecting means, it must have a rating of at least 115% of the rated-load current. If the disconnecting means is to be a switch rated in horsepower, then a horsepower rating must be determined for the hermetic motor-compressor that is equivalent to a conventional motor with comparable values of full-load current and locked-rotor current. The equivalent horsepower rating is first determined by comparing the nameplate rated-load current and the

440-12(a)(2)

locked-rotor current with the values given in the appropriate Code tables for con-

[4]In the example given here it is assumed that the branch-circuit selection current is not given and the nameplate rating is used.

ventional motors. If the results disagree, the larger of the listed horsepower ratings must be used. The equivalent motor that corresponds to the nameplate rated-load of 25 amperes is a 5-horsepower motor with a full-load current listed as 28 amperes. The equivalent motor with a locked-rotor current of 100 amperes or greater is a 3-horsepower motor with a locked-rotor current of 102 amperes. Since the larger value must be used, an equivalent rating of 5 horsepower will be used for the purpose of sizing the disconnecting means according to the Code requirement.

TABLE 430-148
TABLE 430-151

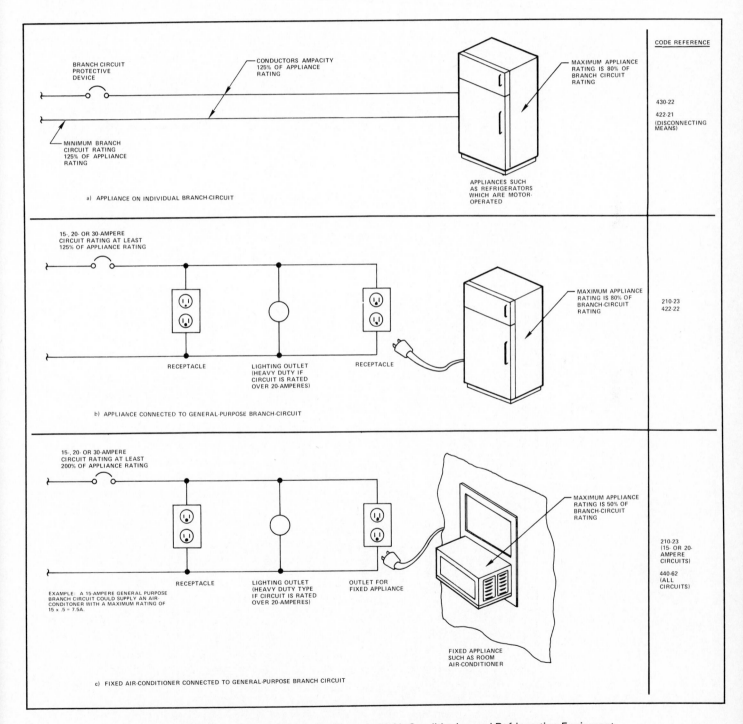

Figure 3-11. Branch Circuits for Household Air Conditioning and Refrigerating Equipment

Table 3-8. Branch-Circuit Calculations For Hermetic Motor-Compressors

Motor	Selection of Disconnecting Means (440-12)	Minimum Rating of Disconnect (440-12)	Maximum Rating of Short-Circuit and Ground Fault Protection (440-22(a))	Minimum Ampacity of Conductors (440-32)	Minimum Rating of Controllers (440-41)	Maximum Rating of Overload Protection (440-52)
Hermetic Refrigerant Motor-Compressor	Nameplate rated-load current and locked-rotor current (If disconnecting means and controller are rated in hp find equivalent hp for motor)	Amperes: 115% of nameplate rated-load current HP:— Larger value determined by either hp or hp equivalent of locked-rotor current	175% of rated load current (225% absolute maximum to start motor)	125% of rated load current	Equal to rated current and locked rotor current	Overlay relay = 140% of nameplate Time-delay or Inverse-time device = 125% of nameplate
240V Motor-Compressor. Nameplate current 25A, Locked rotor current 100A	Equivalent hp 25A ⟶ 5 hp 100A ⟶ 3 hp	Circuit breaker 1.15 × 25 = 28.8A or 5-hp switch	1.75 × 25A = 43.75A (use 40A)	1.25 × 25A = 31.25A	5 hp	Relay 1.40 × 25A = 35A

68

The branch-circuit short-circuit and ground-fault protection device must be **440-22(a)**
rated or set at not more than 175% of the nameplate rated-load current for hermetic
motor-compressors, but it may be increased to a maximum of 225% if the specified
value is insufficient to handle the motor-compressor's starting current. The use of
the next larger standard size circuit breaker is *not* permitted for hermetic motor-
compressors as was the case for conventional motors. The maximum rating that may
be used for the motor-compressor in the example is, therefore, 1.75×25 amperes =
43.75 amperes. A standard 40-ampere circuit breaker would be used for the branch
circuit.

The branch-circuit conductors must have an ampacity of at least 125% of the **440-32**
nameplate rated-load current. In the example the required minimum ampacity of the
conductors would be 1.25×25 amperes = 31.25 amperes for the compressor. No. 8
THW copper conductors are used for the circuit.

The controller for a hermetic motor-compressor must have a continuous duty **440-41(a)**
full-load current rating and locked-rotor current rating of not less than the nameplate rated-load current and the locked-rotor current of the motor-compressor,
respectively. When these ratings are marked on the controller, the selection is based
on the motor-controller's nameplate values. If the controller is rated in terms of
horsepower (one or both current ratings not given), the appropriate full-load and
locked-rotor current tables for conventional motors are used to determine equivalent
horsepower rating, as was done to select the disconnecting means. The motor-
compressor in the example would require a 5-horsepower controller.

Overload protection may be provided by a separate *overload relay* set to trip at **440-52**
not more than 140% of the motor-compressor's nameplate rated-load current. A
fuse or *inverse-time circuit breaker* could be used (and may also serve as the branch-

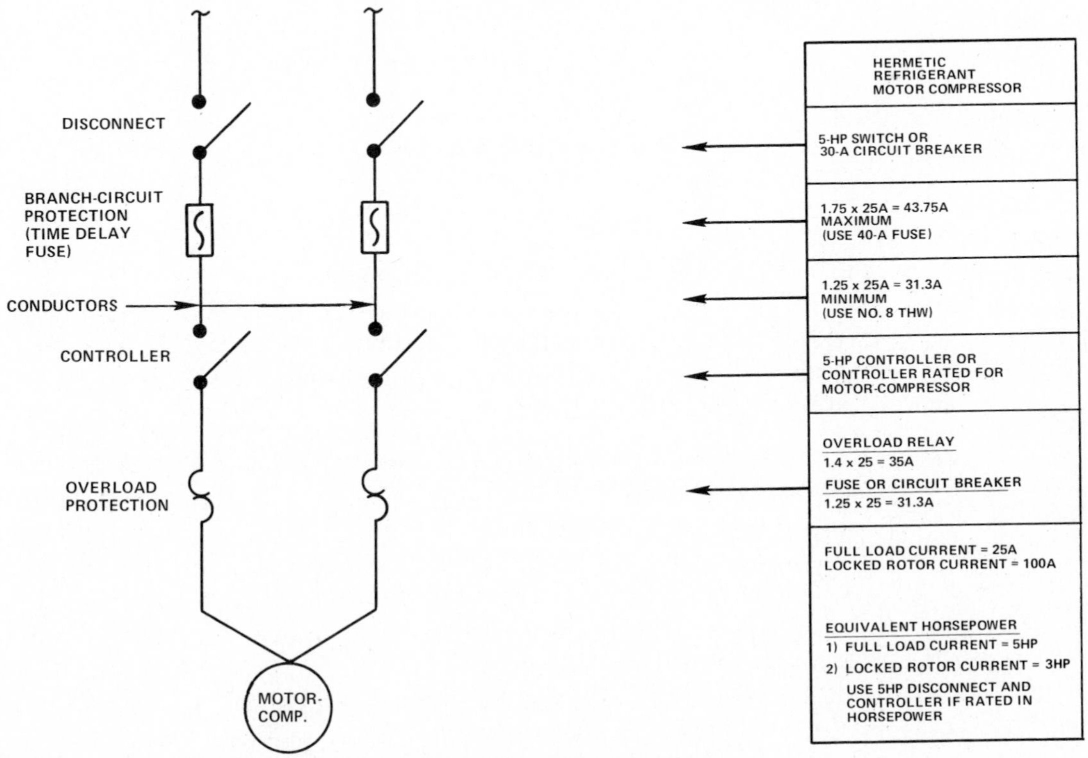

Figure 3-12. Branch Circuit for Motor-Compressor

circuit protective device) if it is rated at not more than 125% of the rated-load current. The motor-compressor in the example has a rated-load current of 25 amperes and would require an overload relay rated or set at not more than 1.4×25 amperes = 35 amperes. A fuse or inverse-time circuit breaker rated or set at not more than 1.25×25 amperes = 31.25 amperes could also be used.

QUIZ

(Closed-Book)

1. A hermetic motor-compressor has a rated-load current of 100 amperes. What are the allowable ratings of the following branch-circuit elements?
   (a) Branch-circuit conductors
   (b) Branch-circuit protective device
   (c) Disconnecting means
   (d) Overload protection if a fuse
2. The locked-rotor current of a hermetic motor-compressor is used to select which of the following items?
   (a) Overload protection
   (b) Motor-compressor controller
   (c) Branch-circuit protective device
3. A room air conditioner is to be attached to an individual branch circuit. If the load is 12 amperes, what is the minimum size branch circuit required?
4. What is the largest rated room air conditioner unit that may be attached to a 15-ampere branch circuit that also serves other receptacles?
5. If a motor-compressor does not have the locked-rotor current marked on the nameplate, what value must be assumed for the locked-rotor current?
6. How must the disconnecting means for a motor-compressor be rated?

(Open-Book)

1. Design the branch circuit for a hermetic motor-compressor with a rated-load current of 20 amperes. Use THW copper conductors and an overload relay. Assume that the controller is supplied with the motor.
2. What are the current rating and locked-rotor current rating of a 460-volt, 10-horsepower controller for a three-phase motor-compressor?
3. Determine the rating of the disconnecting means and the controller for a 480-volt, three-phase motor-compressor with a nameplate full-load current of 124 amperes and a locked-rotor current of 730 amperes.
4. If the motor in problem 3 had a locked-rotor current of 750 amperes, what would the required horsepower rating of the disconnecting means become?

3-1.6 Mixed Loads on Branch Circuits

The discussion in this chapter thus far has been concerned with rules for individual branch circuits supplying a single type of load. The material that follows deals with considerations made necessary when two or more loads of different types are connected to a single branch circuit.

Appliance Loads Supplied by General-Purpose Branch Circuits. When appliances are supplied by a general-purpose branch circuit or if an appliance branch circuit has more than one outlet, the ratings of the appliances that may be connected to the circuit are restricted. The standard 15-, 20-, 30-, 40-, and 50-ampere branch circuits may supply several outlets, but the loads that are permitted to be attached are specified by the Code.[5] The permissible loads for standard branch circuits are summarized in Table 3-9 and are illustrated in Figure 3-13. As indicated, the rating of a cord- and plug-connected appliance supplied by a 15- or 20-ampere branch

210-3
210-23

[5]For purposes of this discussion, the terms "utilization equipment" and "appliance" are considered to be synonymous. Multioutlet circuits of greater than 50-ampere rating are allowed in restricted usage for industrial establishments.

Table 3-9. Permissible Loads on Branch Circuits With Two or More Outlets

Circuit Rating	Permissible Load	Rating of Cord- and Plug-Connected Appliance	Rating of Fixed Appliance
15- or 20-ampere (general-purpose branch circuit)	Lighting units, other utilization equipment, or combinations	80% of branch-circuit rating.	50% of branch-circuit rating
30 ampere	Fixed lighting units with heavy-duty lampholders (not in dwellings)	80% of branch-circuit rating	Not specified
	Utilization equipment		
40 or 50 ampere	Fixed lighting units with heavy-duty lampholders (not in dwellings)	Not specified	Not specified
	Infrared heating units or other utilization equipment (not in dwellings)		
	Fixed cooking appliances		

circuit may not exceed 80% of the branch-circuit rating. For a 20-ampere branch circuit, for example, the maximum permissible load is .8 × 20 amperes = 16 amperes. The load of a fixed appliance connected to 15- or 20-ampere general-purpose branch circuits may not exceed 50% of the rating of the circuit.

The rules for mixed loads on branch circuits are highly dependent on the type of load to be served. As an example, assume that an occupancy has the following continuous loads for which branch circuits are to be designed, as illustrated in Figure 3-14:

a. Two 115-volt, 7-ampere air-conditioning window units
b. Twelve 120-volt receptacles (continuous)

If 20-ampere branch circuits were used for the receptacle load, the number of two-wire circuits required would be

$$\frac{1.25 \times 12 \text{ outlets} \times 180 \text{ VA/outlet}}{20 \text{ A} \times 120 \text{ V}} = 1.125 \qquad \textbf{220-3(c)}$$

or two circuits. The circuit shown in Figure 3-14 is a single three-wire circuit which is equivalent to two two-wire circuits.

The receptacle circuits can accommodate the air-conditioning units if certain restrictions are observed. The window air conditioner units are connected by cord and attachment plug. The load may not exceed 50% of the rating of a branch circuit **440-62(c)** if lighting units or other appliances are also supplied. One window unit can be connected to each of the two 20-ampere receptacle circuits since each 7-ampere load represents less than 50% of the circuit rating. The marked current rating on the nameplate is used to determine the load and no additional 25% is needed to compute the air-conditioning load. **210-7(c)**

In the case of the receptacles, an equipment grounding conductor is required. The required conductor size would be No. 12 for the 20-ampere circuits. **TABLE 250-95**

Motors and Other Loads on One Branch Circuit. In some cases, it is desirable to connect a motor and other loads to a single branch circuit. The Code allows such **430-53**

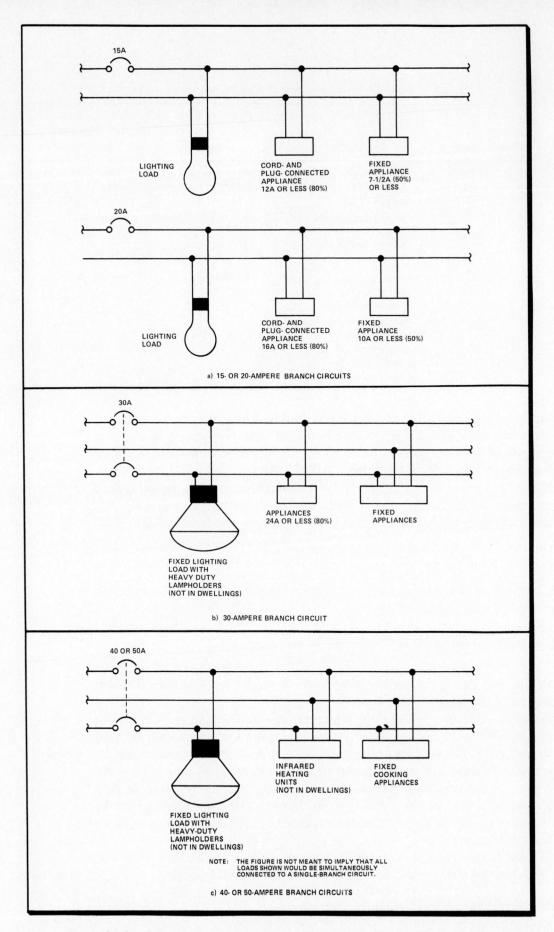

Figure 3-13. Permissible Loads for Branch Circuits with Two or More Outlets

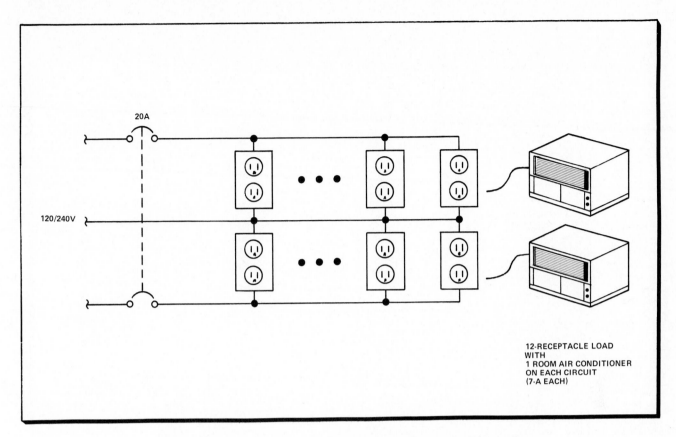

20A

120/240V

12-RECEPTACLE LOAD
WITH
1 ROOM AIR CONDITIONER
ON EACH CIRCUIT
(7-A EACH)

Figure 3-14. Branch Circuits for Room Air Conditioners and Other Loads

combinations of mixed loads under certain specified conditions pertaining to the characteristics and protection of the motors. Many possibilities of mixed motor and nonmotor loads exist, but only the more common situations will be discussed here, including:

a. A motor on a general-purpose branch circuit
b. A motor and other loads on a single branch circuit

A motor may be connected to a general-purpose branch circuit. Such a motor could be included as part of an appliance that would have an overall rating that included the motor load or it could be a separate appliance. If the motor is a separate appliance, such as a ⅓-horsepower, 115-volt disposal that draws 7.2 amperes, **TABLE 430-148** it must have individual running overcurrent protection. A wall switch could act as both the disconnecting means and the controller. The general-purpose branch circuit must be rated at least 20 amperes since the disposal is a fixed appliance and represents a load of 1.25 × 7.2 amperes = 9 amperes. The Code rules are satisfied since the total **210-23** rating of fixed appliances does not exceed 50% of the branch-circuit rating when lighting units or other utilization equipment, or both, are supplied.

If a single branch circuit supplies a motor and other loads but is not a general-purpose branch circuit, the branch-circuit design is based on 125% of the full-load current of the motor plus the current drawn by the other loads on the circuit. Figure 3-15 illustrates the design rules for the circuit elements.

Figure 3-16 illustrates a 240-volt circuit that supplies a 5-horsepower, 230-volt, single-phase motor and a 2400-watt resistive load. The full-load current for the mo- **TABLE 430-148** tor is listed in the appropriate Code table as 28 amperes. The motor requires a

73

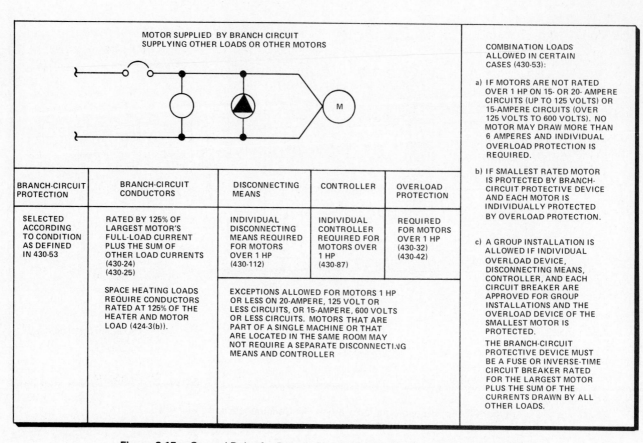

MOTOR SUPPLIED BY BRANCH CIRCUIT
SUPPLYING OTHER LOADS OR OTHER MOTORS

BRANCH-CIRCUIT PROTECTION	BRANCH-CIRCUIT CONDUCTORS	DISCONNECTING MEANS	CONTROLLER	OVERLOAD PROTECTION
SELECTED ACCORDING TO CONDITION AS DEFINED IN 430-53	RATED BY 125% OF LARGEST MOTOR'S FULL-LOAD CURRENT PLUS THE SUM OF OTHER LOAD CURRENTS (430-24) (430-25) SPACE HEATING LOADS REQUIRE CONDUCTORS RATED AT 125% OF THE HEATER AND MOTOR LOAD (424-3(b)).	INDIVIDUAL DISCONNECTING MEANS REQUIRED FOR MOTORS OVER 1 HP (430-112)	INDIVIDUAL CONTROLLER REQUIRED FOR MOTORS OVER 1 HP (430-87)	REQUIRED FOR MOTORS OVER 1 HP (430-32) (430-42)
		EXCEPTIONS ALLOWED FOR MOTORS 1 HP OR LESS ON 20-AMPERE, 125 VOLT OR LESS CIRCUITS, OR 15-AMPERE, 600 VOLTS OR LESS CIRCUITS. MOTORS THAT ARE PART OF A SINGLE MACHINE OR THAT ARE LOCATED IN THE SAME ROOM MAY NOT REQUIRE A SEPARATE DISCONNECTING MEANS AND CONTROLLER		

COMBINATION LOADS ALLOWED IN CERTAIN CASES (430-53):

a) IF MOTORS ARE NOT RATED OVER 1 HP ON 15- OR 20- AMPERE CIRCUITS (UP TO 125 VOLTS) OR 15-AMPERE CIRCUITS (OVER 125 VOLTS TO 600 VOLTS). NO MOTOR MAY DRAW MORE THAN 6 AMPERES AND INDIVIDUAL OVERLOAD PROTECTION IS REQUIRED.

b) IF SMALLEST RATED MOTOR IS PROTECTED BY BRANCH-CIRCUIT PROTECTIVE DEVICE AND EACH MOTOR IS INDIVIDUALLY PROTECTED BY OVERLOAD PROTECTION.

c) A GROUP INSTALLATION IS ALLOWED IF INDIVIDUAL OVERLOAD DEVICE, DISCONNECTING MEANS, CONTROLLER, AND EACH CIRCUIT BREAKER ARE APPROVED FOR GROUP INSTALLATIONS AND THE OVERLOAD DEVICE OF THE SMALLEST MOTOR IS PROTECTED.

THE BRANCH-CIRCUIT PROTECTIVE DEVICE MUST BE A FUSE OR INVERSE-TIME CIRCUIT BREAKER RATED FOR THE LARGEST MOTOR PLUS THE SUM OF THE CURRENTS DRAWN BY ALL OTHER LOADS.

Figure 3-15. General Rules for Branch Circuits Supplying Motors and Other Loads

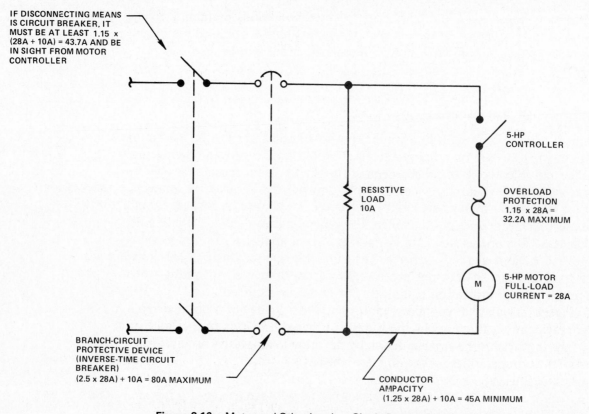

IF DISCONNECTING MEANS IS CIRCUIT BREAKER, IT MUST BE AT LEAST 1.15 x (28A + 10A) = 43.7A AND BE IN SIGHT FROM MOTOR CONTROLLER

5-HP CONTROLLER

RESISTIVE LOAD 10A

OVERLOAD PROTECTION 1.15 x 28A = 32.2A MAXIMUM

5-HP MOTOR FULL-LOAD CURRENT = 28A

BRANCH-CIRCUIT PROTECTIVE DEVICE (INVERSE-TIME CIRCUIT BREAKER) (2.5 x 28A) + 10A = 80A MAXIMUM

CONDUCTOR AMPACITY (1.25 x 28A) + 10A = 45A MINIMUM

Figure 3-16. Motor and Other Load on Single Branch Circuit

74

5-horsepower controller and overload protection rated or set at not more than 1.15 × 28 amperes = 32.5 amperes. The conductor ampacity must be at least 125% of the motor load plus the 10 amperes drawn by the resistive load (2400 ÷ 240 volts = 10 amperes). The required conductor ampacity is, therefore, **430-25**

$$1.25 \times 28\,A + 10\,A = 45\,A$$

The branch-circuit protective device for the branch circuit is designed to allow the motor to start while carrying the current drawn by the other load. When an inverse-time circuit breaker is used, the maximum rating is 250% of the full-load current of the motor plus the current drawn by the other load, or **430-53(c)**

$$2.5 \times 28A + 10\,A = 80\,A \text{ maximum}$$

The disconnecting means to disconnect both motor and resistive load may be a circuit breaker rated at not less than 115% of the full-load current of the motor plus the resistive load, or **430-109** **430-110**

$$1.15 \times (28\,A + 10\,A) = 43.7\,A$$

The branch-circuit circuit breaker could serve this purpose if it were within sight from the motor controller. **430-102**

QUIZ

(Closed-Book)
    1. A fixed appliance connected to a 15- or 20-ampere branch circuit may have what maximum rating?
    2. What loads may a 30-ampere branch circuit supply in (a) any occupancy and in (b) a dwelling?
    3. What size branch circuit is required for a window air conditioner with a nameplate rating of 8 amperes if the circuit also supplies other outlets?
    4. How is the branch-circuit current determined when a motor and another load are supplied by one circuit?
    5. A circuit breaker that acts as a disconnecting means for a combination load must be rated at:
        (a) 115% of the total load
        (b) 125% of the motor load plus the other load
        (c) 115% of the motor load plus the other load

(Open-Book)
    1. Determine the size of THW copper conductors to supply a load consisting of a 10-horsepower motor and a 1000-watt resistive unit on one branch circuit. The circuit operates at 240 volts.

3-1.7 Branch-Circuit Design Summary

This subsection will demonstrate suggested techniques for performing branch-circuit calculations for circuit design. The examples used are intended only to represent types of branch-circuit calculations that might be required; they do not cover all Code rules applicable to branch-circuit design.

Procedure. The chart in Figure 3-17 shows a step-by-step procedure for design of branch circuits. The loads are first separated by types and the load current in

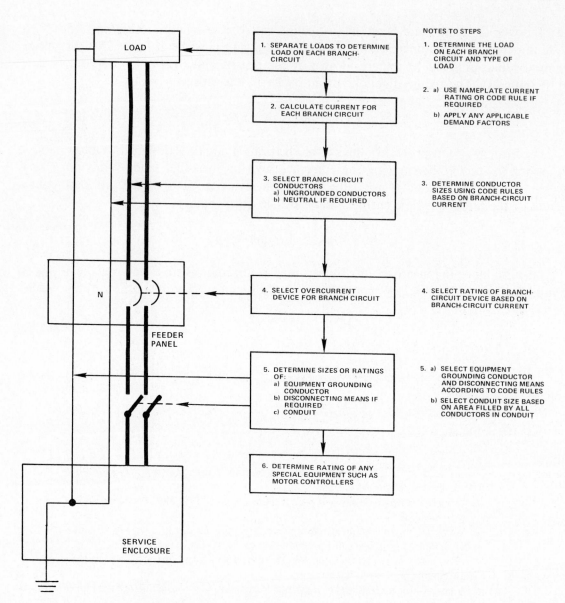

Figure 3-17. Branch-Circuit Calculation Procedure

amperes is calculated for each load. The circuit conductors and the branch-circuit

220-3 protective device are selected next. The Code requires that the branch-circuit rating

210-22 be not less than the computed load including continuous loads. Next, the conductors

210-19(a) are selected to carry the load and be protected by the overcurrent protective device. Since the conductor ampacities given in the Code are for continuous loading, theoretically the conductor ampacity could be 25% less than the rating of the circuit supplying only continuous loads. In practice, this is likely to occur only in a large-capacity industrial circuit.

As an example, assume that a 390-ampere load on an individual branch circuit is considered continuous. The branch-circuit conductors must carry at least 390 amperes. However, if the overcurrent protective device is not listed for operation at 100% of its rating, the required setting cannot be less than 390 × 1.25 = 487.5

240-6 amperes. The next standard setting is 500 amperes. Type THW copper conductors

TABLE 310-16 of size 600 MCM with an ampacity of 420 amperes could be used to supply the load but they cannot be protected at 500 amperes since 450 amperes is the next higher standard size above 420 amperes. Type THW conductors of size 700 MCM (460 amperes) could be protected by the 500-ampere overcurrent device. Thus, conductors with an ampacity of 460 amperes protected by an overcurrent device set at 500 amperes will meet Code requirements to protect the ungrounded conductors and also avoid overheating the overcurrent device. If a neutral conductor is present, it would not have to be increased in size to meet requirements stated as necessary for protection of the *ungrounded* conductors. **240-3**

The equipment grounding conductor is selected from the appropriate Code table based on the rating or setting of the overcurrent device protecting the circuit. The equipment grounding conductor specified in that table could be larger than the circuit conductors in a motor circuit due to the allowed rating of the branch-circuit protective device. The Code recognizes this and stipulates that the equipment grounding conductor is not required to be larger than the circuit conductors. For example, a 25-horsepower, 460-volt, three-phase motor circuit protected by a set of fuses of 110 amperes requires a No. 8 grounding conductor if the circuit conductors are No. 8. If the Code table were used without regard to the exception, a No. 6 equipment grounding conductor would be required. **250-95** **250-95 EXCEPTION 2** **TABLE 250-95**

The disconnecting means for a motor circuit must meet specific Code requirements. The ampere rating of the device must be at least 115% of the full-load current rating of the motor. A rating in horsepower is also required for certain motor disconnecting means. **430-110**

Complete Example. To summarize the technique for branch-circuit design, Table 3-10 presents examples for the selection of the conductors, branch-circuit protective device, and the equipment grounding conductor for various branch circuits. The resulting feeder panel is shown in Figure 3-18.

A 120/240-volt feeder supplies the following load for which branch circuits are to be designed:

    a. A 2-horsepower, 230-volt, squirrel-cage motor
    b. One hundred duplex receptacles
    c. One hundred 200-watt incandescent lamps

Type THW copper conductors are used for the circuits and 20-ampere circuits are selected to supply the lights and receptacles that constitute a continuous load in this example.

The motor circuit is assumed to be protected from short circuits by an instantaneous trip circuit breaker that may be set as high as 700% of the full-load current.[6] The 2-horsepower, 230-volt motor has a full-load current of 12 amperes; therefore, the circuit breaker may be set as high as 84 amperes. The conductor ampacity must be at least 1.25×12 amperes $= 15$ amperes. A circuit of No. 14 conductors could supply the motor. **TABLE 430-152** **TABLE 430-148**

The number of branch circuits required for the lighting and receptacle loads is determined as follows:

---

[6]Special Code rules apply to the use of instantaneous-trip circuit breakers in motor circuits. These breakers are sometimes called magnetic-only breakers. The example here only considers the maximum setting.

Table 3-10. Branch-Circuit Design Examples for Various Loads

Load	Connected Load	Branch-Circuit Load Value	Branch-Circuit Conductor Size (THW Copper)	Branch-Circuit Protective Device	Equipment Grounding Conductor	Code Ref.
230-volt 2-hp Motor	12 amperes	1.25 × 12 amperes = 15 amperes	Two No. 14	7 × 12 amperes = 84 amperes for instaneous-trip breaker. Use standard 90-ampere breaker (maximum allowed)	No. 14 (same as circuit conductors)	Table 430-148 430-22 Table 430-152, 240-6 Equipment Ground: 250-95 Exception No. 2
100 Receptacles (Continuous Load)	$\dfrac{180 \text{ volt-amperes}}{120 \text{ volts}}$ = 1.5 amperes each	1.25 × 1.5 amperes = 1.88 amperes each Number of units per 20-ampere circuit = $\dfrac{20 \text{ amperes}}{1.88 \text{ amperes}}$ = 10.6 or 10 units/circuit Number of circuits = $\dfrac{100 \text{ units}}{10 \text{ units/circuit}}$ = 10 circuits	Use five 3-wire, 120/240-volt, 20-ampere circuits Three No. 12 conductors per circuit	Use five two-pole, 20-ampere circuit breakers	No. 12 with each circuit	220-3(c) 220-3(a)
100 200-watt Lamps (Continuous Load)	$\dfrac{200 \text{ watts}}{120 \text{ volts}}$ = 1.67 amperes each	1.25 × 1.67 amperes = 2.1 amperes each Number of units per 20-ampere circuit = $\dfrac{20 \text{ amperes}}{2.1 \text{ amperes}}$ = 9.5 or 9 units/circuit Number of circuits = $\dfrac{100 \text{ units}}{9 \text{ units/circuit}}$ 11.1 or 12 circuits	Use six 3-wire 120/240-volt, 20-ampere circuits Three No. 12 conductors per circuit	Use six-two pole, 20-ampere circuit breakers	No. 12 with each circuit if required to ground fixture	220-3(a)

220-3

1. Determine the unit load. Each receptacle is rated at 180 volt-amperes and each lighting unit at 200 watts. Twenty five percent is added to these unit loads if the loads are continuous.

2. Compute the number of units per circuit

$$\text{units/circuit} = \frac{\text{rating of circuit (amperes)}}{\text{continuous load/unit (amperes)}}$$

3. The required number of circuits is then

$$\text{number of circuits} = \frac{\text{units}}{\text{units/circuit}}$$

The receptacles require 10 two-wire circuits or 5 three-wire, 20-ampere circuits. The lights are served by 12 two-wire or 6 three-wire circuits.

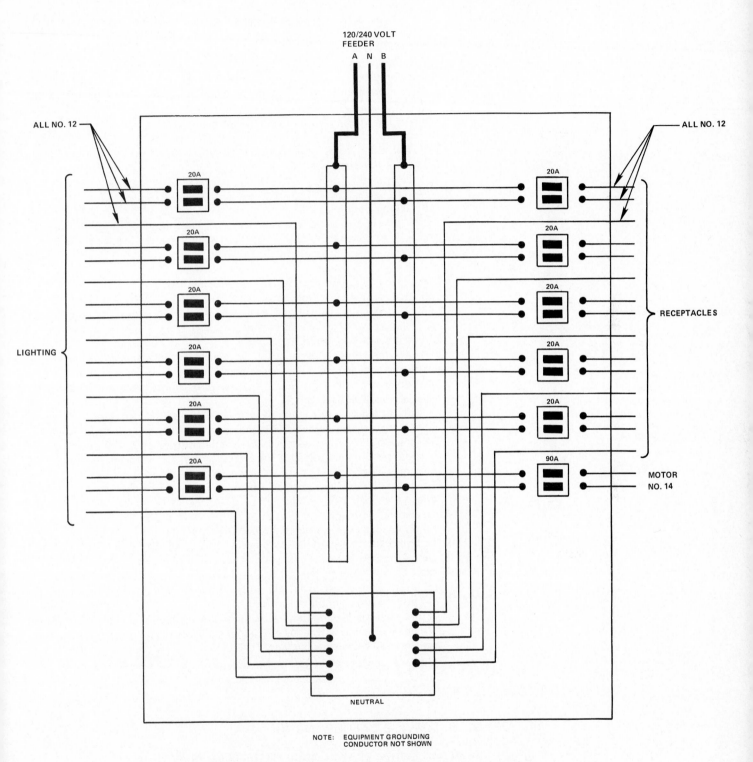

120/240 VOLT
FEEDER

A N B

ALL NO. 12

ALL NO. 12

20A

20A

20A

20A

20A

20A

20A

20A

20A

20A

20A

90A

LIGHTING

RECEPTACLES

MOTOR
NO. 14

NEUTRAL

NOTE: EQUIPMENT GROUNDING
CONDUCTOR NOT SHOWN

Figure 3-18.   Resultant Feeder Panel for Branch-Circuit Design Example

QUIZ

(Closed-Book)

1. A motor that draws 100 amperes at full-load requires supply conductors with an ampacity of
   (a) 100 amperes
   (b) 125 amperes
   (c) 150 amperes
2. A 20-ampere branch circuit supplies a continuous load. What is the maximum load the circuit may supply?
3. How is the full-load current of a 10-horsepower, 230-volt single-phase motor determined?
4. The maximum size of a nontime-delay fuse for a motor circuit with a full-load current of 100 amperes is
   (a) 200 amperes
   (b) 175 amperes
   (c) 300 amperes
5. What is the branch-circuit load for a receptacle outlet in an industrial occupancy?
6. What is the computed branch-circuit load value for each receptacle outlet if it supplies a continuous load?

(Open-Book)

1. What is the full-load current for the following motors?
   (a) ½-horsepower, 115-volt, single-phase
   (b) ½-horsepower, 230-volt, single-phase
   (c) 3-horsepower, 230-volt, three-phase
   (d) 3-horsepower, 480-volt, three-phase
2. What is the required ampacity of the branch-circuit conductors supplying a 10-horsepower, 460-volt, three-phase motor?
3. If the branch circuit in problem 2 is protected by an instantaneous trip breaker, how large may the breaker be?
4. A 240-volt load is rated at 50 kilowatts. If the load is considered continuous, what is the required branch-circuit conductor ampacity and the circuit rating?
5. What size THW copper conductors are required to supply a 240-volt, 50-kilo-volt-ampere continuous load if the protective device is rated for 100% continuous operation?
6. How many three-wire, 20-ampere branch circuits are required to serve a 50-kilovolt-ampere, 240-volt load?
7. Determine the number of 15-ampere, 120-volt, two-wire circuits required to supply 100 duplex receptacles if the load is continuous?
8. Determine the number of receptacles per circuit in problem 7.
9. Design the branch circuits for an occupancy with the following loads:
   (a) 2-kilovolt-ampere, 120-volt lighting load
   (b) 25-horsepower, 460-volt, three-phase squirrel-cage motor (the circuit is protected by a nontime-delay fuse)
   (c) Ten 120-volt lighting fixtures with four 40-watt fluorescent lamps each. Each unit has a nameplate rating of 1.7 amperes
   (d) Ten 120-volt heavy-duty lampholders

   Use THW copper conductors for the circuits. All loads are considered continuous and the lighting circuits are rated 20 or 30 amperes.

3-2 FEEDER OR SERVICE CALCULATIONS

The purpose of this section is to describe the manner in which feeder and service calculations are made when various types of loads are supplied. Unless otherwise noted, the calculations presented here are independent of the type of occupancy involved. The discussions are separated into five major areas covering lighting and receptacle circuits, motor circuits, appliance circuits, mixed loads, and special feeder problems.

The feeder-circuit calculation differs from the branch-circuit calculation in several significant ways. First, although the connected load for the feeder is computed in the same manner as for branch circuits, various demand factors may be applied to feeders. The demand factors generally depend on the type of occupancy involved. For example, two electric ranges of 12 kilowatts each, if used in a commercial occupancy such as a restaurant, would represent a total load of 24 kilowatts; but, according to the Code table "Demand Loads for Household Electric Ranges, etc.," two 12-kilowatt ranges in a home would represent a total feeder demand load of only 11 kilowatts. The Code also provides lighting-load feeder demand factors that differ according to the type of occupancy involved. These factors will be discussed in Chapters 4 and 5 of the Guide. **220-10(a)** **220-20** **TABLE 220-19** **220-11**

In feeder or service calculations the smaller of two dissimilar loads that will not be operated simultaneously (such as heating and air conditioning) may be neglected. This is obviously not possible for branch circuits if the loads are served by separate branch circuits. **220-21**

Finally, feeder-circuit calculations differ from branch-circuit calculations in that a 70% demand factor may be applied to the neutral of the feeder in certain cases for that portion of the unbalanced load in excess of 200 amperes. Only branch circuits supplying electric ranges in dwellings may have the neutral reduced. **220-22**

3-2.1 Feeders for Lighting and Receptacle Circuits

The design of a feeder circuit supplying only lighting outlets and receptacle outlets begins with a determination of the actual connected load. Table 3-11 summarizes the Code requirements for computing lighting and receptacle feeder loads. Note that the requirements are the same as for the equivalent branch-circuit loads.

The feeder conductors must be selected according to the size of the load to be served in volt-amperes. The conductor ampacity determined this way will be equal to the amount of current due to both continuous and noncontinuous loads. However, the feeder overcurrent protective device must be rated to carry the current of the noncontinuous loads plus 125% of that of the continuous loads unless the device is listed for operation at 100% of its rating. The ungrounded feeder conductors must also be selected so that they are protected by the feeder overcurrent protective device. Therefore, in cases where the overcurrent device rating in amperes is larger than the conductor ampacity that would otherwise be necessary to carry the connected load, the conductor size might need to be increased to meet Code rules for conductor protection. This increase would not apply to the neutral (grounded) conductor. **220-10(a)** **240-3**

With respect to services, both the service conductors and the service overcurrent protective device are selected to carry the current of the noncontinuous load plus 125% of the current of the continuous loads when the circuit is protected by fuses or standard circuit breakers. **220-10(b)**

Feeder Design Example. A feeder design example will demonstrate the techniques used to determine feeder ampacity and overcurrent protection for a feeder supplying various lighting and receptacle loads. The feeder circuit supplies the following loads:

a. Floor area of 100 feet by 50 feet with a load of 3.5 volt-amperes per square foot
b. A 50-foot long show window
c. Two hundred 120-volt duplex receptacles for general use. Assume unity power factor and noncontinuous operations.
d. A 100-foot multioutlet assembly along baseboards for appliance loads used continuously.

Table 3-11. Summary of Code Rules for Computing Feeder Lighting and Receptacle Loads

Type of Load	Feeder Load	Code Rules
General illumination	Larger of Volt-amperes per square foot area of listed occupancy or Actual load if known (increase by 25% if continuous)	220-10, 220-3(b) If volt-amperes per square foot is used, see Table 220-3(b)
Heavy-duty lamp-holders in fixed lighting units	Larger of 600 volt-amperes per outlet or Actual load if known (increase by 25% if continuous)	220-10, 220-3(c)
Show window illumination	Minimum of 200 volt-amperes per foot of length	220-12
Receptacles	Larger of 180 volt-amperes per outlet or Actual load if known (increase by 25% if continuous)	220-10, 220-3(c), 220-13
Multioutlet Assemblies	(a) 180 volt-amperes per each 5 feet for general use (b) 180 volt-amperes per foot for appliance loads (increase by 25% if continuous)	220-10, 220-3(c)

220-13 The feeder is assumed to be a 120/240-volt, three-wire circuit and the lighting load is continuous. Type THW aluminum conductors will be used in a raceway (conduit). Demand factors may be applied to the receptacle load and the neutral since the type of occupancy may be assumed to be commercial.

220-3 The required feeder is depicted in Figure 3-19. The feeder loads are computed in the same manner as for branch-circuit loads. The general illumination lighting load is

$$100 \text{ ft} \times 50 \text{ ft} \times 3.5 \text{ VA/ft}^2 \times 1.25 = 21\ 875 \text{ VA}$$

(based on the given volt-amperes/square foot unit load)

220-12 The show window load is

$$50 \text{ ft} \times 200 \text{ VA/ft} = 10\ 000 \text{ VA}$$

220-13 The receptacle load of 200×180 VA/unit $= 36\ 000$ VA is reduced to

$$10\ 000 \text{ VA} + 0.5\ (26\ 000 \text{ VA}) = 23\ 000 \text{ VA}$$

The load for the 100 feet of multioutlet assembly is

$$100 \text{ ft} \times 180 \text{ VA/ft} \times 1.25 = 22\ 500 \text{ VA}$$

The total computed load is then

$$
\begin{array}{r}
21\ 875 \text{ VA} \\
10\ 000 \text{ VA} \\
23\ 000 \text{ VA} \\
\underline{22\ 500 \text{ VA}} \\
77\ 375 \text{ VA}
\end{array}
$$

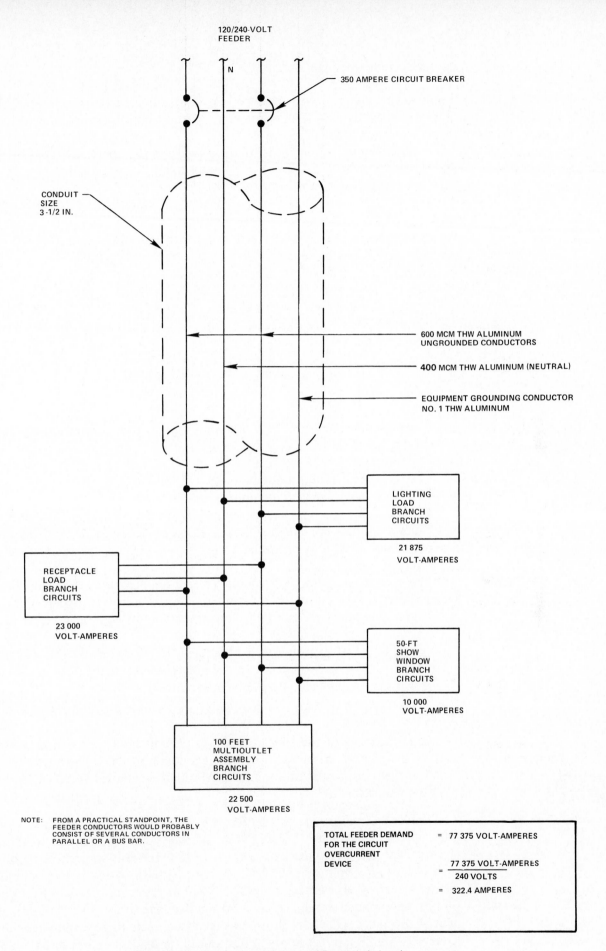

120/240-VOLT
FEEDER

N

350 AMPERE CIRCUIT BREAKER

CONDUIT
SIZE
3-1/2 IN.

600 MCM THW ALUMINUM
UNGROUNDED CONDUCTORS

400 MCM THW ALUMINUM (NEUTRAL)

EQUIPMENT GROUNDING CONDUCTOR
NO. 1 THW ALUMINUM

LIGHTING
LOAD
BRANCH
CIRCUITS

21 875
VOLT-AMPERES

RECEPTACLE
LOAD
BRANCH
CIRCUITS

23 000
VOLT-AMPERES

50-FT
SHOW
WINDOW
BRANCH
CIRCUITS

10 000
VOLT-AMPERES

100 FEET
MULTIOUTLET
ASSEMBLY
BRANCH
CIRCUITS

22 500
VOLT-AMPERES

NOTE: FROM A PRACTICAL STANDPOINT, THE
FEEDER CONDUCTORS WOULD PROBABLY
CONSIST OF SEVERAL CONDUCTORS IN
PARALLEL OR A BUS BAR.

TOTAL FEEDER DEMAND   = 77 375 VOLT-AMPERES
FOR THE CIRCUIT
OVERCURRENT
DEVICE   = $\dfrac{77\ 375\ \text{VOLT-AMPERES}}{240\ \text{VOLTS}}$

                               = 322.4 AMPERES

Figure 3-19.   Feeder for Lighting and Receptacle Loads

The load in amperes used to select the feeder overcurrent device is 77375 volt-amperes ÷ 240 volts = 322.4 amperes. A standard 350-ampere device is the minimum that could be used.

The conductor ampacity is that required for the connected load of the lights and the multioutlet assembly plus the other loads. That value is

$$17500 \text{ VA} + 10000 \text{ VA} + 23000 \text{ VA} + 18000 \text{ VA} = 68500 \text{ VA}$$

for the lights, show window, receptacles, and multioutlet assembly, respectively. The connected load current is

$$68500 \text{ VA} \div 240 \text{ V} = 285 \text{ A}$$

220-22 The neutral load is computed as

$$
\begin{array}{r}
200 \text{ A (first) at } 100\% = 200 \text{ A} \\
(285 - 200) \text{ A (excess) at } 70\% = \underline{60 \text{ A}} \\
260 \text{ A}
\end{array}
$$

The minimum size type THW aluminum conductors to carry the connected load current are

**TABLE
310-16**

Ungrounded conductors: 500 MCM
Neutral conductor: 400 MCM

240-3 Since a 350-ampere overcurrent protective device must be used, the minimum size permissible for the ungrounded conductors is 600 MCM. The neutral does not have to be increased in size.

TABLE 250-95 The equipment grounding conductor size required for a 350-ampere circuit is No. 1 aluminum (based on the 350-ampere overcurrent protective device).

The conduit must enclose the following conductors with the given area:

Two 600 MCM ungrounded conductors $(2 \times 1.0261 \text{ in.}^2)$ = 2.0522 in.²
One 400 MCM neutral   .6969 in.²
One No. 1 equipment grounding conductor $\underline{.2027 \text{ in.}^2}$
Total area of conductors = 2.9518 in.²

**CHAPTER 9
TABLE 1, TABLE 4** The total area of the four conductors (2.9518 square inches) may not exceed 40% of the area of the conduit. A 3 1/2-inch conduit, which has a total area of 9.9 square inches with a 40% fill area of 3.96 square inches (.4 × 9.9 = 3.96), would meet the requirements.

Feeder Neutral Load for Lighting Circuits. If a circuit is a three-wire, single-phase circuit, such as a 120/240-volt circuit, the neutral conductor would carry no current if the load were balanced. This is also true for four-wire, three-phase circuits, such as 480Y/277-volt or 208Y/120-volt circuits. In the worst case, it is unlikely that a large feeder circuit carrying a load of 200 amperes or more would have a full load on one ungrounded conductor and no load on the other. Furthermore, since the **220-22** neutral currents tend to cancel for resistive loads, including incandescent lamps, the Code allows the neutral capacity to be reduced for loads over 200 amperes.

When the feeder supplies electric-discharge lighting, no reduction in feeder ampacity may be made for that portion of the load attributable to this type of lighting. Fluorescent lamps, neon tubes, mercury-vapor lamps, and the like are considered electric-discharge lighting.

Figure 3-20 illustrates three examples of calculations for sizing the feeder neutral for incandescent and electric-discharge lighting loads. The voltages shown are typical for the type of loads considered.

The first example is an incandescent load of 65 kilovolt-amperes on a 240-volt circuit. The feeder circuit requires ungrounded conductors with an ampacity of 65 000 volt-amperes ÷ 240 volts = 271 amperes. A demand factor of 70% may be applied to that portion of the load in excess of 200 amperes for purposes of sizing the neutral conductor. The neutral conductor must be rated at least

220-22

$$
\begin{array}{rcl}
\text{200 A (first) at } 100\% &=& 200 \ \ \text{A} \\
(271 - 200) \text{ A at } 70\% &=& \underline{49.7 \text{ A}} \\
&& 249.7 \text{ A}
\end{array}
$$

The second example is a fluorescent lighting load of 300 kilovolt-amperes on a 480Y/277-volt feeder. The current drawn is 100 000 volt-amperes ÷ 277 volts = 361 amperes since all branch circuits are 277-volt circuits. No reduction in the neutral capacity is permitted for fluorescent lighting loads. Type THHN conductors are used.

220-22

As a third example, a 300-kilovolt-ampere fluorescent lighting load and a 360-kilovolt-ampere incandescent lighting load are supplied by the same feeder. The total feeder demand load is 660 kilovolt-amperes. The ampacity of the ungrounded conductors must be at least 220 000 volt-amperes ÷ 277 volts = 794 amperes.

Since a demand factor of 70% may be applied only to that portion of the incandescent lighting load exceeding 200 amperes, the loads must be computed separately. The results are:

$$
\text{Incandescent load} = \frac{120\ 000 \text{ VA}}{277 \text{ V}} = 433 \text{ A}
$$

$$
\text{Fluorescent load} = \frac{100\ 000 \text{ VA}}{277 \text{ V}} = \underline{361 \text{ A}}
$$

$$
794 \text{ A}
$$

as before. The required neutral ampacity becomes

$$
\begin{array}{rl}
\text{Incandescent load} = 200 + .7 \times (433 - 200) &= 363 \text{ A} \\
\text{Fluorescent load} &= \underline{361 \text{ A}} \\
\text{Total load for neutral} &= 724 \text{ A}
\end{array}
$$

The neutral ampacity is thus reduced by 70 amperes since part of the load is incandescent lighting.

QUIZ

(Closed-Book)   **1.** The feeder load for a 40-foot long show window is

   (a) 4000 volt-amperes
   (b) 8000 volt-amperes
   (c) 10 000 volt-amperes

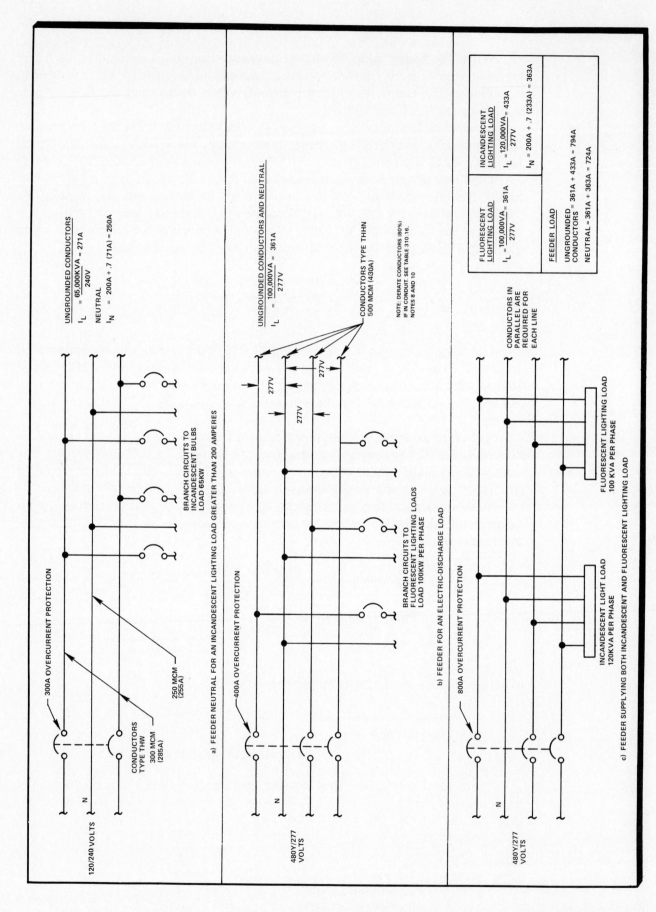

Figure 3-20. Feeder Neutral Calculations for Incandescent and Fluorescent Lighting Loads

86

2. In nondwelling occupancies, what feeder load is required for each receptacle outlet?
3. For what type of circuits is a 70% demand factor allowed for the neutral conductors?
4. No reduction in the feeder neutral is allowed for
   (a) A feeder that supplies any electric-discharge load
   (b) That portion of a feeder load contributed by electric-discharge lighting
   (c) Four-wire, three-phase ac circuits
5. Upon what factors does the size of the feeder overcurrent protective device depend if the feeder supplies only lighting loads?

(Open-Book)
1. Design a service circuit to supply a 10 000-square foot building. The 120/240-volt loads are as follows:
   (a) 3.5 volt-amperes/square foot lighting
   (b) 100 duplex receptacles (continuous)
   Use THW copper conductors.
2. Design a feeder circuit to supply a 52 kilovolt-ampere fluorescent lighting load and a 104 kilovolt-ampere incandescent lighting load. Use type THW aluminum conductors. The feeder voltage is 120/240 volts.
3. Select the ungrounded conductors for a feeder if the loads in Problem 1 are supplied by a feeder rather than a service.

3-2.2 Feeders for Motor Circuits

The most common motor feeder circuit consists simply of the conductors and an overcurrent device to protect these conductors. For continuous duty motors, the **430-24** conductor ampacity must not be less than the sum of the full-load currents of all the motors plus an additional 25% of the full-load current of the largest motor. The **430-62(a)** motor feeder overcurrent protective device may not be larger than the rating or setting of the branch-circuit overcurrent protective device for any motor plus the sum of the full-load currents of the *other* motors. The Code makes no provisions for increasing the rating to the next standard size.

Two Motors Supplied by One Feeder. As an example, the 5-horsepower motor and the 10-horsepower motor shown in Figure 3-21 are both supplied by the same 120-volt feeder. The minimum ampacity for the conductors is

$$
\begin{array}{r}
56 \text{ A (5-hp motor)} \\
+\,100 \text{ A (10-hp motor)} \\
\underline{+\,25 \text{ A (.25} \times \text{ full-load current of 100 A)}} \\
181 \text{ A}
\end{array}
$$

If both branch circuits are protected by nontime-delay fuses, the 5-horsepower **TABLE 430-150** motor circuit may use 3×56 amperes = 168-ampere fuses and the 10-horsepower motor may use 3×100 amperes = 300-ampere fuses. The feeder overcurrent protection cannot be rated or set at greater than

$$
\begin{array}{r}
300 \text{ A (largest branch-circuit protective device)} \\
\underline{+\;\;56 \text{ A (full-load current of other motors)}} \\
356 \text{ A}
\end{array}
$$

If a standard-size device is chosen, it must be rated *lower* than 356 amperes. This would require a 350-ampere standard-size fuse or circuit breaker. The 350-ampere **240-6** fuse would allow the largest motor to start while the other motor is running.

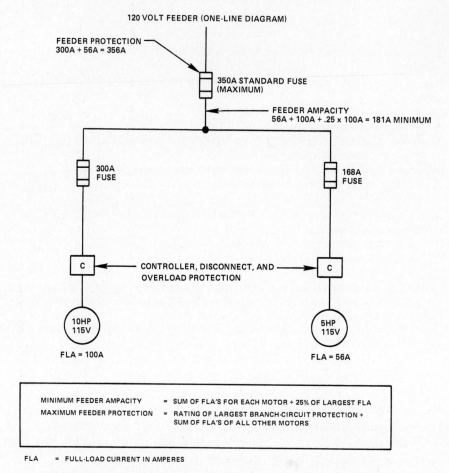

120 VOLT FEEDER (ONE-LINE DIAGRAM)

FEEDER PROTECTION
300A + 56A = 356A

350A STANDARD FUSE
(MAXIMUM)

FEEDER AMPACITY
56A + 100A + .25 x 100A = 181A MINIMUM

300A
FUSE

168A
FUSE

C ← CONTROLLER, DISCONNECT, AND → C
OVERLOAD PROTECTION

10HP
115V

5HP
115V

FLA = 100A

FLA = 56A

| MINIMUM FEEDER AMPACITY | = | SUM OF FLA'S FOR EACH MOTOR + 25% OF LARGEST FLA |
| MAXIMUM FEEDER PROTECTION | = | RATING OF LARGEST BRANCH-CIRCUIT PROTECTION + SUM OF FLA'S OF ALL OTHER MOTORS |

FLA = FULL-LOAD CURRENT IN AMPERES

Figure 3-21. Motor-Feeder Design for Two Motors on One Feeder

Motor-Feeder Design Procedure. Figure 3-22 shows three motors. A 25-horse-power motor with Code letter F and two 30-horsepower wound-rotor motors are supplied by a feeder circuit. The steps in the solution to the problem are indicated in the figure. First, the full-load current in amperes is found from the appropriate **TABLE 430-150** Code tables. According to the table of full-load currents for 460-volt, three-phase motors, the 25-horsepower motor draws 34 amperes and each 30-horsepower motor **430-24** draws 40 amperes. The total load on the feeder is then calculated based on the full-load currents of the motors plus 25% of the full load of the largest motor, one of the 30-horsepower motors represents the "largest motor" in this case or

$$
\begin{aligned}
&34\text{A (25 hp)} \\
+\ &40\text{A (30 hp)} \\
+\ &40\text{A (30 hp)} \\
+\ &10\text{A (.25} \times 40) \\
\hline
&124\text{ A}
\end{aligned}
$$

To determine the feeder protective device rating, the protective devices for the individual branch circuits must be determined. Assuming that nontime-delay fuses **TABLE 430-152** are being used, the 25-horsepower squirrel-cage motor with Code letter F may use a fuse 300% as large as its full-load current or 3 × 34 amperes = 102 amperes. **TABLE 430-152** A standard 110-ampere fuse could be used for the branch-circuit protection. Each

88

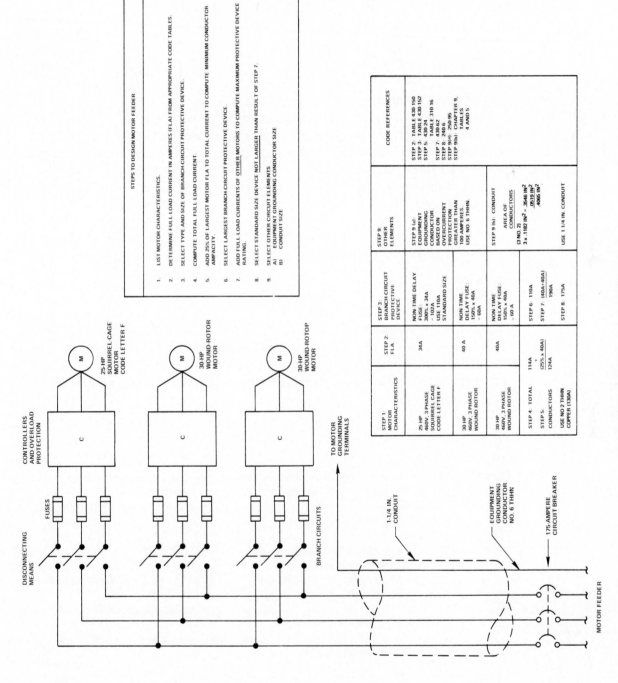

Figure 3-22. Motor-Feeder Design Procedure

STEPS TO DESIGN MOTOR FEEDER

1. LIST MOTOR CHARACTERISTICS.
2. DETERMINE FULL LOAD CURRENT IN AMPERES (FLA) FROM APPROPRIATE CODE TABLES.
3. SELECT TYPE AND SIZE OF BRANCH CIRCUIT PROTECTIVE DEVICE.
4. COMPUTE TOTAL FULL LOAD CURRENT.
5. ADD 25% OF LARGEST MOTOR FLA TO TOTAL CURRENT TO COMPUTE MINIMUM CONDUCTOR AMPACITY.
6. SELECT LARGEST BRANCH CIRCUIT PROTECTIVE DEVICE.
7. ADD FULL LOAD CURRENTS OF OTHER MOTORS TO COMPUTE MAXIMUM PROTECTIVE DEVICE RATING.
8. SELECT STANDARD SIZE DEVICE NOT LARGER THAN RESULT OF STEP 7.
9. SELECT OTHER CIRCUIT ELEMENTS
   A) EQUIPMENT GROUNDING CONDUCTOR SIZE
   B) CONDUIT SIZE

STEP 1 MOTOR CHARACTERISTICS	STEP 2 FLA	STEP 3 BRANCH CIRCUIT PROTECTIVE DEVICE	STEP 9 OTHER ELEMENTS	CODE REFERENCES
25 HP 460V, 3 PHASE SQUIRREL CAGE CODE LETTER F	34A	NON TIME DELAY FUSE: 300% x 34A = 102A USE 110A STANDARD SIZE	STEP 9 (a): EQUIPMENT GROUNDING CONDUCTOR. BASED ON OVERCURRENT PROTECTION GREATER THAN 100 AMPERES. USE NO. 6 THHN.	STEP 2: TABLE 430.150 STEP 3: TABLE 430.152 STEP 5: 430.24 TABLE 310.16 STEP 7: 430.62 STEP 8: 240.6 STEP 9(a): 250.95 STEP 9(b): CHAPTER 9, TABLES 4 AND 5
30 HP 460V, 3 PHASE WOUND ROTOR	40 A	NON TIME DELAY FUSE: 150% x 40A = 60A		
30 HP 460V, 3 PHASE WOUND ROTOR	40A	NON TIME DELAY FUSE: 150% x 40A = 60 A	STEP 9 (b): CONDUIT AREA OF CONDUCTORS (3 NO. 2) 3 x .1182 IN² = .3546 IN² .0519 IN² .4065 IN²	
STEP 4: TOTAL	114A	STEP 6: 110A		
STEP 5: CONDUCTORS	(25% x 40A)	STEP 7: (40A+40A) 190A	USE 1 1/4 IN. CONDUIT	
USE NO 2 THHN COPPER (130A)	124A	STEP 8: 175A		

CONTROLLERS AND OVERLOAD PROTECTION

25-HP SQUIRREL-CAGE MOTOR CODE LETTER F

30-HP WOUND-ROTOR MOTOR

30-HP WOUND-ROTOP MOTOR

TO MOTOR GROUNDING TERMINALS

FUSES

DISCONNECTING MEANS

BRANCH CIRCUITS

1-1/4 IN. CONDUIT

EQUIPMENT GROUNDING CONDUCTOR NO. 6 THHN

175-AMPERE CIRCUIT BREAKER

MOTOR FEEDER

30-horsepower wound-rotor motor is allowed a fuse as large as 150% of its full-load current, or 1.5 × 40 amperes = 60 amperes. In this case, the smallest motor has the largest fuse for branch-circuit protection. For the feeder, the maximum rating of its protective device is

$$
\begin{array}{r}
110 \text{ A (largest branch-circuit protective device)} \\
+ \quad 40 \text{ A (full-load current of other motors)} \\
+ \quad \underline{40 \text{ A}} \\
190 \text{ A}
\end{array}
$$

240-6 The next smaller standard size fuse is 175 amperes.

TABLE 250-95 The equipment grounding conductor must be at least a No. 6 copper wire. If type THHN copper conductors are arbitrarily chosen for the feeder conductors, No. 2 conductors rated at 130 amperes could be used. Thus, a conduit must enclose

CHAPTER 9 three No. 1 THHN conductors and one No. 6 THHN equipment grounding conductor.

TABLE 5 The final step in the solution to the problem is to determine the conduit size based on the area of the conductors. The areas of the conductors are

$$
\begin{array}{ll}
\text{Three No. 2 THHN} = 3 \times .1182 \text{ sq in.}^2 = .3546 \text{ sq in.}^2 \\
\text{One No. 6 THHN} & = \underline{.0519 \text{ sq in.}^2} \\
& \text{Total area} = .4065 \text{ sq in.}^2
\end{array}
$$

CHAPTER 9
TABLE 4 Since there are more than two conductors in the conduit, it may only be 40% filled; therefore, a 1¼-inch conduit with a 40% fill area of .60 square inches is required.

440-22(a) If one or more of the circuits connected to a feeder supply motor-compressor units for air-conditioning equipment, the feeder design problem is not changed. The branch-circuit protective device, however, must follow the rules for motor-compressor circuits.

QUIZ

(Closed-Book) 1. Upon what factors does the size of the feeder overcurrent protective device depend if the feeder supplies several motor branch circuits?
2. What is the conductor ampacity for a feeder that supplies several motors and other loads?

(Open-Book) 1. Design the feeder circuit for three 230-volt, single-phase, 10-horsepower motors. The motors are supplied by individual branch circuits and the branch circuits are protected by instantaneous trip circuit breakers. Use type THW copper conductors.
2. Design a feeder circuit to supply a hermetic motor-compressor rated at 25 amperes and a 5-horsepower motor, both of which are 480-volt, three-phase motors. The motors are protected by inverse-time circuit breakers. Type THW aluminum conductors are used in all circuits.

3-2.3 Feeders for Appliances and Electric Space Heating

The design of a feeder for appliance branch circuits begins with a consideration

220-14 of the type of appliance to be served. The load of one or more motor operated appliances, for instance, is computed in accordance with the rules for motor feeders previously discussed. Other appliances would be considered at their rated values or at 125% of the rated value if a continuous load. The load of fixed electric space- **220-15** heating units is taken to be 100% of the total connected load.

The motor operated appliances and fixed space-heating units are loads which are not subject to demand factors in typical situations. Most nonmotor operated appliances also fall into this category.

Commercial cooking equipment, household cooking equipment, and household electric clothes dryers are appliances for which feeder demand factors may be applied under certain conditions. The application of feeder demand factors to this equipment is discussed in detail.

Feeder Design Example. A feeder design example will demonstrate the techniques used to determine feeder ampacity and overcurrent protection for a feeder supplying several circuits. The feeder load consists of the following 240-volt loads:

    a. Motor operated appliance with 5-horsepower motor
    b. Two motor operated appliances with 3-horsepower motors
    c. Air conditioner unit with 25-ampere full-load current
    d. Three 8-kilovolt-ampere electric space heaters
    e. Five 12-kilovolt-ampere hot water heaters

The feeder design is shown in Figure 3-23 in which an organized form is used to arrange the calculations. Since the motors are rated in amperes, all loads are computed in amperes for convenience. The motor circuits are protected by inverse-time circuit breakers and the conductors are chosen to be type THW copper for the two-wire, 240-volt circuit.

Before the load is calculated, it is necessary to compare the heating and air- **220-21** conditioning load in order to neglect the smaller of the two loads. The three heaters draw a total current of

$$\frac{3 \times 8\,000 \text{ VA}}{240 \text{ V}} = 100 \text{ A}$$

Thus, since the 25-ampere load of the air conditioner is the smaller, it may be neglected. The load of the hot water heaters is

$$\frac{5 \times 12\,000 \text{ VA}}{240 \text{ V}} = 250 \text{ A}$$

The 5-horsepower motor draws 28 amperes and each 3-horsepower motor draws 17 amperes at 240 volts. The feeder load is then

$$
\begin{array}{l}
28 \text{ A} \\
17 \text{ A} \\
17 \text{ A} \\
\underline{7 \text{ A} \ (.25 \times 28)} \\
69 \text{ A}
\end{array}
$$

The total load is $100 + 250 + 69 = 419$ amperes. Size 600 MCM THW copper conductors are required.

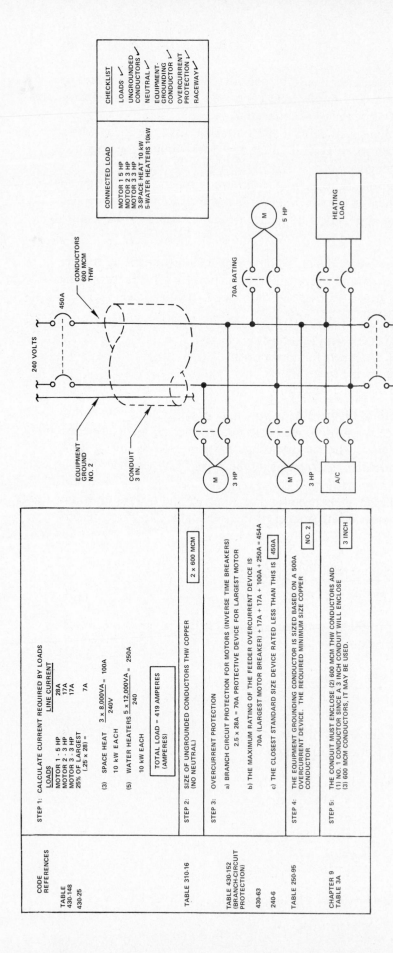

Figure 3-23. Calculation of Feeder for Appliance Loads

92

The contribution to the load made by the motors is the same as that for a motor feeder alone. Since the motor branch circuits are protected by inverse-time breakers, the rating may be as much as 250% of the full-load current of the motor it protects.   **TABLE 430-152** The rating of the branch-circuit protective device is used as the basis for selecting the feeder overcurrent device. If the feeder supplies motor operated appliances and   **430-63** other appliances, the rating of the feeder overcurrent devices is determined as follows:

maximum rating or setting
of feeder over current device = rating or setting of the largest branch-circuit
protective device for any motor + sum of full-
load currents of all other motors + sum of
load currents of all other loads

In this example, the 5-horsepower motor with a full-load current of 28 amperes   **TABLE 430-148** is obviously the largest motor. An inverse-time circuit breaker protecting the branch   **TABLE 430-152** circuit for this motor may be rated at 250% of the 28 amperes or 2.5 × 28 amperes = 70 amperes. To determine the rating of the feeder protection, the other load currents must be added to this value with a result of 454 amperes. Since this is not a standard-   **240-6** size device, the next lower size must be used, resulting in a protective device rated at 450 amperes.

The circuit requires an equipment grounding conductor the equivalent of No. 2   **TABLE 250-95** copper wire since the 450-ampere protective device is more than 400 amperes but less than 500 amperes. The conduit could serve as the equipment grounding conductor   **250-91(b)** from the service equipment to the feeder panel, but the conductors have been in-cluded here to demonstrate the selection method.

The conduit size in this case may be read directly from the Code table specifying   **CHAPTER 9** the "maximum number of conductors in trade sizes of conduit, etc." A 2½-inch   **TABLE 3A** conduit will allow only one 600 MCM THW conductor, but a 3-inch conduit will allow three such conductors. Two 600 MCM conductors and one No. 2 easily fit into the 3-inch conduit and satisfy the Code rules.

Feeder Demand Factors for Appliance Loads in Industrial and Commercial Occupancies. Feeder demand factors are not provided for appliance loads in indus-trial occupancies and in most commercial occupancies. The only exception occurs   **220-20** when three or more kitchen appliances are used in a commercial establishment such as a restaurant. The load for the largest motor operated appliance and any con-tinuous appliance load must be increased to 125% of the rating of the appliance.

Commercial Cooking Equipment. The use of demand factors for commercial kitchen equipment is explained in Table 3-12 in which a shortened table of demand   **TABLE 220-20** factors is given. The Code allows the feeder load for three such units to be reduced to 90% of the connected load. Thus, three 15-kilowatt commercial electric ranges represent a connected load of 3 × 15 kilowatts = 45 kilowatts but a maximum demand load of only .9 × 45 kilowatts = 40.5 kilowatts.[7]

The demand factor depends on the number of units installed and becomes 80% for four units, 70% for five units, and 65% for six or more units. The example in Table 3-12 shows a connected load of four cooking units.

[7]The feeder load must not be less than the sum of the connected load of the two largest appliances.

Barbeque machines	12.5 kW
Fryer	5.9 kW
Heavy-duty range	21.6 kW
Oven	11.0 kW
Total connected load =	51.0 kW

The demand factor taken from the Code table "Feeder Demand Factors for Commercial Electric Cooking Equipment, etc." is 80% for four units. The demand load is then .8 × 51 kilowatts = 40.8 kilowatts. For three-wire, 240-volt loads, the current for the feeder is

$$\frac{40\ 800\ \text{W}}{240\ \text{V}} = 170\ \text{A}$$

TABLE 310-16   instead of 51 000 watts/240 volts = 212.5 amperes if no demand factor were applied.

Table 3-12. Summary of Code Rules for Application of Demand Factors and Demand Loads to Cooking Equipment and Dryers

Commercial Cooking Equipment (220-20)	Dwellings Only	
	Electric Clothes Dryers (220-18)	Electric Ranges (220-19, 220-22)
A demand factor is allowed for three or more units. (Table 220-20)	A demand factor is allowed for five or more units. (Table 220-18)	A reduced load is allowed for ranges rated over 1-3/4 kW. (Table 220-19)

Demand Factors for Commercial Cooking Equipment		Demand Factors for Electrical Clothes Dryer		Demand Loads for Electric Cooking Equipment rated 8-3/4 kW to 12 kW	
Number of Units	Demand Factor	Number of Units	Demand Factor	Number of Units	Maximum Demand
1-2	100%	1-4	100%	1	8 kW
3	90%	5	80%	2	11 kW
4	80%	6	70%	3	14 kW
		10	50%	10	25 kW
		20-24	35%	20	35 kW
				26-40	15 kW plus 1 kW for each range
				41 and over	25 kW plus 3/4 kW for each range

Commercial Cooking Equipment — Example

Load:
Barbecue machine (12.5 kW)
Fryer (5.9 kW)
Heavy-duty range (21.6 kW)
Oven (11 kW)

Connected load:
12.5 kW
5.9 kW
21.6 kW
11.0 kW
51 kW

Feeder load for 4 units:
.8 × 51 kW = 40.8 kW

Electric Clothes Dryers — Example

Load:
20 5-kW clothes dryers

Connected load:
20 × 5 kW = 100 kW

Feeder load for 20 units:
.35 × 100 kW = 35 kW

Feeder Neutral load is (70%):
.7 × 35 kW = 24.5 kW

Electric Ranges — Example

Load:
Ten 12 kW electric ranges

Connected load
10 × 12 kW = 120 kW

Feeder load for 10 units:
25 kW

Feeder Neutral load is (70%):
.7 × 25 kW = 17.5 kW

Size 3/0 THW copper conductors could be used and would require 2-inch conduit if the conduit also were to serve as the equipment grounding conductor.

**CHAPTER 9
TABLE 3A**

Feeder Demand Factors for Electric Clothes Dryers and Electric Ranges in Dwellings. Since a feeder circuit normally supplies a large number of branch circuits for appliances in multifamily dwellings, the Code permits the application of a demand factor to reduce the maximum demand load for certain appliances. The Code recognizes that a large number of similar appliances, such as electric ranges or clothes dryers, are not likely to all be used simultaneously in a large apartment complex. Thus, a reduced demand is applied for more than four electric clothes dryers or any number of electric ranges when the feeder ampacity is calculated. The neutral load is calculated as 70% of the demand load.

**TABLE 220-18
220-22**

The use of a demand factor for a number of electric clothes dryers is shown in Table 3-12. If twenty 5-kilowatt clothes dryers are connected to a feeder, the Code table ''Demand Factors for Household Electric Clothes Dryers'' allows a demand factor of 35%. The maximum demand load for 20 dryers is then

$$.35 \times 20 \text{ dryers} \times 5 \text{ kW/dryer} = 35 \text{ kW}$$

instead of the connected load of 100 kilowatts. The ampacity of the feeder conductors must be at least 35 000 watts/240 volts = 146 amperes.

The Code provides a table to determine the feeder demand load for household electric ranges, wall-mounted ovens, counter-mounted cooking units, and other household cooking appliances. The equipment covered by this table must be rated between 1¾ kilowatts and 27 kilowatts. The standard household range (the so-called 30-inch range) is rated about 8 kilowatts at 208 volts and 10 kilowatts at 240 volts.[8] A small apartment-size model might require 7.8 kilowatts at 208 volts. A large range with two ovens could be rated at 14 kilowatts or more. A range platform or countertop range represents a smaller load of about 5 kilowatts. Thus, the Code range table has three columns, each for ranges of different ratings. Column A, which is used most often, gives the maximum demand, and *not* the demand factor, for ranges rated between 8¾ kilowatts and 12 kilowatts, although it could be used for any range rated over 1¾ kilowatts.

TABLE 220-19

Design Examples for Household Electric Ranges. According to the Code table, ten 12-kilowatt ranges, which represent a connected load of 10 × 12 kilowatts = 120 kilowatts, require the feeder capacity to be designed for a demand load of only 25 kilowatts. In addition, the feeder neutral capacity for a standard 120/240-volt or 208Y/120-volt circuit may be reduced to 70% of this maximum demand load or .7 × 25 kilowatts = 17.5 kilowatts. The required ampacity is then

220-22

$$\text{ungrounded conductor current} = \frac{25\ 000 \text{ W}}{240 \text{ V}} = 104.2 \text{ A}$$

$$\text{neutral current} = .7 \times 104.2 \text{ A} = 72.9 \text{ A}$$

The feeder would require two No. 2 and one No. 4 THW copper conductors.

TABLE 310-16

---

[8]The electric range is a resistance load; increasing the voltage increases the power consumption by the square of the voltage; i.e.,

$$\text{power} = \frac{(\text{voltage})^2}{\text{resistance}}$$

where the voltage is the voltage measured *across* the resistance.

**TABLE 220-19
NOTE 4** The reductions for ranges as given by column A of the Code range table are shown in shortened form in Table 3-12. It should be noted that the use of the range table for a single range is the same as described previously for branch circuits supplying electric ranges.

**TABLE 220-19
COLUMN A** If there are more than 26 ranges, but fewer than 41, a fixed demand of 15 kilowatts is increased by 1 kilowatt for each additional range. For a feeder supplying more than 40 ranges, the demand is 25 kilowatts plus ¾ kilowatt for each range. As shown in Figure 3-24, example E, the maximum demand load for fifty 12-kilowatt ranges is

$$25 \, \text{kw} \; + \; (.75 \, \text{kW/range} \times 50 \, \text{ranges}) = 62.5 \, \text{kW}$$

The neutral demand load is $.7 \times 62.5$ kilowatts = 43.75 kilowatts. This represents a great reduction in feeder ampacity and increased cost-savings over the feeder required to supply the full connected load of 50×12 kilowatts = 600 kilowatts.

The notes to the Code range table detail the method of finding the feeder demand for the following feeder loads:

    a. A number of ranges rated over 12 kilowatts and all of the same rating. Column A of the Code range table can only be used for ranges up to 12 kilowatts in rating.

    b. A number of ranges rated over 12 kilowatts when some of the ranges have unequal ratings.

    c. Smaller ranges rated 1¾ kilowatts through 8¾ kilowatts. In these cases, the use of column A might give a larger demand load than the actual range rating for a smaller number of ranges.

**TABLE 220-19
NOTE 4** Each of the above cases is illustrated in Figure 3-24. The note concerning the branch-circuit load and hence the feeder load for wall-mounted ovens and counter-mounted cooking units was presented previously in the branch-circuit discussions.

**TABLE 220-19
NOTE 1** For ranges from more than 12 kilowatts to 27 kilowatts and all of the same rating, 5% of the demand load given in column A of the Code range table is added to the demand shown in that column for each kilowatt by which the rating exceeds 12 kilowatts. If five 16-kilowatt ranges are connected to a feeder, the maximum demand load becomes

$$
\begin{array}{r}
(\text{value from Code table}) = 20 \, \text{kW} \\
(.05 \times 20 \, \text{kW}) \times (16 - 12) = \underline{4 \, \text{kW}} \\
24 \, \text{kW}
\end{array}
$$

The neutral demand load is $.7 \times 24$ kilowatts = 16.8 kilowatts.

**TABLE 220-19
NOTE 2** If several ranges of unequal ratings over 12 kilowatts are connected to a feeder, the average value of the load is first computed as follows:

$$\text{average rating} = \frac{\text{sum of ratings}}{\text{number of ranges}}$$

Each range rated under 12 kilowatts is treated as a 12-kilowatt range. Thus, for a 16-kilowatt, 17-kilowatt, and a 10-kilowatt range, all connected to the same feeder, the average rating is

$$\frac{16 \, \text{kW} + 17 \, \text{kW} + 12 \, \text{kW}}{3} = \frac{45 \, \text{kW}}{3} = 15 \, \text{kW}$$

TYPE OF RANGES	RULES	EXAMPLE	CODE REF
A. OVER 12-kW RATING THROUGH 27 kW, RANGES ALL OF THE SAME RATING	INCREASE DEMAND IN COLUMN A BY 5% FOR EACH ADDITIONAL kW OVER 12 kW	FIVE 16 kW RANGES — 120/240V FEEDER. FEEDER LOAD: DEMAND LOAD = 20 kW, + .05 x 20 kW x (16-12) = 4 kW, = 24 kW. NEUTRAL LOAD = .7 x 24 kW = 16 kW	TABLE 220-19 (COLUMN A) NOTE 1
B. OVER 12-kW RATING THROUGH 27 kW, RANGES OF UNEQUAL RATINGS	1) COMPUTE AVERAGE VALUE (USING 12 kW FOR ANY RANGE RATED LESS THAN 12 kW) 2) INCREASE DEMAND IN COLUMN A BY 5% FOR EACH kW THE AVERAGE EXCEEDS 12 kW	THREE RANGES OF UNEQUAL RATINGS — 17 kW, 16 kW, 10 kW. NOTE: RATE AT 12 kW. 120/240V FEEDER. CONNECTED LOAD = 43 kW. AVERAGE RATING = $\frac{16\ kW + 17\ kW + 12\ kW}{3}$ = 15 kW. DEMAND LOAD = 14 kW + .05 x 14 kW (15-12) = 2.1 kW = 16.1 kW. NEUTRAL LOAD = .7 x 16.1 kW = 11.3 kW	TABLE 220-19 (COLUMN A) NOTE 2
C. RANGES 3-1/2 kW TO 8-3/4 kW RATING	ADD NAMEPLATE RATINGS TOGETHER AND MULTIPLY BY DEMAND FACTOR IN COLUMN C IF COMBINED ON ONE CIRCUIT	FIFTY 6 kW RANGES — 120/240V FEEDER. CONNECTED LOAD = 50 x 6 kW = 300 kW. MAXIMUM DEMAND LOAD = .2 x 300 kW = 60 kW. NEUTRAL LOAD = .7 x 60 kW = 42 kW	TABLE 220-19 (COLUMN C) NOTE 3
D. RANGES 1-3/4 TO 3-1/2 kW RATING	ADD NAMEPLATE RATINGS TOGETHER AND MULTIPLY BY DEMAND FACTOR IN COLUMN C IF COMBINED ON ONE CIRCUIT	TEN 3 kW RANGES (COUNTERTOP RANGES) — 120/240V FEEDER. CONNECTED LOAD = 10 x 3 kW = 30 kW. MAXIMUM DEMAND LOAD = .49 x 30 kW = 14.7 kW	TABLE 220-19 (COLUMN B) NOTE 3, NOTE 4
E. OVER 41 RANGES RATED 8-3/4 kW TO 12 kW ON ONE FEEDER	THE DEMAND LOAD IS 25 kW PLUS 3/4 kW FOR EACH RANGE	FIFTY 12 kW RANGES — 120/240V FEEDER. CONNECTED LOAD = 600 kW. DEMAND LOAD = 25 kW + 3/4 kW/// RANGE x 50 RANGES = 62.5 kW. NEUTRAL LOAD = 43.75 kW	TABLE 220-19 (COLUMN A)

Figure 3-24. Examples of Application of Demand Loads for Household Electric Ranges

The load is then treated as three 15-kilowatt ranges with a basic demand of 14 kilowatts by the Code range table. As in the previous example, the maximum feeder demand is

$$
\begin{array}{r}
\text{(value from table)} = 14.0\,\text{kW} \\
(.05 \times 14\,\text{kW}) \times (15 - 12) = \underline{2.1\,\text{kW}} \\
16.1\,\text{kW}
\end{array}
$$

The neutral demand load is $.7 \times 16.1$ kilowatts = 11.3 kilowatts.

**TABLE 220-19
NOTE 3**   When ranges connected to a feeder are rated less than 8¾ kilowatts, it is usually a reduction in demand load to use one of the other columns of the Code range table.[9] As a simple example, one 8-kilowatt range has a maximum demand of 8 kilowatts using column A but only a demand of $.8 \times 8$ kilowatts = 6.4 kilowatts if the demand factor in column C of the Code range table is used. Several examples of the use of these demand factors are given in Figure 3-24. Example C in that figure treats fifty 6-kilowatt ranges which allow a demand factor of 20%. For ten 3-kilowatt ranges (example D), the demand factor is 49% since the ranges are rated less than 3½ kilowatts.

Single-Phase Ranges Connected to a Three-Phase Feeder. In certain installations a three-phase feeder, such as a 208Y/120-volt circuit, supplies single-phase **220-19** ranges with each range connected to two phases and the neutral. The Code provides that the total load shall be computed on the basis of twice the maximum number of **CHAPTER 9** ranges connected to any two phases. A Code example, also shown in Figure 3-25, **EXAMPLE 7** demonstrates the technique used to determine the maximum demand load on the feeder.

As can be seen from Figure 3-25, each ungrounded feeder line has twice the number of ranges connected to it than the number of ranges per phase.[10] The number of ranges per phase for thirty 12-kilowatt ranges connected to a 208Y/120-volt feeder is

$$\frac{\text{number of ranges}}{3} = \frac{30\,\text{ranges}}{3} = 10\,\text{ranges/phase}$$

Then, according to the Code rule, the maximum demand is

$$2 \times (\text{number of ranges per phase}) = 2 \times 10\,\text{ranges/phase} = 20\,\text{ranges}$$

TABLE 220-19   According to the Code range table, the demand load for twenty 12-kilowatt ranges is 35 kilowatts. This load is shared by two phases of the three-phase circuit. The load per phase is then 35 kilowatts/2 = 17.5 kilowatts based on the maximum demand. Of course, the total load for the three-phase system is not 35 kilowatts; it is 3×17.5 kilowatts = 52.5 kilowatts. To determine the line current for each phase, the load per phase or the total three-phase load can be used. Both methods are shown in Figure 3-25, but the method using the three-phase load will be given here.

[9]It is prudent to check with the local enforcing agency to be sure that the use of these other Code columns is permitted.

[10]If the number of ranges is not divisible by 3, say 11 ranges, use the maximum number connected to any phase. For 11 ranges, they would be distributed as 4, 4, and 3 per phase, respectively.

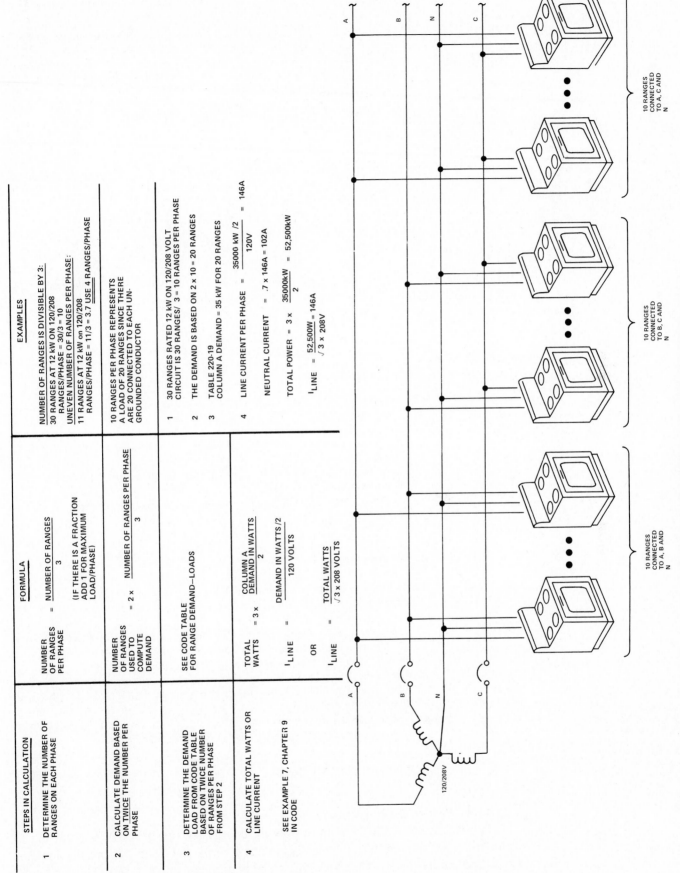

STEPS IN CALCULATION		FORMULA	EXAMPLES
1	DETERMINE THE NUMBER OF RANGES ON EACH PHASE	NUMBER OF RANGES PER PHASE $= \dfrac{\text{NUMBER OF RANGES}}{3}$ (IF THERE IS A FRACTION ADD 1 FOR MAXIMUM LOAD/PHASE)	NUMBER OF RANGES IS DIVISIBLE BY 3: 30 RANGES AT 12 kW ON 120/208 RANGES/PHASE = 30/3 = 10 UNEVEN NUMBER OF RANGES PER PHASE: 11 RANGES AT 12 kW on 120/208 RANGES/PHASE = 11/3 = 3.7 USE 4 RANGES/PHASE
2	CALCULATE DEMAND BASED ON TWICE THE NUMBER PER PHASE	NUMBER OF RANGES USED TO COMPUTE DEMAND $= 2 \times \dfrac{\text{NUMBER OF RANGES PER PHASE}}{3}$	10 RANGES PER PHASE REPRESENTS A LOAD OF 20 RANGES SINCE THERE ARE 20 CONNECTED TO EACH UN-GROUNDED CONDUCTOR
3	DETERMINE THE DEMAND LOAD FROM CODE TABLE BASED ON TWICE NUMBER OF RANGES PER PHASE FROM STEP 2	SEE CODE TABLE FOR RANGE DEMAND—LOADS	1 30 RANGES RATED 12 kW ON 120/208 VOLT CIRCUIT IS 30 RANGES/ 3 = 10 RANGES PER PHASE 2 THE DEMAND IS BASED ON 2 x 10 = 20 RANGES 3 TABLE 220-19 COLUMN A DEMAND = 35 kW FOR 20 RANGES
4	CALCULATE TOTAL WATTS OR LINE CURRENT SEE EXAMPLE 7, CHAPTER 9 IN CODE	TOTAL WATTS $= 3 \times \dfrac{\text{COLUMN A DEMAND IN WATTS}}{2}$ $I_{LINE} = \dfrac{\text{DEMAND IN WATTS}/2}{120 \text{ VOLTS}}$ OR $I_{LINE} = \dfrac{\text{TOTAL WATTS}}{\sqrt{3} \times 208 \text{ VOLTS}}$	4 LINE CURRENT PER PHASE $= \dfrac{35000 \text{ kW } /2}{120\text{V}} = 146\text{A}$ NEUTRAL CURRENT $= .7 \times 146\text{A} = 102\text{A}$ TOTAL POWER $= 3 \times \dfrac{35000\text{kW}}{2} = 52,500\text{kW}$ $I_{LINE} = \dfrac{52,500\text{W}}{\sqrt{3} \times 208\text{V}} = 146\text{A}$

Figure 3-25. Single-Phase Ranges on Three-Phase Feeder

$$\text{line current} = \frac{3 \times \text{(watts per phase)}}{\sqrt{3} \times \text{voltage}} = \frac{3 \times \dfrac{\text{demand from code range tables}}{2}}{\sqrt{3} \times 208\,\text{V}}$$

$$= \frac{3 \times \dfrac{35\,000\,\text{W}}{2}}{1.732 \times 208\,\text{V}} = \frac{3 \times 17\,500\,\text{W}}{360.3\,\text{V}} = 145.7\,\text{A}$$

Note that the basic formula is no different from that for any three-phase load, that is,

$$\text{demand load in amperes} = \frac{\text{total power in watts}}{\sqrt{3} \times \text{voltage}}$$

If the demand load in the Code range table were *not* used, the load would have been 20×12 kilowatts = 240 kilowatts requiring a feeder ampacity of

$$\frac{240\,000\,\text{W}}{\sqrt{3} \times 208\,\text{V}} = 666\,\text{A}$$

QUIZ

(Closed-Book)

1. The feeder demand factor for fixed electric space-heating equipment is 100%. True or false?
2. Motor operated appliances require a feeder circuit designed according to the rules for motor circuits. True or false?
3. The feeder demand factor for four units used for commercial cooking is:
   (a) 100%
   (b) 90%
   (c) 80%
4. How many household electric clothes dryers must be connected to a feeder before a reduced demand factor may be applied?
5. The feeder neutral for a feeder supplying electric ranges or dryers in a dwelling may be reduced by what percentage of the demand load?

(Open-Book)

1. Design the feeder circuit to supply a 7½-horsepower motor-operated appliance, a 10-kilowatt heating load, and a 12-ampere air conditioner motor. The circuit is 240 volts, single-phase and it uses type THW copper conductors.
2. What is the demand load for thirty 5-kilowatt electric clothes dryers in a dwelling?
3. What is the demand load for twenty-five 10-kilowatt electric ranges in a dwelling? What is the neutral load?
4. Design a feeder to supply twenty-five 10-kilowatt electric ranges and thirty 5-kilowatt electric clothes dryers. Use THW copper conductors.

3-2.4 Feeder Design Summary

The feeder or service circuit supplying energy to any occupancy connects to branch circuits for lights, receptacles, heating units, air-conditioning equipment, appliances, and motor loads plus any special equipment to be served. The technique used to calculate the total feeder load begins by separating the individual loads into these categories since different rules may apply to each type of load. As has been shown in previous subsections, each category may have to be further divided, as is the case with fluorescent lights and incandescent lamps on the same feeder. The total feeder load is then simply the sum of the maximum demand load caused by each category or subcategory of loads. If motor loads are present, the feeder protective device and also its disconnecting means, if required, depend on the rating of the largest motor branch circuit.

A feeder one-line diagram is shown in Figure 3-26. Each branch circuit shown may actually represent one or more branch circuits for each type of load. The rules illustrated apply regardless of type of occupancy except as noted in the figure. **TABLE 220-13** If the receptacle loads are not considered continuous, a demand factor may be applied. **TABLE 220-19** When they are used in a dwelling, electric ranges and electric clothes dryers are **TABLE 220-18** subject to the demand factors listed in the appropriate Code tables.

When a feeder supplies motor loads and other loads, the feeder overcurrent protective device must be large enough in amperes to allow the motor with the largest starting current to start as well as carry the current from the other loads. When **220-10** continuous loads are present, the overcurrent protective device will be selected to carry 125% of the current from the continuous loads in addition to that required by any noncontinuous loads and the motor loads; unless the device is listed for operation at 100% of its rating.

If the continuous load is significantly larger than the other loads, the feeder conductors sized according to the connected load plus motor loads could be calculated to have an ampacity about 25% lower than the rating of the overcurrent device. For example, a 400-ampere continuous load and an 8-ampere motor with a 20-ampere branch-circuit overcurrent device would be protected by a feeder overcurrent protective device set at less than

$$(400 \text{ A} \times 1.25) + 20 \text{ A (largest branch-circuit device)} = 520 \text{ A}.$$

Based on the connected load, the feeder conductors require an ampacity of only

$$400 \text{ A} + (8 \text{ A} \times 1.25) = 410 \text{ A}.$$

In this design, the conductors would not be protected in accordance with their am- **240-3** pacities due to the higher capacity of the feeder overcurrent device required by the continuous load. A more conservative design would be to select the conductor ampacity and the overcurrent protective device using 125% of the continuous load. In the example, the conductor ampacity would then be

$$(400 \text{ A} \times 1.25) + (8 \text{ A} \times 1.25) = 510 \text{ A}.$$

A standard 500-ampere overcurrent device could be used and the conductors would be properly protected.

When a feeder or service supplies continuous loads plus motor loads in the examples given in this guide, the design will be based on 125% of the continuous load added to the load represented by the motors.

Summary Examples. The steps required to calculate the feeder or service capacity are outlined in Figure 3-27. The application of these steps and a form in

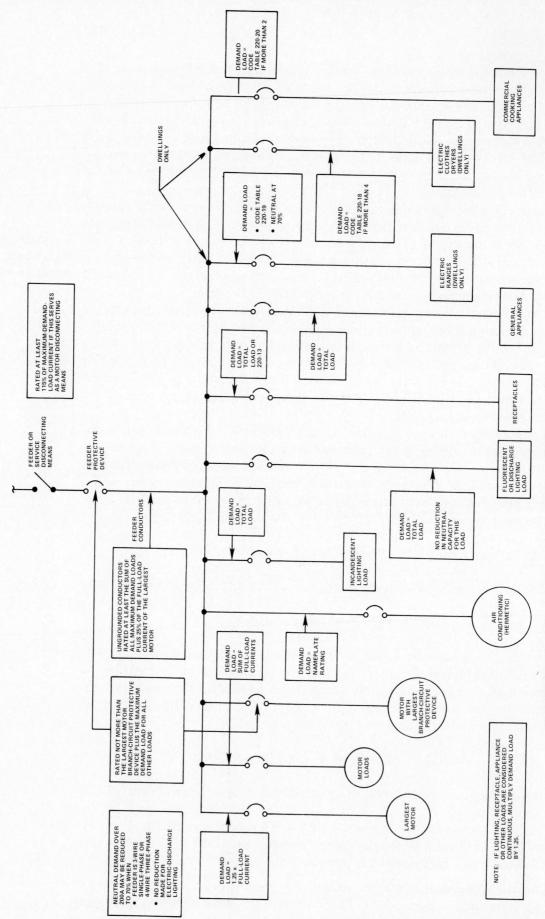

Figure 3-26. Basic Requirements for Feeder Supplying Mixed Loads

102

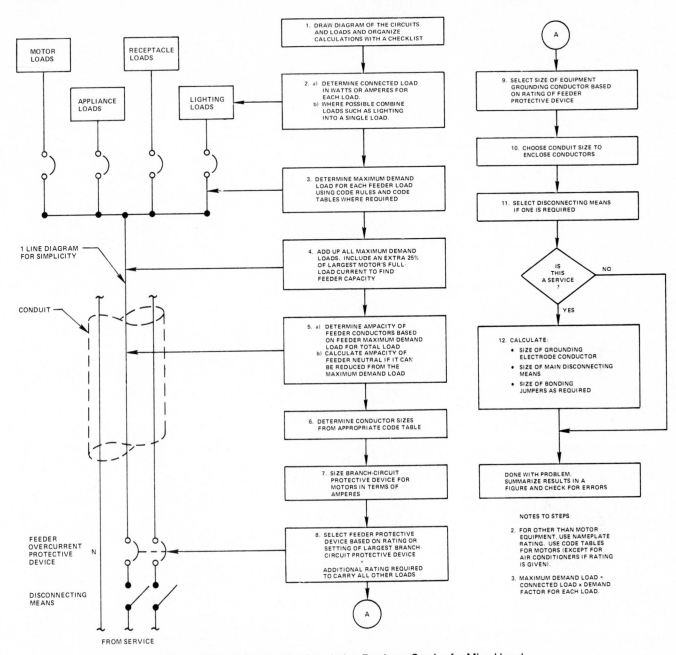

Figure 3-27. Procedure for Calculating Feeder or Service for Mixed Loads

Figure 3-28 for use with the calculations of feeder capacity can be demonstrated with an example that combines several calculations for feeders previously given.

Assume that a 120/240-volt feeder supplies the following continuous loads:

  a. A 12 500-volt-ampere lighting load
  b. Two hundred receptacles as a continuous load
  c. A 5-horsepower, 230-volt motor with circuit breaker protection
  d. A 10-horsepower, 230-volt motor with fuse protection
  e. A 30-kilovolt-ampere (continuous rating), 240-volt heating unit

The conductors used for the circuits are THW copper conductors.

The organization form of Figure 3-28 should be completed to show a diagram, a list of the loads, and a checklist. Code references should be noted for each step of

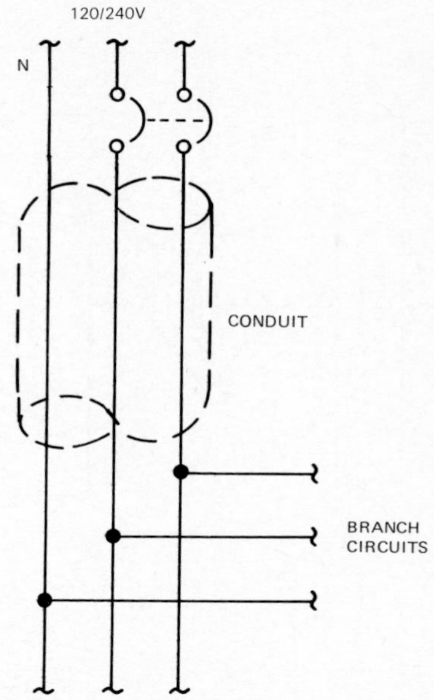

CONNECTED LOAD	CHECKLIST
LIGHTS	LOADS ✓
RECEPTACLES	UNGROUNDED CONDUCTORS ✓
5 HP MOTOR	NEUTRAL ✓
10 HP MOTOR	OVERCURRENT PROTECTIVE DEVICE ✓
HEATING	EQUIPMENT GROUNDING CONDUCTOR ✓
	RACEWAYS ✓

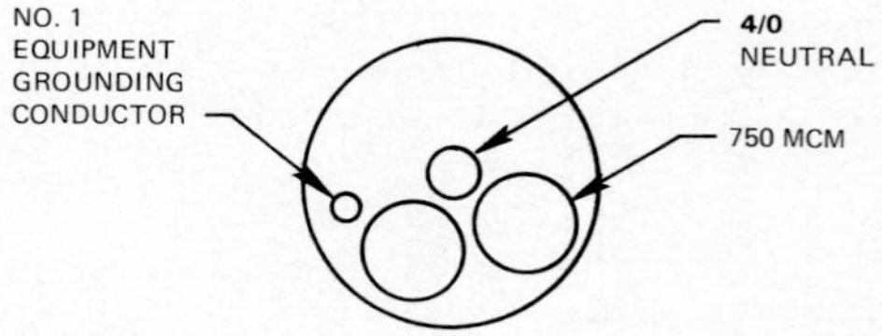

CODE REFERENCE	LOAD	FEEDER DEMAND LINE LOAD	FEEDER DEMAND NEUTRAL LOAD
FEEDER LOADS 220-3(c)(4)	LIGHTING LOAD 12,500VA AT 120/240V $I_L = \frac{12,500VA}{240V} \times 1.25 = 65.1A$	65.1A	52A
	RECEPTACLE LOAD 200 RECEPTACLES AT 180 VA EACH $I_L = \frac{200 \times 180VA}{240V} \times 1.25 = 187.5A$	187.5A	150A
TABLE 430-148	MOTORS 5 HP AT 240V = 28A 10 HP AT 240V = 50A	28A 50A	0 0
430-25	25% OF LARGEST MOTOR .25 × 50A = 12.5A	12.5A	0
	HEATING 30 KVA AT 240V $I_L = \frac{30,000VA}{240V} = 125A$	125A	0
	TOTAL CURRENT	468.1A ✓	202A ✓
CONDUCTOR SIZE TABLE 310-16 NEUTRAL 220-22	UNGROUNDED CONDUCTORS: 468.1 A MINIMUM, THW SIZE = 750 MCM ✓ NEUTRAL CONDUCTOR: MINIMUM AMPACITY = 200A + .7 × (202A - 200A) = 201A THW SIZE = 4/0 ✓		
OVERCURRENT PROTECTION 430-52 430-63	a) FIND RATING OF LARGEST BRANCH-CIRCUIT PROTECTIVE DEVICE FOR MOTOR 5 HP MOTOR WITH CIRCUIT BREAKER SET AT 700% × 28A = 196A 10 HP MOTOR WITH FUSE RATED AT 300% × 50A = 150A USE 196A (OR 200A STANDARD SIZE) b) FEEDER PROTECTION 200A + 65.1A + 187.5 + 50A + 125A = 627.6A, USE 600A ✓		
EQUIPMENT GROUND CONDUCTOR 250-95	SELECT SIZE OF CONDUCTOR BASED ON RATING OF OVERCURRENT PROTECTIVE DEVICE 600A DEVICE REQUIRES NO. 1 CONDUCTOR — NO. 1 ✓		
CONDUIT SIZE CHAPTER 9, TABLES 4 AND 5	a) CALCULATE CONDUCTOR AREA (2) 750 MCM = 2 × 1.2252 IN² = 2.4504 IN² (1) 4/0 = .3904 IN² (1) NO. 1 = .2027 IN² 3.0435 IN² b) SELECT SMALLEST CONDUIT WHOSE 40% FILL AREA IS LARGER THAN TOTAL CONDUCTOR AREA 3-1/2 IN CONDUIT HAS 40% FILL AREA OF 3.96 IN. — 3-1/2 IN ✓		

Figure 3-28. Suggested Organization of Feeder Capacity Calculation

104

the calculation. A decision is first made about the units (volt-amperes or amperes) to be used to calculate the total feeder load. Amperes were chosen in this case since motor loads are given in amperes. The line and neutral loads are calculated separately since the 240-volt loads do not require a neutral.

As outlined in Figure 3-27, the connected load is converted to the maximum demand load for the ungrounded conductors and the neutral. Since the largest motor draws 50 amperes, a value of .25 × 50 amperes = 12.5 amperes is added to the line load. Once the required ampacity of each conductor is known, the conductor size is selected. The 468.1 ampere load requires size 750 MCM THW copper conductors. The neutral conductor size is considerably less than that of the ungrounded conductors because of the 240-volt loads and the fact that its size is based on the connected load. It can be reduced even further for the portion exceeding 200 amperes before the conductor size was chosen. The neutral load of 202 amperes from the circuits is reduced to 201 amperes, which allows a size 4/0 conductor to be used. **TABLE 430-148**

The feeder protective device cannot be chosen until the protective devices for the motor branch circuits have been selected. A 5-horsepower motor circuit may be protected by an instantaneous trip circuit breaker as large as 7 × 28 amperes = 196 amperes. A standard 200-ampere device is selected. A nontime-delay fuse protecting the 10-horsepower motor circuit could be as large as 3 × 50 amperes = 150 amperes. In this case, the circuit breaker of the smaller motor determines the rating of the feeder protective device. The rating of the feeder protective device, then, is 200 amperes plus the additional rating required to carry the other loads. The total of 627.6 amperes requires the feeder protective device to be a standard 600-ampere device. **TABLE 430-152** **430-52** **EXCEPTION 1** **430-63**

The calculations of the equipment grounding conductor size and the conduit size proceed as shown in previous examples. A summary of the solution and a view of the filled conduit are included in Figure 3-28 to complete the problem.

If the feeder is a three-phase circuit, the calculations are similar except that the *line load* is the current in each of three conductors for a balanced load. Again, it is recommended that an organizational form such as the example shown in Figure 3-29 be prepared to minimize the chances of error.

Assume that a 480Y/277-volt feeder supplies the following loads:

a. A 100-kilovolt-ampere lighting load balanced on the feeder
b. Two 10-horsepower, three-phase, 480-volt motors
c. A 50-horsepower, three-phase, 480-volt motor

Type TW copper conductors are used for the feeder and the motor branch circuits are protected by nontime-delay fuses. A feeder disconnecting means is also required, but the conduit serves as the equipment grounding conductor.

The balanced lighting load, if continuous, represents a feeder capacity of **220-10**

$$1.25 \times \frac{100\ 000\ \text{VA}}{\sqrt{3} \times 480\ \text{V}} = 150\ \text{A}$$

and consists of 277-volt lamps connected between a phase and the neutral. The motor currents and 25% of the largest motor current are added to find the total feeder load of 259 amperes. Type TW conductors of size 350 MCM are used for the ungrounded conductors and size No. 1/0 is used for the neutral based on a current of 120.3 amperes. **TABLE 310-16**

The branch-circuit fuse for the 50-horsepower motor with a full-load current of **TABLE 430-150**

CONNECTED LOADS	CHECKLIST	
LIGHTS	LOADS	✓
(2) 10 HP MOTORS	UNGROUNDED CONDUCTORS	✓
(1) 50 HP MOTOR	NEUTRAL	✓
	OVERCURRENT PROTECTIVE DEVICE	✓
	DISCONNECT	✓
	CONDUIT	✓

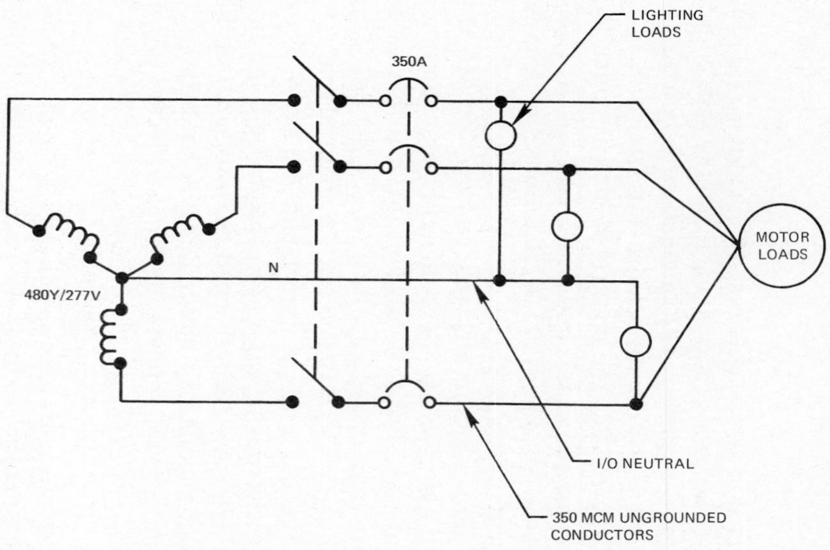

CODE REFERENCE	LOAD	FEEDER DEMAND	
		LINE LOAD	NEUTRAL LOAD
FEEDER LOADS / TABLE 420-150	LIGHTING LOAD 100 KVA AT 227/480V $I_L = \dfrac{100,000VA}{\sqrt{3} \times 480V} \times 1.25 = 150A$	150A	120.3A
	MOTOR LOADS (2) 10 HP AT 480V 2 x 14A = 28A	28A	0
	(1) 50 HP AT 480V = 65A	65A	0
	25% OF LARGEST MOTOR .25 x 65A = 16.3A	16.3A	0
	TOTAL CURRENT	259A ✓	120.3A ✓
CONDUCTOR SIZE / TABLE 310-16	UNGROUNDED CONDUCTORS 259A MINIMUM	TW SIZE = 350 MCM ✓	
	NEUTRAL CONDUCTOR 125A MINIMUM	TW SIZE = I/O ✓	
OVERCURRENT PROTECTION / TABLE 430-152, 240-6, 430-63	a) FIND RATING OF LARGEST BRANCH-CIRCUIT PROTECTIVE DEVICE FOR MOTORS ✓ 50 HP: FUSE SIZE = 300% x 65A = 195A, USE STANDARD 200A ✓ b) FEEDER PROTECTION 200A + 150A + 28A = 378A LOWER STANDARD SIZE = 350A ✓		
DISCONNECTING MEANS	115% OF TOTAL LOAD CURRENT (CONNECTED LOAD) LOAD CURRENT = 120A + 14A + 14A + 65A = 213A RATING OF DISCONNECT = 115% x 213 = 245A ✓		
CONDUIT SIZE / CHAPTER 9, TABLE 4, TABLE 5	AREA OF CONDUCTORS (3) 350 MCM TW = 3 x .6291 IN² = 1.8873 IN² (1) NO. I/O TW = 0.2367 IN² 2.1240 IN² REQUIRES A 3 IN. CONDUIT 3 IN. ✓		

Figure 3-29. Feeder Calculations for Three-Phase Circuit

65 amperes can be as large as 3 × 65 amperes = 195 amperes. The next higher standard-size fuse would be rated 200 amperes. The rating of the feeder protective device may not exceed a value of 200 amperes plus the additional rating required for the other loads, or a total of 378 amperes. The closest standard-size device rated less than 378 amperes is a 350-ampere device.   **430-52 EXCEPTION 1** **430-63**

The size of the disconnecting means is determined by considering the total load as a *single* motor by taking the summation of all loads including the lighting load at the full-load condition for the motors. The rating of the disconnecting means must not be less than 115% of this value. For a total load of 213 amperes, the disconnecting means must be rated at least 1.15 × 213 amperes = 245 amperes. This disconnecting means is used to disconnect the feeder circuit from its source of supply. The individual motors would be required to have their own disconnecting means except in special circumstances.   **430-110**

The conduit must enclose three 350 MCM conductors and one No. 1/0 conductor. A 3-inch conduit is required to satisfy the Code rules.   **CHAPTER 9 TABLE 5, TABLE 4**

QUIZ

(Closed-Book)

1. Convert the following motor loads to a load in volt-amperes:
   (a) A 1-horsepower, 115-volt, single-phase, full-load current = 16 amperes
   (b) A 5-horsepower, 230-volt, single-phase, full-load current = 28 amperes
   (c) A 10-horsepower, 460-volts, three-phase, full-load current = 10 amperes
2. When is the feeder load simply equal to the sum of the branch-circuit loads?
3. Under what circumstances could the feeder disconnecting means disconnect several motors as well as other loads.

(Open-Book)

1. Determine the feeder rating for the following loads served by a 480Y/277-volt, three-phase feeder:
   (a) A 14-kilovolt-ampere, 277-volt, lighting load that is continuous
   (b) A 30-horsepower, 480-volt motor
2. Design the feeder circuit that supplies the following loads:
   (a) Two 4-kilowatt wall-mounted ovens and one 6-kilowatt counter-mounted cooking unit in a dwelling
   (b) A 4.5-kilowatt, 240-volt hot water heater
   (c) A 5-kilowatt, 240-volt electric clothes dryer
   (d) Six 7-ampere, 240-volt air-conditioning units
   (e) A 5-kilowatt, 120-volt lighting load that is not continuous. The feeder is 120/240 volts and uses type THW aluminum conductors.

3-2.5 Special Feeder Problems

This subsection discusses various special cases that may arise when feeder circuits are being designed. These cases are common when very large loads or industrial loads are served. The design approach is basically the same as that presented previously in the Guide. The special cases to be discussed include parallel circuits, transformer circuits, and unbalanced loads on three-phase circuits.

Feeders with Parallel Conductors. The Code allows conductors of size No. 1/0 and larger to be connected in parallel when all of the conductors are identical. This is necessary when the load is so large that a single conductor would not have sufficient ampacity to carry the load. For instance, if a three-wire feeder must carry 1000 amperes and if it is enclosed in a raceway, no copper conductors listed in the Code   **310-4**

TABLE 310-16 table of ampacities is sufficient to carry the load. In this case, two or more smaller conductors are combined in parallel to act as each of the feeder conductors.

TABLE 310-17 Based simply on the area of the conductors, connecting two conductors in parallel doubles the current-carrying capacity. Thus, a 350 MCM THW copper conductor in free air (not in a raceway) can carry 505 amperes and two conductors in parallel would have an ampacity of 2×505 amperes = 1010 amperes as shown in Figure 3-30(a) and (b). If a third conductor were added in parallel, the ampacity would be 3×505 amperes = 1515 amperes, and so on.

TABLE 310-16 In a raceway, a circuit using three 350 MCM THW copper conductors would have an ampere rating of 310 amperes each, as shown in Figure 3-30(c). If each phase

TABLE 310-16 NOTE 8 conductor were paralleled as two 350 MCM conductors, as shown in Figure 3-30(d), the raceway would contain six conductors. Because of the heating effect, the Code requires that the maximum allowable load current of each conductor be reduced to 80% of its normal value. Thus, the six-conductor circuit has a capacity of only

$$.8 \times 620 \, A = 496 \, A$$

TABLE 310-16 NOTE 10 When more than six current-carrying conductors are contained in a raceway, the Code requires a reduction greater than 80%. The table in Figure 3-30(e) lists these Code factors. Notice that only current-carrying conductors are counted. If electric-discharge lighting is not served, the neutral conductor in balanced circuits is not counted.

To carry the 1000-ampere load of the example using THW 500 MCM conductors in a 480-volt circuit would require at least four conductors in parallel for each phase as shown in Figure 3-30(f). This was determined as follows:

1. It is clear that at least three conductors per phase are needed at 380 amperes each, but that would reduce the ampacity of each conductor to 70%, or $.7 \times 380$ amperes = 266 amperes since nine conductors would be in the raceway. Thus, three conductors in parallel can carry only 3×266 amperes = 798 amperes on each phase.
2. The ampacity of four 500 MCM conductors in parallel is 4×380 amperes = 1520 amperes. Since there would be $3 \times 4 = 12$ conductors for the three-phase circuit, this must be reduced to 70% or $.7 \times 1520$ amperes = 1064 amperes. This ampacity is sufficient to carry the load.

300-20(a)

TABLE 310-16 Another solution to the problem of providing a 1000-ampere feeder or service is to use three raceways. This is permitted as long as all phase conductors, the neutral, and equipment grounding conductors for a complete circuit are run together in each conduit. Since each 500 MCM THW circuit has an ampacity of 380 amperes when enclosed in conduit, three such circuits in parallel could serve a load of 3×380 amperes = 1140 amperes. This is sufficient for our example. With only three current-carrying conductors in each conduit, no derating is necessary as is shown in Figure 3-30(g).

TABLE 250-95 If the conduit does not also serve as the equipment grounding conductor, the circuit requires three No. 2/0 copper conductors for grounding based on the 1000-ampere overcurrent device. *Each* conduit must contain one of these grounding conductors and the three must also be connected in parallel.

Feeders with Transformers. If a feeder or service contains a transformer to step up or step down the supply voltage, the connected load in amperes must be converted to the equivalent load in amperes at the primary of the transformer. Thus, a 100-ampere load connected to a 120-volt secondary requires only 50 amperes from the primary circuit if the primary is 240 volts. If both the primary and secondary

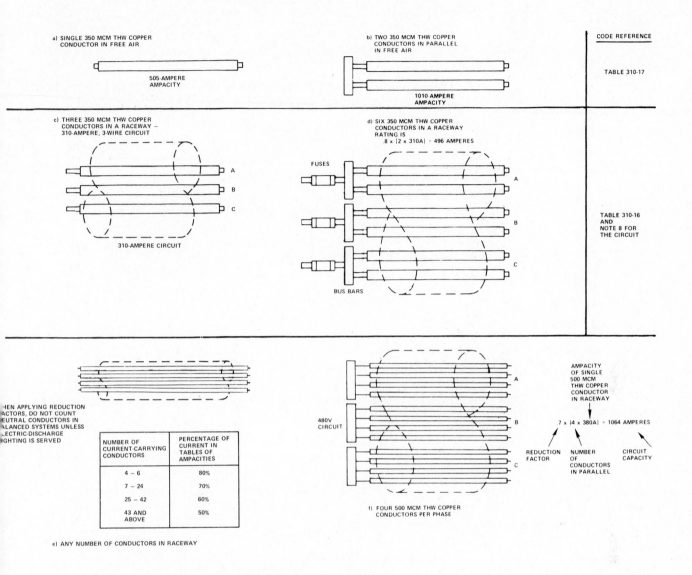

a) SINGLE 350 MCM THW COPPER CONDUCTOR IN FREE AIR

505-AMPERE AMPACITY

b) TWO 350 MCM THW COPPER CONDUCTORS IN PARALLEL IN FREE AIR

1010-AMPERE AMPACITY

CODE REFERENCE

TABLE 310-17

c) THREE 350 MCM THW COPPER CONDUCTORS IN A RACEWAY – 310-AMPERE, 3-WIRE CIRCUIT

A
B
C

310-AMPERE CIRCUIT

d) SIX 350 MCM THW COPPER CONDUCTORS IN A RACEWAY RATING IS .8 x (2 x 310A) = 496 AMPERES

FUSES

A

B

C

BUS BARS

TABLE 310-16 AND NOTE 8 FOR THE CIRCUIT

...HEN APPLYING REDUCTION ...ACTORS, DO NOT COUNT ...EUTRAL CONDUCTORS IN ...ALANCED SYSTEMS UNLESS ...LECTRIC-DISCHARGE ...IGHTING IS SERVED

NUMBER OF CURRENT-CARRYING CONDUCTORS	PERCENTAGE OF CURRENT IN TABLES OF AMPACITIES
4 – 6	80%
7 – 24	70%
25 – 42	60%
43 AND ABOVE	50%

e) ANY NUMBER OF CONDUCTORS IN RACEWAY

480V CIRCUIT

A

B

C

AMPACITY OF SINGLE 500 MCM THW COPPER CONDUCTOR IN RACEWAY

7 x (4 x 380A) = 1064 AMPERES

REDUCTION FACTOR   NUMBER OF CONDUCTORS IN PARALLEL   CIRCUIT CAPACITY

f) FOUR 500 MCM THW COPPER CONDUCTORS PER PHASE

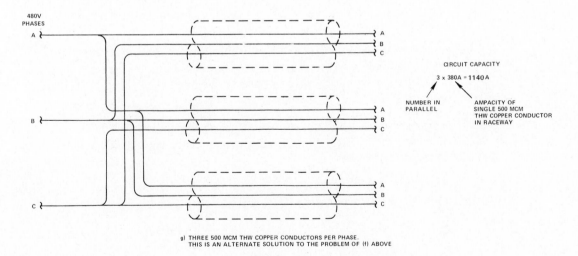

480V PHASES

A

B

C

A
B
C

A
B
C

A
B
C

CIRCUIT CAPACITY

3 x 380A = 1140 A

NUMBER IN PARALLEL   AMPACITY OF SINGLE 500 MCM THW COPPER CONDUCTOR IN RACEWAY

g) THREE 500 MCM THW COPPER CONDUCTORS PER PHASE. THIS IS AN ALTERNATE SOLUTION TO THE PROBLEM OF (f) ABOVE

Figure 3-30. Parallel Conductors

circuits are single-phase or if both are three-phase, the conversion process is a simple one.

The input kilovoltamperes for a transformer is equal to the output kilovoltamperes if very small losses in the transformer are neglected. The basic formula is as follows:

$$kVA_{in} = kVA_{out}$$

Thus, for a single-phase transformer as in the example above

$$I_{primary} \times V_{primary} = I_{secondary} \times V_{secondary}$$

where I is the current and V is the voltage as designated. For the 120/240-volt transformer with a 100-ampere load, the primary current is then

$$I_{primary} = \frac{I_{secondary} \times V_{secondary}}{V_{primary}}$$

$$= \frac{100 \text{ A} \times 120 \text{ V}}{240 \text{ V}} = 50 \text{ A}$$

For three-phase transformers with balanced loads, the relationship between the power input and the power output is

$$\sqrt{3} \times I_{primary} \times V_{primary} = \sqrt{3} \times I_{secondary} \times V_{secondary}$$

and the factor $\sqrt{3} = 1.73$ cancels from both sides of the equation. Thus, a 208Y/120-volt secondary with a load of 100 amperes supplied by a 480-volt primary transformer requires a primary current of

$$\frac{\sqrt{3} \times I_{secondary} \times V_{secondary}}{\sqrt{3} \times V_{primary}} = \frac{100 \text{ A} \times 208 \text{ V}}{480 \text{ V}}$$

$$= 43.3 \text{ A}$$

The fact that the secondary is a four-wire circuit does not matter as long as the load is balanced. The voltage between phases is used as the secondary voltage.

As an example, a 480-volt, three-phase service supplies three transformers as shown in Figure 3-31(a). The loads include the following:

   a. 100 amperes from a 208Y/120-volt transformer secondary
   b. 300 amperes from a 480Y/277-volt transformer secondary
   c. 200 amperes at 480 volts

The size of type THW copper service conductors is to be computed. The loads draw the following primary current at 480 volts:

$$\text{a. } 100 \text{ A} \times \frac{208 \text{ V}}{480 \text{ V}} = 43.3 \text{ A}$$

$$\text{b. } 300 \text{ A} \times \frac{480 \text{ V}}{480 \text{ V}} = 300.0 \text{ A}$$

$$\text{c. } 200\,A \times \frac{480\,V}{480\,V} = \underline{200.0\,A}$$

$$\text{Total} \qquad 543.3\,A$$

Figure 3-31. Sizing Feeder Conductors for Transformer Circuits

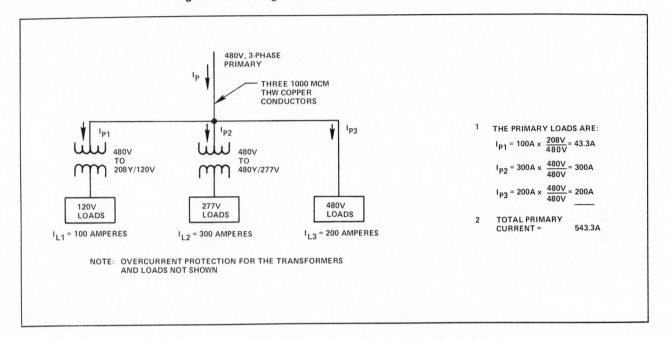

The service conductors must have an ampacity of at least 543.3 amperes. This would require three 1000 MCM THW copper conductors if the circuit is enclosed in a raceway.

If the loads are given in watts, the problem of sizing the conductors is simply one of converting the total load in watts to a load in amperes at the primary voltage. For example, if the loads of Figure 3-31 had been given in watts, they would have been:

  a. A 36-kilowatt load on a 208/Y120-volt transformer secondary
  b. A 249.4-kilowatt load on a 480Y/277-volt transformer secondary
  c. A 166.3-kilowatt load at 480 volts

In that case, the total load would be 36 kilowatts + 249.4 kilowatts + 166.3 kilowatts = 451.7 kilowatts. At 480 volts, the primary load in amperes would be

$$\frac{\text{load in watts}}{\sqrt{3} \times V_{\text{primary}}} = \frac{451\,700\,W}{1.73 \times 480\,V} = 543.3\,A$$

This is the same result as obtained before.

Loads with Power Factors. When the power factor is less than unity, the kilovolt-ampere input is greater than the power in kilowatts required by the load. The kilovolt-ampere input determines the size of the conductors and other equipment in the circuit. For any load,

$$kVA = \frac{\text{load in kilowatts}}{\text{power factor (pf)}}$$

where the power factor is a number less than or equal to 1.0. Thus, a 50-kilowatt load with a power factor of .7 requires a kilovolt-ampere supply of

$$kVA = \frac{50\,kW}{.7} = 71.4\,kVA$$

For a 480-volt, three-phase load, the conductor ampacity must be at least

$$\frac{71\,400\,VA}{\sqrt{3} \times 480\,V} = 85.9\,A$$

even though the power in kilowatts is only 50 kilowatts.

Unbalanced Loads on Three-Phase Feeders. In some cases, the single-phase loads connected to a three-phase feeder may not be equally distributed between the phases in three-wire circuits or between each phase and neutral in four-wire circuits. The standard practice is to add the single-phase load in amperes for each phase to any three-phase loads to compute the total current per phase. A common example occurs when a "high-leg" delta transformer of 120/240 volts supplies both single-phase and three-phase loads. In this case, the 120-volt loads are connected between phases A and C as shown in Figure 3-32. The voltage between phases is 240 volts and the voltage between phase B and the neutral is 208 volts (the "high-voltage leg").

The load for the 120/240-volt circuit shown in Figure 3-32 is as follows:

    a. A 30-kilowatt lighting load (continuous)
    b. A 10-horsepower, 240-volt, three-phase motor
    c. A 7½-horsepower, 240-volt, three-phase motor

If the lighting panel and motor-feeder panel are separate, each would be protected by an overcurrent protective device (selected to be a circuit breaker in the example). Each motor branch circuit is assumed to be protected by a nontime-delay fuse. The size of the type TW copper supply conductors and the feeder and service overcurrent devices must be determined.

The total load in amperes is calculated just as in other examples given previously. Phase B, which supplies *only* the two motors, must have its ampacity calculated separately. The motor loads contribute 57 amperes to each phase as shown in the example. The lighting load contributes 30 000 watts/240 volts × 1.25 = 156 amperes (if the load is continuous) to phases A and C but only 125 amperes to the neutral conductor. The line load contributed by each load is simply added to find the total line load for each conductor. Since lines A and C must carry 213 amperes, they require 250 MCM TW copper conductors. Line B requires an ampacity of only 57 amperes; therefore, a No. 4 TW conductor is sufficient. The neutral requires a No. 1/0 TW conductor to carry a load of 125 amperes.

TABLE 310-16

240-6

220-10

240-3

The overcurrent protective devices must be selected next. With respect to the lighting panel, the overcurrent protective device rating must be at least 156 amperes due to the continuous lighting load. The feeder conductors to this panel could be selected to carry 125 amperes based only on the connected lighting load. However, these conductors must be sized according to the rating of the overcurrent protective

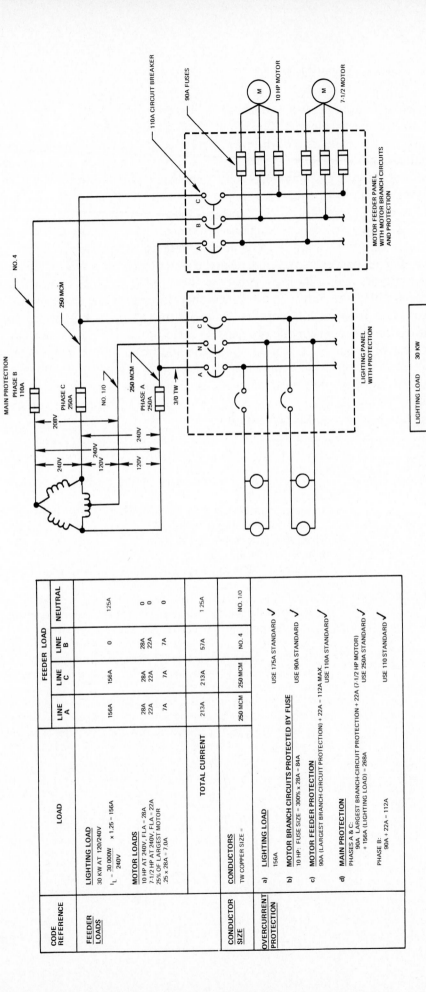

Figure 3-32. Unbalanced Loads on Three-Phase Feeders

113

device. The lighting panel load is 156 amperes. Therefore, a 175-ampere standard circuit breaker is allowed. The panel could be supplied by 3/0 TW conductors. The motor-feeder panel overcurrent protective device depends on the size of the largest branch-circuit protective device for any motor. The 10-horsepower motor with a full-load current of 28 amperes may be protected by a nontime-delay fuse set at not more than 300% of this value, or 84 amperes. The next larger standard-size fuse is 90 amperes. The motor feeder panel protective device can be as large as 90 amperes plus the full-load current of the other motor. The maximum size is 112 amperes, which requires a standard 110-ampere, three-phase circuit breaker.

The main feeder overcurrent protective device for a motor feeder that supplies both motors and other loads must conform to the rules for motor circuits. Phase B supplies only the motor loads and is therefore rated the same as the motor-feeder protective device of 110 amperes. Phases A and C supply both types of loads and the protection for these conductors must be rated no larger than the largest branch-circuit protective device for any motor plus the additional rating to carry the other loads. This is 90 amperes + 22 amperes (7½-horsepower motor) + 156 amperes (lighting) = 268 amperes. The next smaller standard-size device is rated at 250 amperes. Thus, the feeder could be protected by two 250 ampere fuses for phases A and C and one 110-ampere fuse for phase B.

Margin references:
Table 430-150
Table 430-152
430-63
430-63
240-6

QUIZ
(Closed-Book)

1. If two conductors of the same size are connected in parallel, the ampacity _____.
2. How many current carrying conductors must be enclosed in a raceway before their ampacity must be reduced?
3. What is the basic rule for transformers?
4. Define power factor.
5. What are the line-to-line and line-to-neutral voltages for a "high-leg" delta.

(Open-Book)

1. A transformer with a 1000-volt primary has a 10-ampere, 100-volt load on the secondary. What are the primary current and the kilovolt-amperes?
2. What is the current drawn by a 50-kilowatt, single-phase load at 240 volts if the power factor is .8?
3. A 40-kilovolt-ampere fluorescent lighting load is supplied by a 480Y/277-volt, three-phase circuit. If the power factor of the load is .85, calculate (a) the feeder current and (b) the power per phase.
4. A 277-volt fluorescent lighting load on a 480Y/277-volt circuit is 831.38 kilovolt-amperes. If the circuit is in conduit, what ampacity is required of the feeder conductors since the circuit has four current-carrying conductors? How many 500 MCM THW copper conductors are required in parallel for each phase?

TEST CHAPTER 3

1. True or False *(Closed-Book)*

		T	F
1.	The branch-circuit rating is determined by the conductor ampacity.	[ ]	[ ]
2.	Heavy-duty type lampholders are required on branch-circuits having a rating in excess of 20 amperes.	[ ]	[ ]
3.	Continuous loads on branch circuits may not exceed 75% of the rating of the branch circuit.	[ ]	[ ]
4.	A 30-ampere branch circuit may supply fixed lighting units in a dwelling.	[ ]	[ ]
5.	An individual branch circuit may supply only one outlet.	[ ]	[ ]
6.	The feeder neutral may be smaller than the ungrounded conductors in some cases.	[ ]	[ ]
7.	The branch-circuit load for a 12-kilowatt electric range in a dwelling is 8 kilowatts.	[ ]	[ ]
8.	An appliance always requires a disconnecting means.	[ ]	[ ]
9.	A 40-ampere branch circuit may supply several fixed electric space heaters in a dwelling.	[ ]	[ ]
10.	The rating of the overcurrent protective device for a circuit supplying a hermetic motor-compressor may not exceed 225% of the rated load current under any circumstances.	[ ]	[ ]

II. Multiple Choice *(Closed-Book)*

1. The maximum continuous load connected to a 20-ampere branch circuit must not exceed:
   - (a) 20 amperes
   - (b) 16 amperes
   - (c) 14 amperes

   1. _____

2. The minimum size conductors required for a 15-ampere branch circuit in a cable are:
   - (a) No. 12 THW copper
   - (b) No. 14 THW copper
   - (c) No. 10 UF aluminum

   5. _____

3. The standard classification of branch circuits applies only to those circuits with:
   - (a) Two or more outlets
   - (b) More than two outlets
   - (c) More than three outlets

   3. _____

4. The load for heavy-duty lampholders on a branch circuit shall be at least:
   - (a) 180 voltamperes
   - (b) 600 voltamperes
   - (c) 10 amperes

   4. _____

5. A 10-foot long show window constitutes a feeder load of at least:
   - (a) 1500 volt-amperes
   - (b) 1000 volt-amperes
   - (c) 2000 volt-amperes

   5. _____

6. The branch-circuit conductors supplying a motor that draws 40 amperes must have an ampacity not less than:
   - (a) 40 amperes
   - (b) 50 amperes
   - (c) 60 amperes

   6. _____

7. If no other information is given, a motor that draws 100 amperes must be protected by an overload device rated at not more than:
   - (a) 125 amperes
   - (b) 115 amperes
   - (c) 140 amperes

   7. _____

8. The ampacity of the feeder conductors supplying two motors with full-load currents of 28 amperes and 12 amperes respectively must be:
   (a) 47 amperes
   (b) 40 amperes
   (c) 50 amperes

8. _____

9. If a motor draws 50 amperes and its branch circuit is protected by a 150-ampere nontime-delay fuse, the feeder overcurrent protective device may not be larger than:
   (a) 87.5 amperes
   (b) 150 amperes
   (c) 300 amperes

9. _____

10. The smallest size conductors that may be connected in parallel are:
    (a) No. 14
    (b) No. 10
    (c) No. 1/0

10. _____

III. Problems (Open-Book)

1. How many duplex receptacles are allowed on a 20-ampere branch circuit?

2. Design the branch circuits for the following appliances using THW copper conductors and nontime-delay fuses:
   (a) A blower with a ¼-horsepower, 115-volt motor
   (b) A 2000-watt, 240-volt fixed heater

3. What is the minimum size THW aluminum conductors required for a feeder that supplies the following load?
   (a) 99.72-kilovolt-ampere lighting load at 277 volts (balanced)
   (b) Two motors, one 50 horsepower and one 30 horsepower at 480 volts
   The feeder is a 480Y/277-volt circuit.

4. A commercial kitchen has more than six units which constitute a load of 303.3 kilowatts at 480 volts, three-phase. What is the feeder demand?

5. What is the feeder load for the following electric ranges in a dwelling?
   (a) One 16.6-kilowatt range
   (b) Five 11-kilowatt ranges

6. Design the service to supply a 40-kilowatt lighting demand load and ten 12-kilowatt electric ranges in a dwelling. The service is 120/240 volts and uses THW copper conductors in conduit. Use the conduit for equipment grounding. (*Note:* Since this circuit is a service, a grounding electrode conductor and a main disconnecting means are required.)

7. What is the rating of a 480-volt, three-phase circuit with four No. 3 THW copper conductors in parallel for each phase? The entire circuit is in one conduit.

8. Find the size of a 208Y/120-volt service, using THW aluminum conductors, that supplies the following load:
   (a) 10 000 watts of continuous lighting at 120 volts (balanced)
   (b) A 15-horsepower, 208-volt, three-phase motor
   (c) A 5-horsepower, single-phase motor between phases A and B
   (*Note:* The full-load current for a 208-volt motor is determined by increasing the current for a 230-volt motor by 10%.) The THW aluminum conductors are enclosed in conduit.

9. Design the branch circuits and the feeder circuit for the following 480-volt, three-phase motors:
   (a) 25-horsepower squirrel-cage motor
   (b) 30-horsepower wound-rotor motor
   Use THW copper conductors in conduit and nontime-delay fuses to protect the circuits. Include all elements of the circuits.

IV. Special Problems (Open-Book)

1. What is the primary current of a 100-kilovoltampere transformer with a 480-volt primary?

2. A four-wire circuit of No. 12 THW copper conductors supplies fluorescent lighting. If the conductors are in conduit, what values are required for the following?
   (a) The conductor ampacity
   (b) The branch-circuit overcurrent device
   (*Hint:* There are four current-carrying conductors.)

3. A balanced 15-kilowatt, three-phase load with a power factor of .75 is connected to a 208-volt supply. Calculate the current in each line and the total kilovolt-amperes.

4. A three-phase, 480-volt feeder supplies a transformer with a secondary voltage of 240 volts, three-phase. If the load is 207 846 watts at 240 volts, what is the feeder current?

5. A 5-horsepower, 230-volt, three-phase motor operates at a power factor of .80. Determine the size of THW copper conductors needed for the feeder.

4

Calculations
for Dwelling Type
Occupancies

The discussions and calculations presented in previous chapters of the Guide have concentrated on basic design techniques for premises wiring systems. The techniques were generally applicable to the design of branch circuits, feeders, and services for any occupancy. Wiring design calculations presented in this chapter demonstrate the use of the Code rules that apply only to dwelling occupancies.

For purposes of wiring system design, the Code classifies dwelling types as one-family dwellings, two-family dwellings, and multifamily dwellings. A one-family dwelling or an individual apartment unit in a two-family or multifamily dwelling is designated as a *dwelling unit*.

ARTICLE 100*

The premises wiring system for dwelling occupancies includes branch circuits in a dwelling unit, feeders to individual dwelling units in a multifamily dwelling, and the service. The design of each type of circuit is based on specific rules that apply only to dwellings when so stated in the Code.

The first section of this chapter presents calculations for the design of branch circuits, feeders, services, and complete wiring systems for one-family dwellings and individual dwelling units. Branch-circuit design for dwellings is, for the most part,

*References in the margins are to the specific applicable rules and tables in the *National Electrical Code*.

the same as for any type of occupancy. The design of branch circuits for household electric ranges and the requirement for special circuits in the kitchen and laundry are examples of branch-circuit design rules that apply *only* to dwellings.

ARTICLE 220
PART B
220-30

The Code presents two calculation methods for computing the service or feeder for dwelling units. The first, referred to in the Guide as the *standard calculation method,* may be applied to any dwelling unit. The other, an *optional calculation method,* may be used to compute the service or feeder load for a dwelling unit supplied by a three-wire circuit when the ampacity of the feeder or service-entrance conductors will be 100 amperes or more. When the feeder or service load is large, the optional method usually results in a smaller feeder or service capacity than the standard method.

The second section of this chapter deals with multifamily dwellings. Multifamily dwellings, such as apartment houses, contain three or more individual dwelling units supplied by a single feeder or service. Both a standard and an optional calculation method are presented in the Code for computing the feeder or service load for multifamily dwellings.

ARTICLE 220
PART B

220-32

TABLE 220-32

The standard calculation method applies to any dwelling. This method specifies that a demand factor be applied to the load of certain required branch circuits as well as for the load of electric clothes dryers and electric ranges.

An optional calculation method may be used under certain conditions for multifamily dwellings. Each unit must have a single feeder supplying electric cooking equipment and electric space heating or air conditioning, or both. In the optional method, a single demand factor is applied to the total load of the multifamily dwelling. The demand factor is a percentage of the connected load and decreases with the number of individual units. In this case, the individual demand factors for electric ranges and other loads that are allowed using the standard method do not apply.

Table 4-1 summarizes the conditions for use and gives the Code references that apply to standard and optional calculations for dwellings. Examples of each of these cases are given in this chapter.

The service load for two-family dwellings is generally computed by means of the standard calculation in exactly the same manner as for a multifamily dwelling. The optional calculation for multifamily dwellings may be applied to a two-family dwelling under certain special conditions which will be discussed in this chapter.

The final section of this chapter considers the wiring system design for large apartment complexes with both three-phase and single-phase loads.

4-1 ONE-FAMILY DWELLINGS AND INDIVIDUAL DWELLING UNITS

This section presents examples that apply Code rules to the design of the premises wiring system for one-family dwellings and individual dwelling units. The complete design problem for a one-family dwelling or an individual dwelling unit consists of determining the load to be served, determining the number and ratings of required branch circuits, computing the feeder or service load and, finally, selecting the feeder or service conductors and equipment.

Branch-circuit design is independent of the type of dwelling unit; that is, the branch circuits for an individual dwelling unit in a multifamily dwelling are designed by using the same rules as for one-family dwellings. The branch circuits for general lighting and receptacles, small appliances, laundry, specific fixed and stationary appliances, motors, and other equipment are designed by applying the appropriate Code rules to determine the load for each circuit.

An individual dwelling unit in a multifamily dwelling is normally supplied by a feeder circuit that originates at the service for the multifamily dwelling. The Code rules for feeder design in dwellings determine the rating of the feeder and its overcurrent protective device. In a one-family dwelling the branch circuits begin at the service enclosure. The *service must conform to the rules for feeder circuits as well as to those for services* when the rating of the service is calculated.

The most commonly used method of determining the feeder or service load is the standard calculation method. The optional calculation method may be applied if the conditions for use stated in the Code are met. Both calculation methods are demonstrated by means of separate examples of feeder or service load calculations as well as part of the complete wiring system design examples presented at the end of this section.

The Code also provides an optional calculation method for *additional* loads in an existing dwelling unit. The purpose of this calculation as will be discussed in detail in this section, is to determine the maximum load that may be added to an existing service.

220-31

Table 4-1. Calculation Methods for Computing Feeder or Service Loads in Dwellings

a) Service for One-Family Dwelling
               or
Feeder for Individual Dwelling Unit of Multifamily Dwelling

Method	*Conditions For Use*	*Code Reference*
Standard	Any Case	Article 220, Part B
Optional	• Supplied by single 3-wire, 120/240-volt* or 208Y/120-volt circuit • Ampacity of service-entrance or feeder conductors 100 amperes or more	220-30

b) Service or Feeder for Multifamily Dwelling

Method	*Conditions For Use*	*Code Reference*
Standard	Any Case	Article 220, Part B
Optional	Each individual dwelling must meet the following conditions: 1. Supplied by single feeder 2. Equipped with electric cooking equipment 3. Equipped with either electric space heating, or air conditioning, or both	220-32

*This is a nominal voltage. The actual circuit voltage should be used in load calculations.

ARTICLE 100
110-4

4-1.1 Branch-Circuit Design for Dwellings

The general principles of branch-circuit design presented previously in the Guide are applicable to the design of branch circuits for dwellings. The specific rules presented in this section, however, pertain *only* to dwelling occupancies unless note is made to the contrary. For example, motor branch circuits are designed independently of occupancy.

For a one-family dwelling or an individual dwelling unit, the first design problem is to determine the number and rating of the branch circuits. The Code specifies a

220-3(b)

minimum general lighting load. The size of the load is based on the outside square-foot area of the dwelling unit. These branch circuits supply *both* lighting fixtures and receptacles. A unit load of 3 volt-amperes of capacity are required for each square foot of living area. Thus, a 2500-square foot dwelling unit would require general lighting circuit capacity of

$$3 \text{ VA/ft}^2 \times 2500 \text{ ft}^2 = 7500 \text{ VA}$$

210-23(a)

This capacity must be supplied by a number of 15- or 20-ampere branch circuits. The circuits have a capacity at 120 volts as follows:

Capacity of
15-ampere circuit = 15 A × 120 V = 1800 W

Capacity of
20-ampere circuit = 20 A × 120 V = 2400 W

To supply 7500 volt-amperes, the number of 20-ampere, two-wire circuits required is

$$\frac{\text{required capacity in volt-amperes}}{\text{capacity of 20-A circuit}} = \frac{7500 \text{ VA}}{2400 \text{ VA}} = 3.125$$

or four circuits.

220-4(b)
220-4(c)

In addition to the general lighting and receptacle branch circuits, the following circuits are required in a one-family dwelling or an individual dwelling unit:

210-52(e)
EXCEPTIONS 1 AND 2

a. Two 20-ampere small appliance branch circuits
b. One 20-ampere branch circuit for the laundry if laundry facilities are to be installed
c. Individual branch circuits for specific appliances such as clothes dryers
d. Branch circuits for special loads such as motors or space-heating equipment.

As shown in Figure 4-1, even a small dwelling unit having few appliances requires the general lighting and receptacle branch circuits, two small appliance circuits, and a laundry circuit. Other circuits are provided as needed according to the load to be

TABLE 220-19
NOTE 4

served. In the case of an electric range, the *branch-circuit* ampacity may be determined by the load given in the Code range table.

Figure 4-2 illustrates the branch circuits required in a dwelling unit for various types of household loads. The applicable Code references and the branch-circuit rating is given in each case.

CHAPTER 9
EXAMPLE 1

A simple example will serve to introduce the method of determining the required number and rating of the branch circuits for a small one-family dwelling.

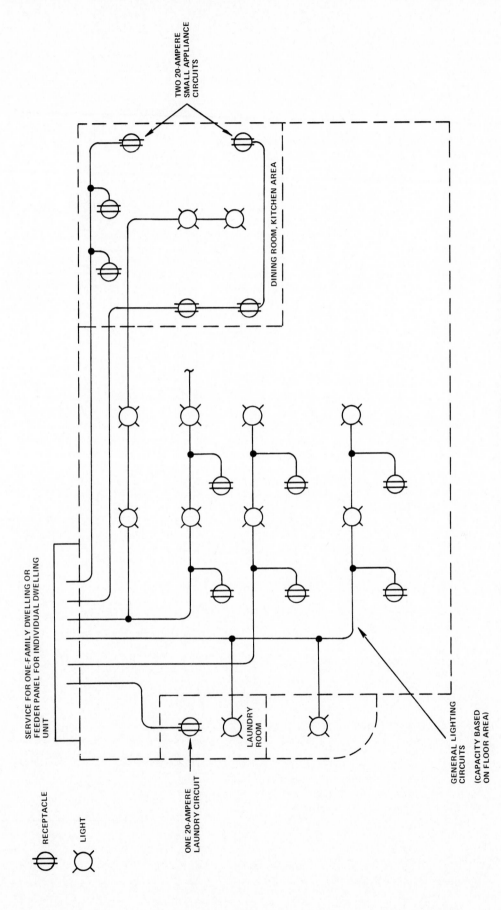

Figure 4-1. Required Branch Circuits for One-Family Dwelling or Individual Dwelling Unit

121

TYPE OF CIRCUIT	BRANCH-CIRCUIT LOAD	CODE REF.
GENERAL LIGHTING AND RECEPTACLES	3VA/FT^2 FOR EACH SQUARE FOOT OF OUTSIDE AREA NUMBER OF CIRCUITS REQUIRED $= \dfrac{3VA/FT^2 \times AREA}{CAPACITY\ OF\ CIRCUIT}$	220-3(b)
INDIVIDUAL APPLIANCE	RATING OF APPLIANCE	220-3(c) 422-5
MOTOR LOAD OR HERMETIC AIR CONDITIONING	125% OF FULL-LOAD CURRENT	430-22 (440-32)
SMALL APPLIANCE CIRCUITS (TWO) REQUIRED	20 AMPERES	220-4 (b)
LAUNDRY CIRCUIT (REQUIRED)	20 AMPERES	220-4(c)
ELECTRIC RANGE	VALUE IN TABLE 220-19	220-19
WATER HEATER (LESS THAN 120 GALLON CAPACITY)	125% OF RATING	422-14 (b)
SPACE HEATING UNITS	125% OF RATING	424-3

Figure 4-2. Branch Circuits in Dwelling Occupancies

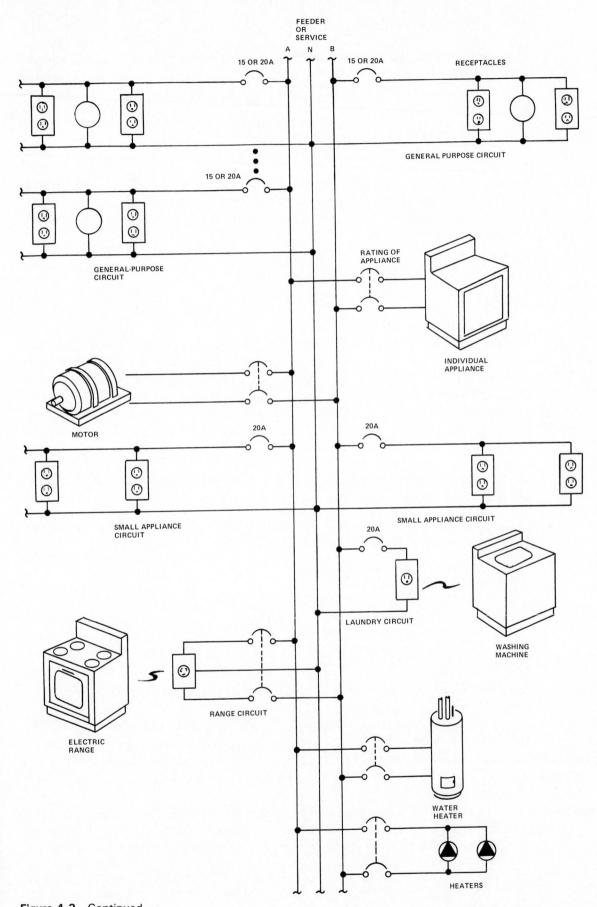

Figure 4.2 Continued.

In this case, a home has a 1500-square foot area and the only appliance is a 12-kilowatt electric range. Figure 4-3 shows the necessary calculations for the dwelling. The number of 20-ampere general lighting and receptacle circuits is

$$\frac{3 \text{ VA/ft}^2 \times 1500 \text{ ft}^2}{120 \text{ V} \times 20 \text{ A}} = 1.9$$

TABLE 220-19

or two circuits. The two small appliance circuits and the laundry circuit must be 20-ampere circuits with a No. 12 equipment grounding conductor.

NOTE 4

The range branch-circuit load is 8 kilowatts for the ungrounded conductors. The load in amperes is

$$\frac{8000 \text{ W}}{240 \text{ V}} = 33.33 \text{ A}$$

210-19(b)
TABLE 310-16
210-19(b)
EXCEPTION 1

220-22

This circuit would require at least a 40-ampere rated branch circuit. Ungrounded conductors of either size No. 8 type TW copper with an ampacity of 40 amperes or size No. 8 THW copper with an ampacity of 50 amperes would be adequate. The minimum neutral ampacity would be .7 × 40 = 28 amperes. A No. 10 conductor may be used for the neutral. Note that this method of determining minimum neutral ampacity based on the rating of the branch circuit applies only to branch circuits. The minimum neutral ampacity for range feeder circuits must be based on the range load.

250-60

A separate equipment grounding conductor is not necessary for this circuit since the range may be grounded to the neutral conductor.

Dwelling Unit with Several Appliances. A more complicated design example is shown in Figure 4-4. The dwelling unit has the following loads:

a. 40 feet by 50 feet of living area
b. 12-kilowatt electric range
c. 1500-watt, 120-volt dishwasher
d. 5-kilowatt, 240-volt electric clothes dryer
e. 30-ampere, 240-volt air conditioner
f. ¹/₃-horsepower, 240-volt air blower
g. 4.5-kilowatt, 240-volt space-heating unit

Type THW copper conductors are used for all circuits and 20-ampere circuits are chosen for general lighting.

The lighting load is 3 volt-amperes/square foot × 40 feet by 50 feet = 6000 volt-amperes which requires three 20-ampere circuits. The two small appliance cir-

TABLE 310-16
TABLE 250-95

cuits and the laundry circuit require three additional 20-ampere circuits. Each circuit would consist of two No. 12 copper conductors and one No. 12 copper equipment grounding conductor.

TABLE 220-19
250-60

The range load is considered to be 8000 watts. Two No. 8 and one No. 10 conductors are used for the circuit. The neutral may serve as the equipment grounding conductor. The dishwasher can be supplied with a 20-ampere circuit with a No. 12

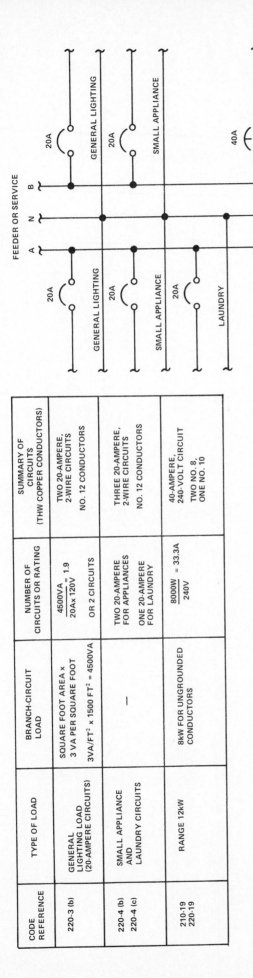

CODE REFERENCE	TYPE OF LOAD	BRANCH-CIRCUIT LOAD	NUMBER OF CIRCUITS OR RATING	SUMMARY OF CIRCUITS (THW COPPER CONDUCTORS)
220-3 (b)	GENERAL LIGHTING LOAD (20-AMPERE CIRCUITS)	SQUARE FOOT AREA x 3 VA PER SQUARE FOOT 3VA/FT² x 1500 FT² = 4500VA	$\frac{4500VA}{20A \times 120V} = 1.9$ OR 2 CIRCUITS	TWO 20-AMPERE, 2-WIRE CIRCUITS NO. 12 CONDUCTORS
220-4 (b) 220-4 (c)	SMALL APPLIANCE AND LAUNDRY CIRCUITS	—	TWO 20-AMPERE FOR APPLIANCES ONE 20-AMPERE FOR LAUNDRY	THREE 20-AMPERE, 2-WIRE CIRCUITS NO. 12 CONDUCTORS
210-19 220-19	RANGE 12kW	8kW FOR UNGROUNDED CONDUCTORS	$\frac{8000W}{240V} = 33.3A$	40-AMPERE, 240-VOLT CIRCUIT TWO NO. 8, ONE NO. 10

NOTE: THE 20-AMPERE CIRCUITS REQUIRE AN EQUIPMENT GROUNDING CONDUCTOR

Figure 4-3. Branch Circuits Required for 1500-Square-Foot One-Family Dwelling with Electric Range

CODE REFERENCE	LOAD	BRANCH-CIRCUIT LOAD	NUMBER OF CIRCUITS OR RATING	SUMMARY OF CIRCUITS (THW COPPER CONDUCTORS)
220-3 (b)	GENERAL LIGHTING	40FT x 50FT = 2000 FT²; 3VA/FT² x 2000FT² = 6000VA	$\frac{6000VA}{20A \times 120V}$ = 2.5 OR 3 CIRCUITS	THREE 20A, 2-WIRE NO. 12 CONDUCTORS
220-4 (b) 220-4 (c)	SMALL APPLIANCE AND LAUNDRY	—	TWO 20A FOR APPLIANCES ONE 20A FOR LAUNDRY	THREE 20A, 2-WIRE NO. 12 CONDUCTORS
220-19 210-19(b)	RANGE 12kW	8kW FOR UNGROUNDED CONDUCTORS	$\frac{8000W}{240V}$ = 33.33A	40A, 120/240V TWO NO. 8, ONE NO. 10 CONDUCTORS
220-4	DISWASHER 1500W	$\frac{1500W}{120V}$ = 12.5A	20-A CIRCUIT	20A, 2-WIRE NO. 12 CONDUCTORS
220-18	CLOTHES DRYER	5000W	$\frac{5000W}{240V}$ = 20.8A USE 25- OR 30-A CIRCUIT	25 OR 30A, 120/240V NO. 10 CONDUCTORS
440-32 440-22	AIR CONDITIONER	1.25 x 30A = 37.5A	PROTECTIVE DEVICE RATING = 1.75 x 30A = 52.5A MAXIMUM STANDARD SIZE IS 50A CONDUCTOR RATING = 37.5A MINIMUM	50A MAXIMUM NO. 8 CONDUCTORS
TABLE 430-148, 430-22	1/3 HP MOTOR 230 VOLTS	1.25 x 3.6A = 4.5 A	15-A CIRCUIT	15A NO. 14 CONDUCTORS
424-3(b)	SPACE HEAT 4.5kW	1.25 x 4500W = 5625W	$\frac{5625W}{240V}$ = 23.4A USE 25- OR 30-A CIRCUIT	25 OR 30A NO. 10 CONDUCTORS

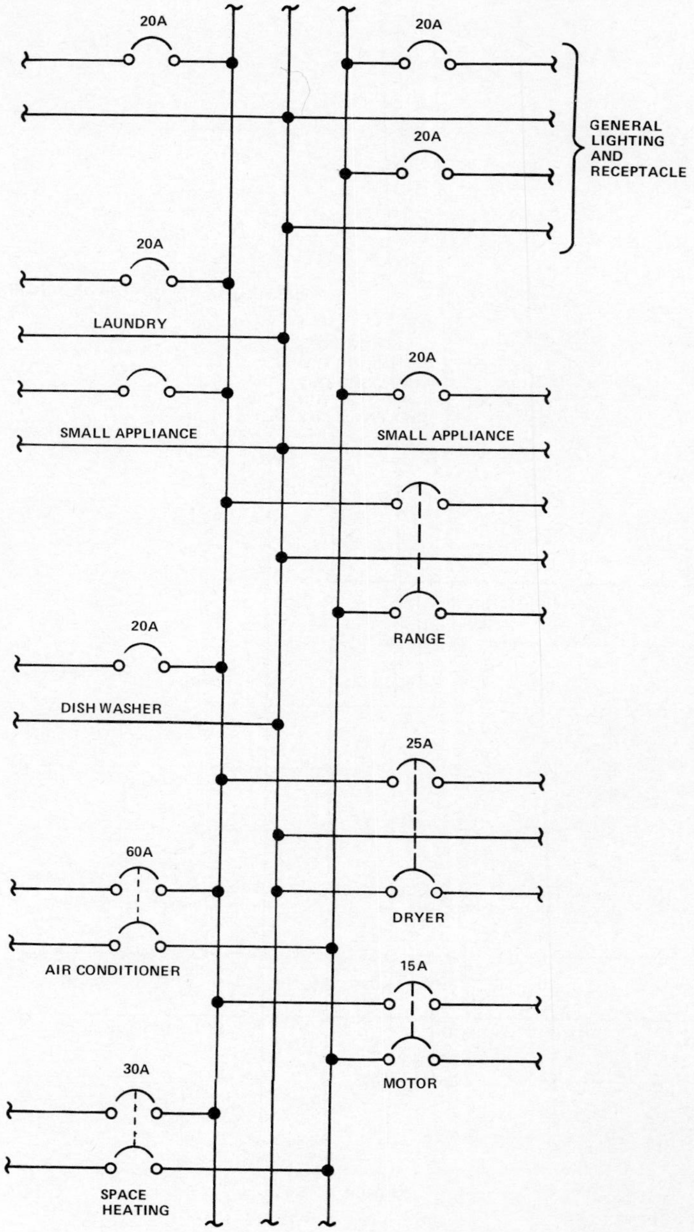

NOTE: ALL CIRCUITS EXCEPT THE RANGE CIRCUIT REQUIRE AN EQUIPMENT GROUNDING CONDUCTOR

Figure 4-4. Branch Circuits Required for 2000-Square-Foot Dwelling with Range, Dishwasher, Dryer, Air Conditioner, Air Blower, and Space Heater

equipment grounding conductor. The dishwasher is a motor-operated appliance, but the motor is not the major load in these appliances and the rating of 1500 watts is assumed to represent the entire load. The clothes dryer is also a motor-operated device, but the same comments for the dishwasher also apply here. The frame of the clothes dryer may be grounded by a No. 10 equipment grounding conductor or by the neutral in some cases. **250-60**

The branch-circuit conductors for the air conditioner must have an ampacity of at least 1.25 × 30 amperes = 37.5 amperes. The branch-circuit protective device may be set at 175% of the rated-load current, or 52.5 amperes, if the air conditioner is assumed to be a hermetically sealed unit. A standard 50-ampere device could be used. The equipment grounding conductor must be a No. 10 copper conductor. The air blower is a ⅓-horsepower motor that draws a current of 3.6 amperes at 230 volts. The small motor could be connected to a 15-ampere circuit which might also supply other loads. **440-32** **440-22(a)** **240-6** **TABLE 250-95** **TABLE 430-148** **TABLE 430-52**

The space-heating equipment branch circuit is rated at 125% of the load current or 1.25 × 4500 watts/240 volts = 23.4 amperes. A standard 25- or 30-ampere branch circuit could supply this load. **424-3**

Dwelling with Space-Heating and Hot Water Heater. A final example will further illustrate the Code rules for individual fixed space-heating units and an electric hot water heater. A dwelling unit has the following loads:

a. 1500 square feet of living area
b. 4.5-kilowatt, 240-volt, 80-gallon water heater
c. 15 kilowatts of 240-volt electric space heating installed in six rooms

Branch circuits for the general lighting load are selected to be 20-ampere circuits and THW copper conductors are used for all circuits. The conductor ampacity must correspond to the circuit rating as the results in Figure 4-5 show. Each circuit requires an equipment grounding conductor (not shown in the figure).

The basic load consists of the general lighting load of 3 volt-amperes/square foot × 1500 square feet = 4500 volt-amperes. This requires three 20-ampere circuits. The small appliance circuits and the laundry circuit add three more 20-ampere circuits.

The hot water heater is considered a continuous load of 1.25 × 4.5 kilowatts = 5625 watts or a load in amperes of 23.4 amperes. A standard 25- or 30-ampere branch circuit would supply the unit. **422-14(b)**

The space-heater units have individual ratings of 15 kilowatts ÷ 6 = 2500 watts if one heater is connected in each room. The branch-circuit load is 1.25 × 2500 watts ÷ 240 volts = 13.0 amperes. Two heaters could be connected to a 30-ampere circuit, in which case three such circuits would be required.

QUIZ
(Closed-Book)

1. Which dimensions are used in measuring the square-foot area of a dwelling?
2. Which branch circuits are always required in a single-family dwelling?
3. What is the branch-circuit load for an electric range rated at 10 kilowatts?
4. The lighting load for a dwelling expressed in terms of a unit load in volt-amperes per square foot must be at least
   (a) 5 volt-amperes
   (b) 3 volt-amperes
   (c) 2 volt-amperes
5. What rating is required for the small appliance and laundry branch circuits?

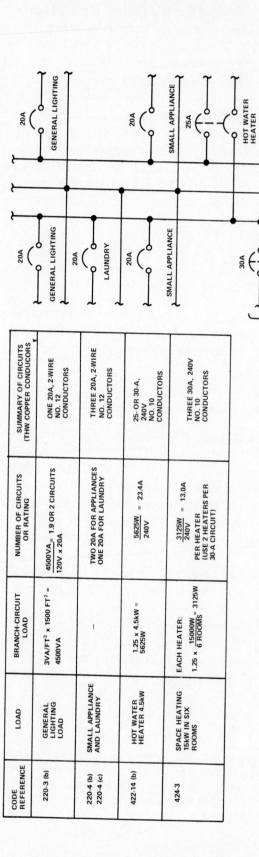

CODE REFERENCE	LOAD	BRANCH-CIRCUIT LOAD	NUMBER OF CIRCUITS OR RATING	SUMMARY OF CIRCUITS (THW COPPER CONDUCORS)
220-3 (b)	GENERAL LIGHTING LOAD	3VA/FT² × 1500 FT² = 4500VA	4500VA = 1.9 OR 2 CIRCUITS, 120V × 20A	ONE 20A, 2-WIRE NO. 12 CONDUCTORS
220-4 (b) 220-4 (c)	SMALL APPLIANCE AND LAUNDRY	–	TWO 20A FOR APPLIANCES ONE 20A FOR LAUNDRY	THREE 20A, 2-WIRE NO. 12 CONDUCTORS
422-14 (b)	HOT WATER HEATER 4.5kW	1.25 × 4.5kW = 5625W	$\frac{5625W}{240V}$ = 23.4A	25- OR 30-A, 240V NO. 10 CONDUCTORS
424-3	SPACE HEATING 15kW IN SIX ROOMS	EACH HEATER: $1.25 \times \frac{15000W}{6\ ROOMS}$ = 3125W	$\frac{3125W}{240V}$ = 13.0A PER HEATER (USE 2 HEATERS PER 30-A CIRCUIT)	THREE 30A, 240V NO. 10 CONDUCTORS

Figure 4-5. Branch-Circuit Calculations for Dwelling Unit With Hot Water Heater and Several Space Heaters

128

(Open-Book)   1. If the calculations indicate that a 45-ampere circuit is required to supply an electric range but only a 50-ampere circuit breaker is available, is it permissible to use a 50-ampere circuit?

2. Design the branch circuits for a dwelling unit with 1600 square feet of floor area and with a 12-kilowatt electric range installed. Use THW aluminum conductors and 20-ampere circuits whenever possible.

3. Design the branch circuits for a dwelling with the following loads:
   (a) 40 foot by 40 foot outside area with a 16-foot by 16-foot extra room attached
   (b) 12-ampere, 230-volt motor-compressor
   (c) 12-kilowatt electric range
   (d) 5-kilowatt 240-volt space-heating unit.
   Use THW copper conductors.

4-1.2 Standard Calculation for Computing Feeder or Service Loads for Dwelling Units

The discussion that follows explains various Code requirements for computing feeder or service loads for dwelling units. The calculation technique used is referred to as the *standard method*. The Code rules to be discussed here apply only to one-family dwellings or to individual dwelling units in a multifamily dwelling unless otherwise noted.   **ARTICLE 220 PART B**

Standard Calculation Techniques. In each dwelling unit the feeder load for each small appliance branch circuit and the laundry circuit is 1500 volt-amperes per circuit. The basic feeder load is then 1500 volt-amperes for each of these circuits plus the required general lighting load calculated at 3 volt-amperes per square foot.   **220-16** **220-11**

A demand factor is applied to the total load caused by the general lighting and receptacle circuits, the small appliance circuits, and the laundry circuit when the standard calculation method is used. The connected general lighting feeder load for a 2000-square foot area dwelling would be as follows: **220-16**

$$
\begin{array}{rl}
3 \text{ volt-amperes/ft}^2 \times 2000 \text{ ft}^2 = & 6000 \text{ VA} \\
\text{Small appliance circuits } (2 \times 1500 \text{ volt-amperes}) = & 3000 \text{ VA} \\
\text{Laundry circuit} \qquad\qquad\qquad = & \underline{1500 \text{ VA}} \\
\text{Total connected load} = & 10\,500 \text{ VA}
\end{array}
$$

The maximum demand load, however, is calculated by using the demand factors given in the Code table "Lighting Load Feeder Demand Factors." This table requires that the first 3000 volt-amperes be included at 100% and the remainder above this value at 35%. In the example the maximum demand load is   **TABLE 220-11**

$$
\begin{array}{rl}
\text{First 3000 volt-amperes at 100\%} & = 3000 \text{ VA} \\
\text{Remaining 7500 volt-amperes at 35\%} & = \underline{2625 \text{ VA}} \\
\text{Demand load} & = 5625 \text{ VA}
\end{array}
$$

This is considerably less than the actual connected load of 10.5 kilovolt-amperes.

The feeder load for specific fixed and stationary appliances is determined according to the Code rules for the type of load involved.[1] The feeder load for motors is not   **220-14**

---

[1] The Code allows a demand factor of 75% for a feeder load of four or more fixed appliances not including electric ranges, clothes dryers, space-heating equipment, and air-conditioning equipment. Check with local code enforcing agency before applying this provision.   **220-17**

220-15
220-18
220-19

dependent on occupancy. Any fixed electric space-heating load is computed as 100% of the total connected load. The load for an electric clothes dryer is 5000 watts (volt-amperes) or the nameplate rating, whichever is *larger*. An electric range is treated as discussed previously in the Guide and the load is calculated according to the value given in the Code range table.

The feeder load is computed in volt-amperes when the feeder or service supplies a balanced load. The required ampacity is the total demand load divided by the circuit voltage.

Example of Standard Calculation. As an example, an individual dwelling unit is considered that has the following loads:

a. 1600-square foot living area
b. 120-volt, 350-watt garbage disposal
c. 12-kilowatt electric range
d. 5-kilowatt, 240-volt electric clothes dryer
e. 35-ampere, 240 volt air conditioner protected by 70-ampere circuit breaker
f. 1/3-horsepower, 240-volt blower motor
g. 10.5-kilowatt, 240-volt space-heating unit

The feeder is 120/240-volt circuit that uses type TW copper conductors. The design problem includes calculations to select all elements of the feeder equipment.

A suggested format for performing the calculations is shown in Figure 4-6. First,

---

CODE RULE	FEEDER LOAD CALCULATION				
		LINE	NEUTRAL		CHECKLIST
220-3 (b) 220-16(a)	GENERAL LIGHTING LOAD = 3VA/FT² X 1600FT² SMALL APPLIANCE CIRCUITS = 2 x 1500VA LAUNDRY CIRCUIT	4,800 3,000 1,500	4,800 3,000 1,500		LTG, SMALL APPL, AND LAUNDRY ✓
	TOTAL LTG, SMALL APPL, AND LAUNDRY CKTS	9,300	9,300		DISPOSAL ✓
					DISHWASHER ✓
220-11	APPLICATION OF DEMAND FACTORS FIRST 3000VA @ 100% REMAINDER (9300 - 3000) @ 35%	3,000 2,205	3,000 2,205		RANGE ✓
	GENERAL LIGHTING DEMAND LOAD	5,205	5,205		DRYER ✓
					HEATING OR A/C ✓
220-18 220-19	5-KW CLOTHES DRYER, NEUTRAL = .7 x 5000W = 3500W 12-KW ELECTRIC RANGE = 8000W; NEUTRAL = .7 x 8000W = 5600W	5,000 8,000	3,500 5,600		BLOWER ✓
220-15, 220-21	10.5-KW SPACE HEATING (NEGLECT AIR CONDITIONING) 120-VOLT, 350-WATT DISPOSAL	10,500	–0–		
TABLE 430-148 430-25	1/3-HP BLOWER MOTOR = 240V x 3.6A 25% OF LARGEST MOTOR = .25 x 864 = 216VA	350 864 216	350 –0– –0–		
		30,135 VA	14,655 VA		

SELECTION OF FEEDER EQUIPMENT

TABLE 310-16 240-3, 240-6 CHAPTER 9 250-91	1. FEEDER AMPACITY = 30,135VA ÷ 240V = 125.5 AMPERES; NEUTRAL = 14,655VA ÷ 240V = 61.6 AMPERES 2. CONDUCTORS: TWO NO. 2/0 AND ONE NO.4 TW COPPER 3. OVERCURRENT PROTECTIVE DEVICE: USE 150-AMPERE STANDARD SIZE BASED ON CONDUCTOR AMPACITY 4. CONDUIT SIZE: AREA OF TWO NO. 2/0 = 2 x .2781 IN.² = .5562 IN.² AREA OF ONE NO. 4 = .1087 IN.² TOTAL AREA OF CONDUCTORS .6649 IN.² REQUIRES 1-1/2 INCH CONDUIT 5. GROUNDING: CONDUIT SERVES AS EQUIPMENT GROUNDING CONDUCTOR

Figure 4-6. Calculation of Feeder Circuit Ampacity for 1600-Square-Foot Dwelling Unit

the general lighting, small appliance, and laundry circuit loads are considered. Their feeder demand load is calculated by using the demand factors allowed by the Code. The total connected load of 9300 volt-amperes for general lighting, small appliance, and laundry circuits contributes only 3000 volt-amperes + .35 (9300 − 3000) volt-amperes = 5205 watts to the feeder demand load. **TABLE 220-11**

The 230-volt clothes dryer adds 5000 volt-amperes to the line load and 3500 volt-amperes to the neutral load. A 12-kilowatt range represents a line load of 8 kilowatts and a neutral load of .7 × 8 kilowatts = 5600 watts (volt-amperes). **220-18 TABLE 220-19 220-22**

Space heating and air conditioning are noncoincident loads and the smaller of the two may be neglected. The air conditioner is neglected in this case because its load of 240 volts × 35 amperes = 8400 volt-amperes is less than the load of the 10.5-kilowatt heater. **220-21**

The 1/3-horsepower blower motor has a basic load of 240 volts × 3.6 amperes = 864 watts. A value of 25% of the blower's load must be added to the feeder load because it is the largest motor. The disposal is a 120-volt, 350-watt load and contributes to the line and the neutral load. **TABLE 430-148 430-25**

The total feeder load is 30 135 volt-amperes which results in an ungrounded conductor ampacity of 30 135 ÷ 240 volts = 125.5 amperes. The required neutral ampacity is 48 amperes. This results in a minimum size type TW conductor of No. 2/0 for the ungrounded conductors and No. 4 for the neutral conductor. **TABLE 310-16**

Although the air-conditioning load can be neglected in the load computation, the rating of the air conditioner's branch-circuit overcurrent protective device must be considered in determining the required rating of the feeder overcurrent protective device. Since the 70-ampere circuit breaker for the air conditioner is the largest rated branch-circuit protective device, the feeder overcurrent protective device must be rated as follows: **440-2(a) 430-63**

```
Air conditioner protective device                    70.0
General lighting = 5202 volt-amperes ÷ 240 volts     21.7
Dryer = 5000 watts ÷ 240 volts                       20.8
Range = 8000 watts ÷ 240 volts                       33.8
Disposal = 350 watts ÷ 240 volts                      1.5
Blower motor                                           3.6
                              Maximum rating         150.9
```

(Refer to the note about computation accuracy concerning the 120-volt disposal.) The space-heating load is not included since it is assumed that the air conditioner and space heating are not used simultaneously. A standard 150-ampere overcurrent protective device would be used to protect the feeder circuit.

A 1½-inch conduit would be required to enclose the conductors. The conduit could also serve as the equipment grounding conductor from the service to the feeder panel. **250-91(d)**

Service Calculation. In the example just considered the feeder circuit supplied a dwelling unit in a multifamily dwelling. If the dwelling unit were instead a one-family dwelling, the load would be supplied from a service. A main disconnecting means, a grounding electrode, and a grounding electrode conductor would then be required. If the load were a 126-ampere load, the service disconnecting means would be rated at least 126 amperes. A 150-ampere circuit breaker used as the over- **230-79** current protective device could also serve as the disconnecting means since the switch rating of such devices is usually considerably higher than the trip setting. The grounding electrode conductor would be required to be not less than size No. 4 for a copper conductor based on size 2/0 service-entrance conductors.

TABLE 250-94

250-23(b)

In the case of a service, the neutral is required to be at least as large as the grounding electrode conductor, or, a No. 4 copper conductor in this case. This requirement does not apply to the feeder neutral in the example just considered.

Note About Computation Accuracy. When 120-volt loads are exactly balanced on a 120/240-volt circuit, it is correct to add the loads in watts and then divide by 240 volts to determine the line and neutral load in amperes. The neutral load represents the maximum unbalanced load at 120 volts. For balanced 120-volt loads, the neutral load is one-half the total load; therefore, dividing the total load in watts by 240-volts yields the correct value for the neutral load in amperes.

If a 120-volt load is not balanced by an identical load, the neutral load should not be determined by dividing by the 240-volt circuit voltage if an exact result is desired. For example, a 120-volt, 350-watt garbage disposal has a neutral load in amperes at 120 volts of

$$\frac{350 \text{ W}}{120 \text{ V}} = 2.92 \text{ A}$$

If the total neutral load is determined by dividing the total load in watts by 240 volts, a value of only 350 watts ÷ 240 volts = 1.46 amperes, which is one-half of the disposal's correct neutral load, will be included in the total.

Standard practice allows the neutral load in amperes to be determined by summing the loads that require a neutral and dividing the total watts by 240 volts. It should be remembered, however, that this method is slightly inaccurate when unbalanced loads are present on three-wire circuits. The exact results can be obtained in such cases by calculating the loads for both ungrounded conductors and the neutral in amperes instead of watts or volt-amperes.

QUIZ

(Closed-Book)

1. What is the service or feeder load for the small appliance branch circuits and the laundry circuit in a dwelling?
2. What is the feeder load for a 4-kilowatt electric clothes dryer in a dwelling?
3. If the standard method of calculation is used, the portion of the general lighting load exceeding 3000 volt-amperes is permitted what demand factor?
   (a) 30%
   (b) 35%
   (c) 40%
4. If the service load for a dwelling is greater than 10 kilovolt-amperes, what is the minimum allowable service rating?

(Open-Book)

1. Design the service for a 1500-square foot home with a 12-kilowatt electric range. Use THW copper conductors and include *all* permissible reductions in the service load and conductors.
2. Design the service for a 2800-square foot dwelling with the following equipment installed:
   (a) 5-kilowatt counter-mounted cooking unit
   (b) Two 4-kilowatt wall-mounted ovens (on the same circuit with cooking unit).

    (c) Washing machine
    (d) 4500-watt, 240-volt clothes dryer
    (e) ¼-horsepower disposal unit
    (f) 1200-watt dishwasher
    (g) Three 9-kilowatt, 240-volt space-heating units
    (h) 15-ampere air conditioner

4-1.3 Optional Calculations for Computing Feeder or Service Loads

The Code provides an optional method for computing feeder and service loads **220-30** for a dwelling unit. This method usually results in a smaller feeder or service rating for the dwelling unit to which it applies than would be calculated by the standard method.

In order for the optional calculation to be used, the dwelling unit must be supplied by a single three-wire, 120/240-volt or 208Y/120-volt feeder or service. The **SEE 230-41** ampacity of the conductors must be 100 amperes or greater. This method is therefore applied to dwelling units that have a relatively large electric load, usually including electric space-heating and air-conditioning equipment.

The neutral conductor of the feeder or service may be calculated by the standard method. The method that yields the smallest neutral load consistent with Code rules should be used.

Optional Calculation Techniques. The optional method of computing feeder or service loads separates the various loads into air conditioning, electric space heating, and all "other loads." A Code table lists demand factors to be applied to the loads **TABLE 220-30** in each category.

The feeder demand for the air conditioning and electric space heating is taken as the largest of either load calculated according to the procedure outlined in the Code. The loads are compared as follows:

1. The air-conditioning load compared to 40% of the connected load of *four or more* separately controlled electric space-heating units.
2. The air-conditioning load compared to 65% of the load of a central heating unit or 65% of the load of *fewer than four* separately controlled space-heating units.

For example, if the air-conditioning load is 10 kilowatts and the connected load **220-21** of six separately controlled heating units is 30 kilowatts, the air-conditioning load, being smaller than the heating load of .4 × 30 kW = 12 kW, would be omitted from the calculations. If, however, a 12-kilowatt heating load is from only three separately controlled units, the demand load is only .65 × 12 kilowatts = 7.8 kilowatts. In that case, the space-heating load would be omitted and the 10-kilowatt air-conditioning load would be used in the calculation.

The "other load" includes the general lighting load at 3 watts per square foot and the required small appliance and laundry branch circuits at 1500 watts each. All fixed appliances, ranges, cooking units, and motors are also included and represent a load according to their nameplate ratings. No demand factors or increase in load rating for the motors need be applied. The total of all "other load" over 10 kilowatts is subject to a 40% demand factor.

Example of Optional Calculation. Figure 4-7 illustrates the results of the optional calculation for a 1500-square foot one-family dwelling with the following equipment:

a. 9 kilowatts of 240-volt electric space heating in five rooms, each room separately controlled
b. 6-ampere, 240-volt air conditioner connected to a 15-ampere circuit
c. 12-kilowatt electric range
d. 2.5-kilowatt, 240-volt water heater
e. 120-volt, 1.2-kilovolt-ampere dishwasher
f. 5-kilowatt, 240-volt clothes dryer

The service equipment elements are to be sized with THW copper conductors used for the service-entrance conductors.

The "other load" includes the 4500-volt-ampere general lighting load plus 3000 volt-amperes for the two small appliance branch circuits and 1500 volt-amperes for the laundry circuit. Each appliance included in the "other load" represents a load according to its nameplate rating. The total "other load" then becomes 29 700 volt-amperes.

TABLE 220-30   Applying the demand factors given in the table, the demand load for the "other load" becomes

First 10 kilovolt-amperes at 100% = 10 000 volt amperes
Remainder (29 700 − 10 000) at 40% = 7880 volt-amperes

The air conditioning or heating demand load is then included after applying the appropriate demand factor. The air conditioning load in this case is only 240 volts × 6 amperes = 1440 volt-amperes. Since it is smaller than 40% of the space-heating nameplate rating, it is neglected. Adding the 3600-watt (volt-ampere) space-heating load produces a total service load of 21 480 volt-amperes.

TABLE 310-16   A 100-ampere service is required and No. 3 THW copper conductors are chosen
(NOTE 3 IGNORED) for the ungrounded service-entrance conductors.[2]

ARTICLE 220,   When calculated by the optional method, the neutral load is 16 880 volt-
PART B amperes or 70.3 amperes. If the standard method is used, the neutral load is only
TABLE 310-16 15 400 volt-amperes or 64.2 amperes. A No. 6 THW copper conductor would be
CHAPTER 9 used. A 1-inch conduit is required to enclose the conductors. The grounding elec-
250-94 trode conductor must be at least a No. 8 copper conductor.

QUIZ
(Closed-Book)   **1.** If the optional method of calculation is used, what is the feeder demand load for electric dryers and electric ranges?
2. A demand factor may be applied to the neutral load of an electric range when the optional method is used. True or false?
3. Under the optional calculation, a 10-kilovolt-ampere central heating load is considered a demand load of how many volt-amperes?
(a) 10 000 volt-amperes
(b) 6500 volt-amperes
(c) 4000 volt-amperes
4. State the conditions under which the optional calculation method may be used.
5. What is the demand load for four or more separately controlled space-heating units if the optional calculation method is used?

[2]A note to the Code tables of conductor ampacities allows the use of smaller conductors in this case. Check with local Code enforcing agency before applying this provision.

SERVICE LOAD CALCULATION

CODE RULE		LINE	NEUTRAL	NEUTRAL BY STANDARD CALCULATION	
220-3	GENERAL LIGHTING LOAD = 3VA/FT² × 1500 FT²	4,500	4,500	4,500	
220-30	SMALL APPLIANCE = 2 × 1500VA.	3,000	3,000	3,000	
	LAUNDRY CIRCUIT	1,500	1,500	1,500	
				9,000	(TOTAL LTG, SMALL APPL, AND LAUNDRY)
				3,000	(FIRST 3000 @ 100%)
				2,100	(REMAINDER @ 35%)
				5,100	(GENERAL LIGHTING DEMAND LOAD)
220-30	12-KW ELECTRIC RANGE	12,000	12,000	5,600	(.7 × 8000W)
	240-VOLT, 2.5-KW WATER HEATER	2,500	–0–	–0–	
	120-VOLT, 1.2-KVA DISHWASHER	1,200	1,200	1,200	
	240-VOLT, 5-KW CLOTHES DRYER	5,000	5,000	3,500	(.7 × 5000W)
	TOTAL "OTHER LOAD"	29,700	27,200		
	APPLICATION OF DEMAND FACTORS				
	FIRST 10KVA OF "OTHER LOAD" @ 100%	10,000	10,000		
	REMAINDER @ 40%	7,880	6,880		
220-21	HEATING = .4 × 9000W	3,600	–0–		
220-30(3)	(AIR CONDITIONING NEGLECTED)				
	SERVICE LOAD	21,480 VA	16,880 VA	15,400 VA	(NEUTRAL LOAD)

SELECTION OF SERVICE EQUIPMENT

CODE RULE		
	1.	SERVICE RATING = 21,480VA ÷ 240V = 89.5 AMPERES; NEUTRAL = 15,400VA ÷ 240V = 64.2 AMPERES
TABLE 310-16, 250-23(b)	2.	SERVICE–ENTRANCE CONDUCTORS: TWO NO. 3 AND ONE NO. 6 THW COPPER CONDUCTORS
230-90(a), 230-79	3.	OVERCURRENT PROTECTIVE DEVICE AND DISCONNECT: USE 100-AMPERE DEVICE
CHAPTER 9	4.	CONDUIT SIZE: AREA OF TWO NO. 3 = 2 × .1263 IN.² = .2526 IN.²
		AREA OF ONE NO. 6 = .0819 IN.²
		TOTAL AREA OF CONDUCTORS .3345 IN.² REQUIRES 1-INCH CONDUIT
250-94	5.	GROUNDING: NO. 8 GROUNDING ELECTRODE CONDUCTOR REQUIRED BASED ON NO. 3 SERVICE–ENTRANCE CONDUCTORS

CHECKLIST

	✓
LTG, SMALL APPL, AND LAUNDRY	✓
RANGE	✓
WATER HEATER	✓
DISHWASHER	✓
DRYER	✓
HEATING OR A/C	✓
SERVICE	✓
NEUTRAL	✓
OVERCURRENT DEVICE	✓
DISCONNECT	✓
CONDUIT	✓
GROUNDING	✓

Figure 4-7. Optional Calculation for 1500-Square-Foot One-Family Dwelling

(Open-Book)   **1.** Design the feeder for a 2000-square foot apartment with the following equipment installed:

(a) 10 kilowatt, 240-volt air conditioner
(b) 12 kilowatt, 240-volt central space-heating unit
(c) Two 4.5-kilowatt wall-mounted ovens
(d) 5-kilowatt counter-mounted cooking unit (on same circuit as ovens)

Use the optional method for the ungrounded conductors and the standard method for the neutral conductor. Use THW copper conductors.

4-1.4 Optional Calculation for Additional Loads in Existing Dwelling Units

220-31   Additional loads may, in some cases, be added to an existing service without enlarging the service rating. The Code provides an optional load calculation method that may be used to determine the largest possible load that may be added to an existing three-wire service when it becomes necessary to add loads to a dwelling.

The calculation method is typically used to determine whether or not an added load will exceed the capacity of a 60-ampere service. When air-conditioning or space-heating equipment is not to be added, the calculation considers any load over 8 kilovolt-amperes to be subject to a 40% demand factor.

Air-conditioning and space-heating equipment is treated differently from other types of appliances. The Code provides a formula that must be used to determine if such equipment may be added to the existing service.

Load Computation without Air Conditioning or Heating Load. The total connected load is computed as follows:

a. 3 volt-amperes per square foot for general-purpose circuits
b. 1500 volt-amperes for each small appliance circuits
c. Fixed or stationary appliances at nameplate rating

The demand load is then determined by applying the demand factors given in terms of percent of *connected* load as follows:

First 8 kilovolt-amperes of load at 100%
Remainder of load at 40%

If the existing load is greater than 8 kilovolt-amperes, the demand load of any added appliance is only 40% of its nameplate rating.

The maximum *demand* load for a 60-ampere, 240-volt service may not exceed 240 volts × 60 amperes = 14 400 volt-amperes. If, however, the demand factors permitted under the optional calculation are applied, the actual *connected* load may exceed that value.

The total *connected* load that may be served by an existing three-wire, 60-ampere service is 24 000 volt-amperes, which is shown by solving for X in the following:

$$8000 \text{ VA} + .4\,(X - 8000)\text{ VA} = 14\ 400 \text{ VA}$$
$$8000 + .4X - 3200 = 14\ 400$$
$$.4X = 14\ 400 - 8000 + 3200$$
$$X = 24\ 000 \text{ VA}$$

where X is the total connected load and $(X - 8000)$ is the excess of the connected load over 8 kilovolt-amperes.

Thus, the difference between 24 000 volt-amperes and the computed existing connected load is the maximum rating of any load that may be added to the existing 60-ampere service.

As an example, a 1500-square foot one-family dwelling with a 12-kilowatt electric range represents a connected load of 19 500 volt-amperes determined as follows:

```
Lighting = 3 volt-amperes/square foot × 1500 square feet  =   4 500
Small appliance circuits = 2 × 1500 volt-amperes          =   3 000
Range                                                     =  12 000
                                                 Total  =  19 500 VA
```

Since the existing connected load exceeds 8 kilovolt-amperes, any additional load would be subject to the 40% demand factor, but its nameplate rating cannot exceed

$$24\ 000\ \text{VA} - 19\ 500\ \text{VA} = 4500\ \text{VA}$$

According to the result first derived.

Load Calculation for Adding Air Conditioning or Space Heating. When air conditioning or space heating is to be added on an existing service, such equipment must be added on the basis of a 100% demand factor. The Code provides a detailed formula for determining whether or not the existing service can handle the added air-conditioning or heating load. The formula requires that the entire load be computed by adding the loads as follows: **220-31**

Larger of—
   air conditioning at 100%
                 or
   (central electric space heating at 100%, or less than four separately controlled space heating units at 100%)
plus
   first 8 kilovolt-amperes of all other load at 100%
plus
   remainder of all other load at 40%

All ''other load'' includes lighting circuits, small appliance circuits, and all fixed appliances at nameplate rating. If the space-heating load consists of four or more separately controlled units, it would be included in the ''other load'' total.

The maximum amount of air-conditioning or space-heating load that can be added on an existing 60-ampere service can be determined by using the following equation:

$$L + 8000\ \text{VA} + .4\ (X - 8000)\ \text{VA} = 14\ 400\ \text{VA}$$

where L is the maximum permissible amount of added air conditioning or space-heating load, X is the total existing connected load; and $(X - 8000)$ is the excess of the existing load over 8 kilovolt-amperes.

The maximum amount of air-conditioning or space-heating load that can be added to the 1500-square foot dwelling in the previous example is found by substituting the 19 500 volt-amperes of the existing connected load for X in the formula and solving for L, which becomes

$$
\begin{aligned}
L + 8000 \text{ VA} + .4(19\ 500 - 8000) \text{ VA} &= 14\ 400 \text{ VA} \\
L + 8000 + .4(11\ 500) &= 14\ 400 \\
L + 8000 + 4600 &= 14\ 400 \\
L &= 14\ 400 - 8000 - 4600 \\
L &= 1800 \text{ VA}
\end{aligned}
$$

Thus, an air conditioner or a central space-heating unit, or fewer than four separately controlled space-heating units may be added providing the nameplate rating of such equipment does not exceed 1800 volt-amperes. As shown previously, some other type of appliance could be added with up to a 4500-volt-ampere nameplate rating since it would be added on the basis of a 40% demand factor.

QUIZ

(Closed-Book)
1. What is the maximum demand load for a 60-ampere, 208-volt service?
2. What is the total possible connected load to a 60-ampere, 208-volt service if no air-conditioning or heating loads are present?

(Open-Book)
1. Compute the rating of the hot water heater that may be added on an existing 60-ampere, 240-volt service in a 2000-square foot, one-family dwelling. The house contains a 10-kilowatt electric range.
2. What is the maximum rating of the air conditioner that could be added to the dwelling of problem 1?

4-1.5 Complete Design Examples for One-Family Dwellings

ARTICLE 220
PART B

220-30

The complete design of the wiring system for a one-family dwelling requires the design of each branch circuit, calculation of the service load, and selection of the service equipment elements. The service load may always be computed by using the standard calculation method. When the service load is 100 amperes or more, a smaller service rating will usually result if the optional calculation method is used to compute the load.

Design Example Using Standard Method. A one-family dwelling with a 120/240-volt service has the following loads:

a. Living area of 800 square feet
b. 5-kilowatt wall-mounted oven
c. 6-kilowatt counter-mounted cooking unit supplied from a separate branch circuit
d. 15-ampere, 240-volt air conditioner protected by a 30-ampere circuit breaker

Type THW copper conductors are used for all conductors in the design problem summarized in Figure 4-8.

The size and number of branch circuits are determined first. The lighting load is

$$
\frac{3 \text{ VA/ft}^2 \times 800 \text{ ft}^2}{120 \text{ V}} = 20 \text{ A}
$$

CODE RULE	BRANCH CIRCUITS REQUIRED		CHECKLIST
220-4	GENERAL LIGHTING LOAD = $\dfrac{3VA/FT^2 \times 800\ FT^2}{120V}$ = 20A	TWO 15- OR ONE 20-AMPERE, 2-WIRE CIRCUIT(S)	LTG, SMALL APPL, AND LAUNDRY ✓
220-4 (b)(c)	SMALL APPLIANCE AND LAUNDRY CIRCUITS	THREE 20-AMPERE, 2-WIRE CIRCUITS	OVEN ✓
TABLE 220-19, NOTE 4	COOKING UNITS: 1) 5000W ÷ 240V = 20.8A 2) 6000W ÷ 240V = 25A	25- OR 30-AMPERE, 240-VOLT CIRCUIT 25- OR 30-AMPERE, 240-VOLT CIRCUIT	COOKING UNIT ✓
440-32, 440-22	AIR CONDITIONING = 1.25 × 15A = 18.75A	30-AMPERE, 240 VOLT CIRCUIT	AIR CONDITIONER ✓

SERVICE LOAD CALCULATION

CODE RULE		LINE	NEUTRAL
220-3 (b)	GENERAL LIGHTING = 3VA/FT² × 800 FT²	2,400	2,400
220-16(a)	SMALL APPLIANCE = 2 × 1500VA	3,000	3,000
220-16(b)	LAUNDRY	1,500	1,500
	TOTAL LTG, SMALL APPL, AND LAUNDRY CKTS	6,900	6,900
220-11	APPLICATION OF DEMAND FACTORS		
	FIRST 3000VA @ 100%	3,000	3,000
	REMAINDER (6900-3000) @ 35%	1,365	1,365
	GENERAL LIGHTING DEMAND LOAD	4,365	4,365
	OVEN: OVEN NEUTRAL = .7 × 5000W = 3500W	5,000	3,500
	COOKING UNIT: COOKING UNIT NEUTRAL = .7 × 6000W = 4200W	6,000	4,200
	AIR CONDITIONER = 1.25 × 240V × 15A	4,500	-0-
	SERVICE LOAD	19,865 VA	12,065 VA

SELECTION OF SERVICE EQUIPMENT

CODE RULE	
	1. SERVICE RATING = 14,685VA ÷ 240V = 82.8 AMPERES; NEUTRAL = 12,065W ÷ 240V = 50.3 AMPERES (USE 100 AMPERE SERVICE)
TABLE 310-16,	2. SERVICE ENTRANCE CONDUCTORS: TWO NO. 3 AND ONE NO. 6 THW COPPER CONDUCTORS (IGNORE NOTE 3)
	3. OVERCURRENT PROTECTIVE DEVICE AND DISCONNECT: USE 100-AMPERE DEVICE
CHAPTER 9	4. CONDUIT SIZE: AREA OF TWO NO. 3 = 2 × .1263 IN.² = .2526 IN.² AREA OF ONE NO. 6 = .0819 IN.² TOTAL AREA OF CONDUCTORS = .3345 IN.² REQUIRES 1-INCH CONDUIT
250-95	5. GROUNDING ELECTRODE CONDUCTOR REQUIRED TO BE NO. 8 COPPER

Figure 4-8. Wiring Design for 800-Square-Foot One-Family Dwelling

and would require two 15- or one 20-ampere, two-wire circuit(s). Three more 20-ampere circuits are required for the two small appliance circuits and the laundry circuit.

TABLE 220-19
NOTE 4 Since the wall-mounted oven and counter-mounted cooking unit are supplied from separate branch circuits, they must be considered as separate units at their nameplate rating for the purposes of branch-circuit calculations. The 5-kilowatt oven represents a branch-circuit load of

$$\frac{5000 \text{ W}}{240 \text{ V}} = 20.8 \text{ A}$$

A 25- or 30-ampere circuit with No. 10 THW copper conductors would be used. The counter-mounted cooking unit load is

$$\frac{6000 \text{ W}}{240 \text{ V}} = 25 \text{ A}$$

Another 25- or 30-ampere circuit would be used to supply this unit.

The air conditioner circuit must supply a load of

440-32

$$1.25 \times 15 \text{ A} = 18.75 \text{ A}$$

The branch-circuit rating must be at least 20 amperes, but, as stated in the problem, the overcurrent protective device is set at 30 amperes.

TABLE 220-19
NOTE 3 The service load was computed by using the standard calculation method as shown in Figure 4-8. For the purposes of computing the service load, no demand factors may be applied to the oven and cooking unit loads. (Note that demand factors may be applied to small "ranges" in some cases.) The total service load of 19 865 volt-amperes requires service-entrance conductors with an ampacity of at 230-41 least 82.8 amperes. The service rating, however, must be at least 100 amperes since there are more than six branch circuits in the dwelling.

430-63 A Code provision allows an increase in the rating permitted for the service overcurrent protective device based on the rating permitted for the largest motor circuit protective device plus the additional rating required by the other loads. This provision is not appropriate here because the required 100-ampere service is still considerably larger than the maximum rating that would result from using the provision.

TABLE 310-16
(NOTE 3 IGNORED)
TABLE 250-94 The ungrounded service-entrance conductors are No. 3 THW copper conductors with a No. 6 neutral. The grounding electrode conductor is required to be a No. 8 copper conductor based on the No. 3 service-entrance conductors.

Design Example Using Optional Method. A large one-family dwelling contains the following loads:

a. 3800 square feet of living area
b. 12-kilowatt electric range
c. 1.0-kilovolt-ampere, 120-volt garbage disposal
d. 10-kilowatt, 240-volt strip heaters (central heating)
e. 6-kilowatt, 240-volt, 130-gallon hot water heater
f. Two 28-ampere, 240-volt air conditioner motor-compressors with circuits protected by inverse-time circuit breakers.
g. Four $1/2$-horsepower, 120-volt blower motors for heating and cooling systems

The branch circuits and service use THW copper conductors. The design problem is summarized in Figure 4-9 on the following page.

The lighting, small appliance, and laundry load requires eight 20-ampere circuits. The 12-kilowatt range is considered as a branch-circuit load of 8 kilowatts and requires a 40-ampere circuit. The disposal and the four ¹/₂-horsepower motors must be supplied by conductors rated at 125% of the full-load current of the motors. **TABLE 220-19, NOTE 4** **430-22**

The conductors for the central electric space heating must be rated at least **424-3(b)**

$$1.25 \times \frac{10\ 000\ \text{W}}{240\ \text{V}} = 52.1\ \text{A}$$

The 6-kilowatt hot water heater represents a load of 25 amperes at 240 volts. The hot water heater is not considered a continuous load since its capacity is more than 120 gallons. **422-14(b)**

The branch-circuit ampacity of the motor-compressors must be 125% of the nameplate rated-load current, or 1.25 × 28 amperes = 35 amperes. The rating of the inverse-time circuit breaker serving as the branch-circuit overcurrent protective device must be no larger than 1.75 × 28 amperes = 49 amperes, or a 45-ampere standard size. **440-32** **440-22**

The computation of the service load is straightforward after the noncoincident heating and air-conditioning loads are compared. The load of the two motor-compressors is 2 × 240 volts × 28 amperes = 13 440 volt-amperes. The central electric space heating is considered at the 65% diversified demand, or .65 × 10 000 watts = 6500 watts (volt-amperes). Since the heating load is the smaller load, it is neglected. The four blower motors, however, are part of the heating and cooling system and their load of 4704 volt-amperes must be included at the 100% demand factor. **TABLE 220-30**

The "other load" includes the lighting, small appliance, and laundry circuits as well as the nameplate rated-load of all fixed appliances for a total of 34 900 volt-amperes. The applicable demand factors reduce the demand load to 19 960 volt-amperes. The air-conditioning equipment represents an additional 13 440 + 4704 = 18 144 volt-amperes for which no reduction is permitted.

The total service load of 38 104 volt-amperes requires a minimum service rating of

$$\frac{38\ 104\ \text{VA}}{240\ \text{V}} = 158.8\ \text{A}$$

A neutral load of 19 113 volt-amperes, or 79.6 amperes, is contributed by the unbalanced load from the 120-volt and 120/240-volt loads.

No. 2/0 type THW copper conductors are required for the ungrounded conductors and a No. 4 is required for the neutral conductor.

The rating of the feeder overcurrent protective device may not exceed the following: **430-63**

Air Conditioner protective device	45.0
Other motor loads [28+ (4704 ÷ 240)]	47.6
"Other load" (19 960 ÷ 240)	83.2
Maximum rating	175.8 Amperes

A 175-ampere rating is the next smaller standard size for the feeder overcurrent protective device. **240-6**

CHECKLIST

LGT, SMALL APPL, LAUNDRY	✓
RANGE	✓
DISPOSAL	✓
HEATING OR A/C	✓
WATER HEATER	✓
BLOWER MOTORS	✓

BRANCH CIRCUITS REQUIRED

CODE RULE		BRANCH CIRCUITS REQUIRED
220-4	GENERAL LIGHTING LOAD = $\dfrac{3VA/FT^2 \times 3800\ FT^2}{120V \times 20A}$ = 4.75 OR 5 CIRCUITS	FIVE 20-AMPERE, 2-WIRE CIRCUITS
220-4	SMALL APPLIANCE AND LAUNDRY =	THREE 20-AMPERE, 2-WIRE CIRCUITS
TABLE 220-19 NOTE 4	12-KW RANGE = 8000W ÷ 240V = 33.33A	40-AMPERE, 240-VOLT CIRCUIT
210-22	DISPOSAL = 1.25 x 1000VA ÷ 120 = 10.4A	15- OR 20-AMPERE, 2-WIRE CIRCUIT
424-3(b)	STRIP HEATERS = 1.25 x 10000W ÷ 240V = 52.1A	60-AMPERE, 240-VOLT CIRCUIT
	HOT WATER HEATER = 6000W ÷ 240V = 25A	30-AMPERE, 240-VOLT CIRCUIT
440-32, 440-22	AIR CONDITIONER MOTOR-COMPRESSORS = 1.25 x 28A = 35A MAXIMUM RATING OF PROTECTIVE DEVICE = 1.75 x 28 = 49A (45A STANDARD)	TWO 45-AMPERE, 240-VOLT CIRCUITS
430-22 TABLE 430-148	FOUR 1/2-HP, 115-VOLT MOTORS = 1.25 x 9.8A = 12.25A	FOUR 15- OR 20-AMPERE, 2-WIRE CIRCUITS

SERVICE LOAD CALCULATION

CODE RULE		LINE		NEUTRAL BY STANDARD CALCULATION
220-30	GENERAL LIGHTING = 3VA/FT² x 3800 FT²	11,400	11,400	
	SMALL APPLIANCE = 2 x 1500VA	3,000	3,000	
	LAUNDRY	1,500	1,500	
	TOTAL LTG, SMALL APPL, AND LAUNDRY CKTS	15,900	15,900	(TOTAL LTG, APPL, AND LAUNDRY CKTS)
			3,000	(FIRST 3000 @ 100%)
			4,515	(REMAINDER @ 35%)
			7,515	(GENERAL LIGHTING DEMAND LOAD)
	RANGE	12,000	5,600	(.7 x 8000W)
	DISPOSAL	1,000	1,200	
	HOT WATER HEATER	6,000	-0-	
	TOTAL "OTHER LOAD"	34,900		

APPLICATION OF DEMAND FACTORS

		LINE	NEUTRAL
FIRST 10KVA OF "OTHER LOAD" @ 100%		10,000	
REMAINDER (34,900-10,000) @ 40%		9,960	
TOTAL "OTHER LOAD" DEMAND		19,960	-0-
MOTOR-COMPRESSORS @ 100% = 2 x 240V x 28A		13,440	4,998
BLOWER MOTORS @ 100% = 4 x 120V x 9.8A		4,704	$[\,4{,}704 + (.25 \times 120V \times 9.8A)\,]$
SERVICE LOAD		38,104 VA	19,113 VA (NEUTRAL LOAD)

SELECTION OF SERVICE EQUIPMENT

CODE RULE	
	1. SERVICE RATING = 38,104VA ÷ 240V = 158.8 AMPERES; NEUTRAL = 19,113VA ÷ 240V = 79.6 AMPERES
TABLE 310-16	2. SERVICE-ENTRANCE CONDUCTORS: TWO NO. 2/0 AND ONE NO. 4 THW COPPER CONDUCTORS
430-63	3. OVERCURRENT PROTECTIVE DEVICE AND DISCONNECT: USE 175-AMPERE STANDARD SIZE DEVICE BASED ON AIR CONDITIONER DEVICE
CHAPTER 9	4. CONDUIT SIZE: AREA OF TWO NO. 2/0 = 2 x .2781 IN² = .5562 IN² AREA OF ONE NO. 4 = .1087 IN² TOTAL AREA OF CONDUCTORS = .6649 IN² REQUIRES 1-1/2 INCH CONDUIT
250-94	5. GROUNDING ELECTRODE CONDUCTOR REQUIRED TO BE NO. 4 COPPER

Figure 4-9. Design Example for 3800-Square-Foot Dwelling Using Optional Calculation

A 1½-inch conduit is adequate to enclose the conductors and not exceed the 40% fill limitation. A No. 4 copper grounding electrode conductor would be used   **250-94** at the service.

QUIZ

(Closed-Book)   1. When is it reasonable to apply the optional method of calculation to determine the service rating?

                     2. What are the minimum number of branch circuits and the minimum service rating for a 1000-square foot one-family dwelling? There are no electric appliances permanently installed.

(Open-Book)   1. Design the branch circuits and 120/240-volt service for a one-family dwelling with the following load:
   (a) 1500 square feet of area
   (b) 4.5-kilowatt, 240-volt clothes dryer
   (c) 14-kilowatt, 240-volt electric heating units with five separate controls
   (d) 12-kilowatt range
   (e) 3-kilowatt, 240-volt hot water heater
   (f) 13-ampere, 240-volt air conditioner
   Use the optional method and type THW copper conductor for all circuits.

                     2. Design the branch circuits and 120/240-volt service for a one-family dwelling with the following load:
   (a) 2800 square feet of area
   (b) 5-kilowatt counter-mounted cooking unit
   (c) Two 4-kilowatt wall-mounted ovens (on same circuit as cooking unit)
   (d) 4.5-kilowatt, 240-volt clothes dryer
   (e) ¼-horsepower, 120-volt garbage disposal unit
   (f) 1.2 kilovolt-ampere, 120-volt dishwasher
   (g) Three 9-kilowatt, 240-volt space-heating units
   (h) 15-ampere, 120-volt air conditioner
   Use both the standard and optional methods and compare the results.

4-2 MULTIFAMILY DWELLING CALCULATIONS

This section presents examples that apply Code rules to the design of services for multifamily dwellings. The design of the branch circuits and the feeder for an individual dwelling unit in a multifamily dwelling was covered in Section 4-1.

The service load for the multifamily dwelling is *not* simply the sum of the individual dwelling unit loads because of demand factors that may be applied when either the standard calculation or the optional calculation is used to compute the service load.

When the standard calculation is used to compute the service load, the total lighting, small appliance, and laundry loads as well as the total load from all electric ranges and electric clothes dryers are subject to the application of demand factors. In addition, further demand factors may be applied to the portion of the neutral load contributed by electric ranges and the portion of the total neutral load greater than 200 amperes.

When the optional calculation is used, the *total connected load* is subject to the application of a demand factor that varies according to the number of individual units in the dwelling.

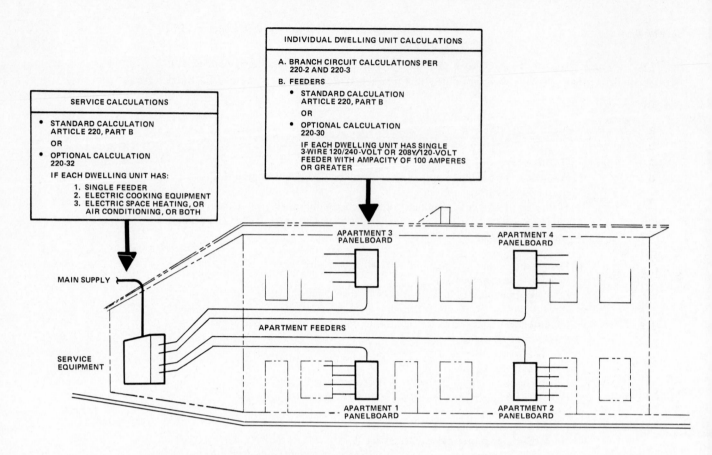

SERVICE CALCULATIONS

- STANDARD CALCULATION
  ARTICLE 220, PART B
  OR
- OPTIONAL CALCULATION
  220-32
  IF EACH DWELLING UNIT HAS:
      1. SINGLE FEEDER
      2. ELECTRIC COOKING EQUIPMENT
      3. ELECTRIC SPACE HEATING, OR
         AIR CONDITIONING, OR BOTH

INDIVIDUAL DWELLING UNIT CALCULATIONS

A. BRANCH CIRCUIT CALCULATIONS PER
   220-2 AND 220-3
B. FEEDERS
   - STANDARD CALCULATION
     ARTICLE 220, PART B
     OR
   - OPTIONAL CALCULATION
     220-30
     IF EACH DWELLING UNIT HAS SINGLE
     3-WIRE 120/240-VOLT OR 208Y/120-VOLT
     FEEDER WITH AMPACITY OF 100 AMPERES
     OR GREATER

MAIN SUPPLY

SERVICE
EQUIPMENT

APARTMENT 3
PANELBOARD

APARTMENT 4
PANELBOARD

APARTMENT FEEDERS

APARTMENT 1
PANELBOARD

APARTMENT 2
PANELBOARD

Figure 4-10. Calculation Methods for Designing Wiring Systems in Multifamily Dwellings

A summary of the calculation methods for designing wiring systems in multi-family dwellings and the applicable Code references are shown in Figure 4-10. The selection of a calculation method for computing the service load is not affected by the method used to design the feeders to the individual dwelling units.

The rules for computing the service load are also used for computing a main feeder load when the wiring system consists of a service that supplies main feeders which, in turn, supply a number of subfeeders to individual dwelling units.

In addition to the examples of the calculation methods described above, this section also discusses the use of the optional calculation for two-family dwellings and provides complete design examples for multifamily dwellings.

4-2.1 Standard Calculation for Computing Feeder of Service Loads in Multifamily Dwellings

ARTICLE 220,
PART B

The standard calculation for computing feeder or service loads may be used for any dwelling, whether it is a one-family, two-family, or multifamily dwelling. When the standard calculation is used for a two-family dwelling or for a multifamily dwelling, the total connected load for each type of load is first computed for the entire dwelling. Then any applicable demand factors are applied. The demand factors depend on the size of the load, or the number of appliances in some cases, but *not* on the number of dwelling units.

144

Example of Standard Calculation. Figure 4-11 summarizes the required calculations for designing the service for a 20-unit apartment building. Each apartment unit has the following loads:

a. 850 square feet of living area
b. 1.2-kilovolt-ampere, 120-volt dishwasher
c. 600-volt-ampere, 120-volt garbage disposal
d. 12-kilowatt electric range

The service is a 120/240 volt, three-wire circuit and uses type THW copper service-entrance conductors.

The general lighting load of 3 volt-amperes per square foot, two small appliance circuits, and a laundry circuit are included for each unit. The total general lighting, small appliance, and laundry load for 20 units is 141 000 volt-amperes. The demand factors for lighting loads reduce this load to a demand load of 49 200 volt-amperes. **TABLE 220-11**

The dishwashers and garbage disposals are taken at nameplate rating. The total loads are 24 000 volt-amperes and 12 000 volt-amperes, respectively. A Code provision for applying a demand factor to these appliances is ignored as explained previously. **220-10(a)** **220-17**

The total demand load for 20 ranges of 12-kilowatt rating is given in the Code as 35 kilowatts. The neutral load can be reduced to 70% of the *load* on the ungrounded conductors, or .7 × 35 kilowatts = 24 500 watts (volt-amperes). **TABLE 220-19** **220-22**

Figure 4-11. Standard Calculation for 20-Unit Apartment Building

SERVICE LOAD CALCULATION

CODE RULE		LINE	NEUTRAL
220-3 (b)	GENERAL LIGHTING LOAD = 3VA/FT^2 x 850 FT^2 x 20 UNITS	51,000	51,000
220-16(a)	SMALL APPLIANCE CIRCUITS = 2 x 1500VA x 20 UNITS	60,000	60,000
220-16(b)	LAUNDRY CIRCUITS = 1500VA x 20 UNITS	30,000	30,000
	TOTAL LTG, SMALL APPL, AND LAUNDRY CKTS	141,000	141,000
220-11	APPLICATION OF DEMAND FACTORS		
	FIRST 3000 @ 100%	3,000	3,000
	NEXT (120,000 - 3,000) = 117,000 @ 35%	40,950	40,950
	REMAINDER (141,000 - 120,000) = 21,000 @ 25%	5,250	5,250
	GENERAL LIGHTING DEMAND LOAD	49,200	49,200
220-10(a)	DISHWASHERS = 1.2 KVA x 20 UNITS	24,000	24,000
220-10(a)	DISPOSALS = 600VA x 20 UNITS	12,000	12,000
220-19, 220-22	ELECTRIC RANGES: 20 UNITS (NEUTRAL AT 70%)	35,000	24,500
430-25	25% OF LARGEST MOTOR = .25 x 600VA	150	150
	SERVICE LOAD	120,350 VA	109,850 VA

CHECKLIST	
LTG, SMALL APPL, AND LAUNDRY	✓
DISHWASHERS	✓
DISPOSALS	✓
RANGES	✓
SERVICE EQUIPMENT	✓

SELECTION OF SERVICE EQUIPMENT

CODE RULE	
220-22	1. SERVICE RATING = 120,350VA ÷ 240V = 501 AMPERES; NEUTRAL = 109,850W ÷ 240V = 458 AMPERES FURTHER DEMAND FACTOR FOR NEUTRAL FIRST 200 AMPERES @ 100% = 200 AMPERES BALANCE (458-200) = 258 AMPERES @ 70% = 181 AMPERES 381 AMPERES
TABLE 310-16	2. SERVICE ENTRANCE CONDUCTORS: TWO 900 MCM, ONE 600MCM THW COPPER CONDUCTORS
230-90, 430-63 240-6	3. OVERCURRENT PROTECTIVE DEVICE: USE 600-AMPERE STANDARD SIZE
230-79	4. DISCONNECTING MEANS: RATING REQUIRED TO BE 501 AMPERES OR GREATER (600-AMPERE STANDARD)
250-95, 250-94	5. GROUNDING: EQUIPMENT GROUNDING CONDUCTOR REQUIRED TO BE NO. 1 COPPER; GROUNDING ELECTRODE CONDUCTOR REQUIRED TO BE NO. 2/0 COPPER
CHAPTER 9	6. CONDUIT SIZE: AREA OF TWO 900 MCM = 2 x 1.4208 IN.^{2} = 2.8416 IN.^{2} AREA OF ONE 600 MCM = 1.0261 IN.^{2} AREA OF ONE NO. 1 = .2027 IN.^{2} 4.0704 IN.^{2} REQUIRES 4-INCH CONDUIT

430-25   If it is assumed that the disposals represent the largest rated motor, an additional 25% of the load of one disposal, or .25 × 600 volt-amperes = 150 volt-amperes, is added to the service load.

    The total service load is 120 350 volt-amperes for the ungrounded conductors and 109 850 volt-amperes for the neutral load. The line load in amperes at 240 volts

220-22   is 501 amperes. The calculated neutral load of 458 amperes may be reduced to a load of 200 amperes + 70% of the amount in excess of 200 amperes, or 200 + .7 (458 − 200) = 381 amperes. The minimum sizes for the THW copper service-en-

430-63   trance conductors are 900 MCM for the ungrounded conductors and 600 MCM for the neutral. A standard-size 600-ampere overcurrent protective device could be selected since it is the nearest standard size above the ampacity of the conductors. An

250-95   equipment grounding conductor would be required to be at least a No. 1 copper
230-79   conductor based on the size of the overcurrent protective device. The disconnecting means must be rated not less than the load current of 501 amperes.

250-94   A grounding electrode conductor at the service must be at least a size 2/0 copper conductor. The conduit encloses the two 900 MCM conductors: one 600 MCM con-

CHAPTER 9   ductor and one No. 1 equipment grounding conductor. A 4-inch conduit is needed in order to meet the allowable fill limitation of 40% for four conductors.

QUIZ

(Closed-Book)

1. In a dwelling unit the lighting load of 3001 volt-amperes to 120 000 volt-amperes is subject to what demand factor?
   (a) 25%
   (b) 35%
   (c) 40%

2. The minimum feeder load for an electric clothes dryer in a dwelling unit is _____ watts?

3. If _____ or more electric clothes dryers are served by a feeder, a demand factor of less than 100% is allowed for the load.

(Open-Book)

1. A seven-unit town house complex is served by a 120/240-volt service. Each unit contains the following:
   (a) 1800 square feet of living space
   (b) 12-kilowatt electric range
   (c) 6-ampere, 120-volt garbage disposal
   (d) 12-ampere dishwasher
   (e) Washing machine
   (f) 240-volt electric clothes dryer
   (g) Two 12-ampere, 240-volt air-conditioning compressors
   (h) 1-horsepower, 240-volt air handler motor
   Calculate the service rating and the size of the THW copper service-entrance conductors.

4-2.2 Optional Calculation for Multifamily Dwellings

220-32   The optional calculation for multifamily dwellings may be used to compute the service load when each individual dwelling unit is supplied by a single feeder, is equipped with electric cooking equipment, and is equipped with either electric space heating or air conditioning, or both. The primary feature of the optional calculation is that a total connected load is computed for the entire dwelling and then a demand factor based on the number of units is applied.

The size of the neutral may be computed by using the standard calculation. **220-32(a)** In many cases, this results in a smaller neutral than would be calculated by the optional method.

The optional calculation for multifamily dwellings should not be confused with **220-32** the optional calculation for computing the feeder load for an individual dwelling **220-30** unit. If the stated conditions for its use are satisfied, the optional calculation may be used to compute the service load for a multifamily dwelling regardless of the method used for designing the feeders to the individual dwelling units.

Any loads that are supplied from the service or a main feeder but are not from individual dwelling units are called *house loads*. The load from outside lighting, central laundry, and similar facilities are considered house loads and are computed by using the standard calculation method. The result is added to the service load computed by the optional method.

Example of Optional Calculation. An optional calculation for a 20-unit apartment building is shown in Figure 4-12. Each apartment contains the following loads:

    a. 850 square feet of living area
    b. 12-kilowatt electric range
    c. 1.8-kilowatt, 240-volt space heating
    d. 15-ampere, 120-volt air conditioner

The 120/240-volt service uses 250 MCM THW copper conductors in parallel in multiple conduits.

The connected load on the ungrounded conductors is simply the load of each apartment multiplied by 20. The total connected load is 417 000 volt-amperes. A **TABLE 220-32** demand factor of 38% is applied to reduce this load to 158 460 volt-amperes.

The line load is amperes at 240 volts is 660 amperes and the minimum number **TABLE 310-16** of 250 MCM conductors needed to carry this load is

$$\frac{660 \text{ A}}{255 \text{ A/conductor}} = 2.6 \text{ conductors}$$

or, three per line. When the circuit is run in three conduits, each conduit must contain **300-20** a complete circuit including the neutral conductor.

The neutral ampacity is computed by the standard calculation. The demand **220-19** load for twenty 12-kilowatt ranges is 35 kilowatts. The neutral load for the ranges **220-22** is .7 × 35 kilowatts = 24 500 watts. The total neutral load of 73 700 volt-amperes represents a load in amperes of 307 amperes. A further demand factor may be ap- **220-22** plied to the neutral load over 200 amperes which reduces the total neutral load to 275 amperes. The neutral conductor in each of the three conduits must be capable of carrying 275 ÷ 3 = 92 amperes; therefore, No. 3 THW copper conductors would ordinarily be selected. However, the smallest conductor size permitted for parallel **310-4** use is size 1/0. A 2-inch conduit would be required to enclose the two ungrounded conductors and the neutral of each circuit. The grounding electrode conductor must **250-94** be at least a size 2/0 copper conductor based on the equivalent size of the parallel service-entrance conductors.

SERVICE LOAD CALCULATION

CODE RULE		LINE	NEUTRAL BY STANDARD CALCULATION

220-32(c)(2)	GENERAL LIGHTING LOAD = 3VA/FT² x 850 FT² x 20 UNITS	51,000	51,000
220-32(c)(1)	SMALL APPLIANCE CIRCUITS = 2 x 1500VA x 20 UNITS	60,000	60,000
220-32(c)(1)	LAUNDRY CIRCUITS = 1500VA x 20 UNITS	30,000	30,000
			141,000

TOTAL LTG, SMALL APPL, AND LAUNDRY
APPLICATION OF DEMAND FACTORS

	FIRST 3000 @ 100%		3,000
	NEXT 117,000 @ 35%		40,950
	REMAINING 21,000 @ 25%		5,250
			49,200

220-32(c)(3)	ELECTRIC RANGE = 12 KW x 20 UNITS	240,000	24,500
220-32(c)(5)	ELECTRIC SPACE HEATING (NEGLECT AIR CONDITIONING) = 1800W x 20 UNITS	36,000	GENERAL LIGHTING DEMAND LOAD
			.7 X 35 KW DEMAND LOAD
	TOTAL CONNECTED LOAD	417,000	–0–

APPLICATION OF DEMAND FACTORS

TABLE 220-32	417,000VA @ 38% (20 UNITS)	158,460	–0–
	HOUSE LOAD	–0–	–0–
	SERVICE LOAD	158,460 VA	73,700 VA (NEUTRAL LOAD)

SELECTION OF SERVICE EQUIPMENT

1. 220-22 — SERVICE RATING = 158,460VA ÷ 240V = 660 AMPERES; NEUTRAL = 73,700VA ÷ 240V = 307 AMPERES
   FURTHER DEMAND FACTOR FOR NEUTRAL
   FIRST 200 AMPERES @ 100% = 200 AMPERES
   BALANCE (307-200) @ 70% = 75 AMPERES
   NET COMPUTED NEUTRAL LOAD = 275 AMPERES

2. TABLE 310-16 / 310-4 — SERVICE—ENTRANCE CONDUCTORS: THREE 250-MCM THW COPPER CONDUCTORS IN PARALLEL PER LINE FOR TOTAL AMPACITY OF 765 AMPERES (SIX TOTAL) NEUTRAL CONDUCTOR REQUIRED TO BE NO. 1/0 OR LARGER THW COPPER CONDUCTOR PER CIRCUIT (THREE TOTAL)

3. 230-90 — OVERCURRENT PROTECTIVE DEVICE: USE 700 AMPERE STANDARD SIZE

4. 230-79 — DISCONNECTING MEANS: NOT LESS THAN 660-AMPERE RATING

5. CHAPTER 9 — CONDUIT SIZE: THREE REQUIRED; EACH TO CONTAIN TWO UNGROUNDED CONDUCTORS AND ONE NEUTRAL –
   AREA OF TWO 250-MCM = 2 x .4877 = .9754 IN.²
   AREA OF ONE NO. 1/0 = .2367 IN.²
   1.2121 IN.² REQUIRES 2-INCH CONDUIT

6. 250-94 — GROUNDING ELECTRODE CONDUCTOR: REQUIRED TO BE NO. 2/0 COPPER

CHECKLIST

LTG, SMALL APPL, AND LAUNDRY	✓
RANGES	✓
HEATING OR A/C	✓

Figure 4-12. Example of Optional Calculation for Multifamily Dwelling

Design Example with House Loads. Assume that the 20-unit apartment complex of the previous example contained the following facilities:

a. Central laundry facility with five 5-kilowatt electric clothes dryers and five 3-kilo-volt-ampere, 240-volt washing machines
b. 10-horsepower, 240-volt elevator motor with a 60-minute rating.

The additional load computed by the standard calculation would increase the required ampacity of the service-entrance conductors as follows:

$$\text{Dryer load}^3 = 5 \times 5\,\text{kW} \qquad\qquad\quad = 25\ 000\ \text{VA}$$
$$\text{Washer load}^3 = 5 \times 3\,\text{kVA} \qquad\qquad = 15\ 000\ \text{VA}$$
$$\text{Motor load} = 0.90 \times 240\,\text{V} \times 50\,\text{A} = \underline{10\ 800\ \text{VA}}$$
$$\text{Total house load}\qquad 50\ 800\ \text{VA}$$

TABLE 430-148
TABLE 620-13(a),
430-22(a) exception

The additional load is 50 800 watts ÷ 240 volts = 211.7 amperes. Using the results from the previous example, we see that the service load in amperes is 660 + 211.7 = 871.7 amperes. At least four 250 MCM conductors in parallel would be required to carry the total service load. If laundry receptacles are not installed in the apartment units, the laundry circuit load can be deleted from the service calculation in Figure 4-12.

210-52(e)
EXCEPTION 1

QUIZ
(Closed-Book)
1. The optional method of calculation for multifamily dwellings may be applied when _____ or more dwelling units are served by the feeder or service.
2. House loads may be computed by the optional method of calculation. True or false?

(Open-Book)
1. Using the optional method, design the 120/240-volt service for a 50-unit apartment complex. Each apartment contains the following:
   (a) 1000 square feet of living space
   (b) 10-kilowatt electric range
   (c) 9.6-kilowatt, 240-volt space heater
   (d) 5-kilowatt, 240-volt electric clothes dryer
   (e) 6-ampere, 120-volt garbage disposal
   (f) 12-ampere, 120-volt dishwasher
   The house load consists of ten 200-watt outdoor lights that burn continuously. Compute the neutral by the standard method and use 220-17.

4-2.3 Optional Calculation for Two-Family Dwellings
The optional calculation for multifamily dwellings is intended for dwellings having three or more dwelling units. The Code, however, allows the optional calculation to be used when two dwelling units are supplied by a single feeder or service; the calculated load, however, must be based on *three* units, not two. The load should be computed first by using the optional method for three units and then by the standard method for two units. The lesser of the two loads is then used to determine service rating.

ARTICLE 100
220-33

[3]This load could be reduced by a demand factor when allowed by a local electrical board. However, some authorities consider the central laundry to be commercial in nature and allow no reduction in load.

220-17
220-18

4-2.4 Complete Design Examples
for Multifamily Dwellings

The complete electrical system for a multifamily dwelling includes the branch circuits and feeder for each dwelling unit and the main service. The feeder to each unit may be designed by using the standard calculation method or the optional calculation method if the dwelling units satisfy the conditions stated in the Code. When the required feeder ampacity is 100 amperes or more, the optional calculation method is normally chosen to compute the feeder load for each dwelling unit.

The standard calculation method may be applied to any dwelling, but it is usually advantageous to use the optional calculation method for multifamily dwellings when the units in the dwelling contain large heating, air-conditioning, and appliance loads other than electric ranges.

In a practical design problem both calculation methods would be used and the result that yielded the smaller service rating would be selected. In either case, however, good design practice dictates that sufficient extra capacity over the minimum required by the Code should be provided to allow for future expansion.

Design Example with Comparison Between Standard and Optional Calculations. A 100-unit apartment complex is served by a 120/240-volt circuit. Each dwelling unit contains the following loads:

a. 1000 square feet of living area
b. 10-kilowatt electric range
c. 9.6-kilowatt central space heating
d. 10-ampere, 240-volt air conditioner
e. 5-kilowatt, 240-volt clothes dryer
f. 6-ampere, 120-volt garbage disposal
g. 12-ampere, 120-volt dishwasher

Design the branch circuits and compute the feeder and service loads by using both standard and optional calculation methods and then compare the results.

Branch-Circuit and Feeder Calculations. The branch circuits may be designed as shown in Figure 4-13. A 15-ampere inverse-time circuit breaker was selected for the air-conditioning circuit.

The feeder load for the dwelling units is computed by both calculation methods, but the neutral ampacity is calculated by the standard calculation method only. Figure 4-13 shows the results of the feeder load calculation.

ARTICLE 220
PART B
220-11
220-19
(220-17 NOT USED)
220-15
220-21
430-25

By using the standard method, the general lighting, small appliance, and laundry load of 7500 volt-amperes is reduced to a demand load of 4575 volt-amperes. The 10-kilowatt electric range represents a demand load of only 8 kilowatts. The other appliances and space heater must be computed on the basis of 100% of connected load. The garbage disposal is selected as the largest motor since the load of the air conditioners is neglected when compared to the heating load. An additional 25% of the load of one garbage disposal is thus added to the total load, with a resulting computed load on the ungrounded conductors of 29 515 volt-amperes, or 122.9 amperes.

220-30

The optional calculation for dwelling units requires the larger of the air-conditioning load or 65% of the central space-heating load to be used in the calculation. The demand load for the 9.6-kilowatt heating unit is .65 × 9.6 kilowatts = 6240 watts (volt-amperes); thus, the air-conditioning load of 240 volts × 10 amperes = 2400 volt-amperes is neglected. The other loads in excess of 10 kilovolt-amperes are

CHECKLIST

LTG, SMALL APPL, AND LAUNDRY	✓
RANGE	✓
DRYER	✓
DISPOSAL	✓
DISHWASHER	✓
HEATING OR A/C	✓

BRANCH CIRCUITS REQUIRED

CODE RULE	BRANCH CIRCUITS REQUIRED	SUMMARY
220-4, 220-3 (b)	GENERAL LIGHTING LOAD = $\dfrac{3VA/FT^2 \times 1000\ FT^2}{120V \times 20A}$ = 1.25 OR 2 CIRCUITS	TWO 20-AMPERE, 2-WIRE CIRCUITS
220-4 (b), (c)	SMALL APPLIANCE AND LAUNDRY CIRCUITS	THREE 20-AMPERE, 2-WIRE CIRCUITS
TABLE 220-19 NOTE 4 424-3(b)	RANGE CIRCUIT = 8000W ÷ 240V = 33.3 A	40-AMPERE, 240-VOLT CIRCUIT
	SPACE HEATING = 1.25 x 9600W ÷ 240V = 50A	60-AMPERE, 240-VOLT CIRCUIT
440-32 440-22	AIR CONDITIONER = 1.25 x 10A = 12.5A MAXIMUM BREAKER RATING = 1.75 x 10A = 17.5A	USE 15-AMPERE, 240-VOLT CIRCUIT
430-22	DRYER = 5000W ÷ 240V = 20.8A	25- OR 30-AMPERE, 120/240-VOLT CIRCUIT
	DISPOSAL = 1.25 x 6A = 7.5A	15-AMPERE, 2-WIRE CIRCUIT
	DISHWASHER	15-AMPERE, 2-WIRE CIRCUIT

FEEDER LOAD CALCULATION

CODE RULE		OPTIONAL CALCULATION	STANDARD CALCULATION LINE	NEUTRAL	
220-11	GENERAL LIGHTING LOAD = 3VA/FT² + 1000 FT²	3,000	3,000	3,000	
	SMALL APPLIANCE CIRCUITS = 2 x 1500VA	3,000	3,000	3,000	
	LAUNDRY CIRCUIT	1,500	1,500	1,500	
			7,500	7,500	TOTAL LTG, SMALL APPL, AND LAUNDRY
					APPLICATION OF DEMAND FACTORS
			3,000	3,000	FIRST 3000 @ 100%
			1,575	1,575	REMAINDER (7500-3000) @ 35%
			4,575	4,575	GENERAL LIGHTING DEMAND LOAD
220-19, 220-22	10-KW RANGE	10,000	8,000	5,600	NEUTRAL = .7 x 8000W
	DRYER	5,000	5,000	3,500	NEUTRAL = .7 x 5000W
	DISPOSAL = 120V x 6A		720	720	
	DISHWASHER = 120 x 12A	1,440	1,440	1,440	
			180	180	25% OF LARGEST MOTOR
	TOTAL OTHER LOAD	24,660	9,600 →	-0- →	
			29,515 VA	16,015 VA	FEEDER LOAD
TABLE 220-30	APPLICATION OF DEMAND FACTORS				
	FIRST 10KVA OF OTHER LOAD @ 100%	10,000			
	REMAINDER (24,660-10,000) @ 40%	5,864			
220-15	HEATING = 9.6 KW @ 65%	6,240			
	FEEDER LOAD	22,104 VA			FEEDER LOAD

SELECTION OF FEEDER EQUIPMENT

	CODE RULE		
1.		FEEDER AMPACITY:	LOAD BY OPTIONAL CALCULATION = 22,104VA ÷ 240V = 92.1 AMPERES; LOAD BY STANDARD CALCULATION = 29,515VA ÷ 240V = 122.9 AMPERES
	TABLE 310-16		NEUTRAL = 16,015VA ÷ 240V = 66.7 AMPERES; USE 100-AMPERE SIZE FEEDER BASED ON OPTIONAL CALCULATION.
2.		CONDUCTOR SIZES:	TWO NO. 3 AND ONE NO. 4 THW COPPER CONDUCTORS
3.	240-3	OVERCURRENT PROTECTIVE DEVICE:	USE 100-AMPERE STANDARD SIZE DEVICE
4.	CHAPTER 8	CONDUIT SIZE:	REQUIRES 1¼ INCH CONDUIT
5.	250-91	GROUNDING:	CONDUIT MAY SERVE AS EQUIPMENT GROUNDING CONDUCTOR

Figure 4-13. Branch Circuit and Feeder Calculations for One Unit of 100-Unit Apartment Complex

subject to a demand factor of 40%. The resulting net feeder ampacity is 92.1 amperes which requires a 100-ampere size feeder. The neutral ampacity is 66.7 amperes as computed by the standard method.

Main Service Calculation. The main service for the 100-unit apartment complex is calculated as shown in Figure 4-14. The example demonstrates the large reduction in service rating that can result from using the optional calculation for multifamily dwellings.

220-11   The general lighting, small appliance, and laundry load for 100 units is 750 kilovolt-amperes. The standard method reduces this load to a demand load of 201.45 **TABLE 220-19** kilovolt-amperes. The demand load for one hundred 10-kilowatt ranges is given by **TABLE 220-18** the Code as 25 kilowatts + ³/₄ kilowatt for each range for a total of 100 kilowatts. The demand factor for the 100 electric clothes dryers is 25% of the total connected **430-25** load. The demand factor for the disposals and dishwashers is 75%. The heaters are computed at 100% of the connected load. Although negligible, 25% of the load of the disposal, or 180 volt-amperes, is added to bring the net computed service load to 1548.6 kilovolt-amperes, or a load in amperes of 6453 amperes. The neutral load **220-22** is 521.1 kilovolt-amperes or 2171 amperes. A further demand factor is allowed for the portion of the neutral load over 200 amperes, which reduces the net computed neutral load to 1580 amperes.

TABLE 220-32   In the optional calculation method the total connected load for 100 dwelling units is calculated and then a demand factor of 23% is applied. Because of the large

CODE RULE	SERVICE LOAD CALCULATION				
			STANDARD CALCULATION		
		OPTIONAL CALCULATION	LINE	NEUTRAL	
220-32	GENERAL LIGHTING LOAD = 3VA/FT² x 1000 FT² x 100 UNITS	300,000	300,000	300,000	
	SMALL APPLIANCE CIRCUITS = 2 x 1500VA x 100 UNITS	300,000	300,000	300,000	
	LAUNDRY CIRCUITS	150,000	150,000	150,000	
			750,000	750,000	TOTAL LTG, SMALL APPL, AND LAUNDRY
					APPLICATION OF DEMAND FACTORS
			3,000	3,000	FIRST 3000 @ 100%
			40,950	40,950	NEXT 117,000 @ 35%
			157,500	157,500	REMAINDER (750,000 - 120,000) @ 25%
			201,450	201,450	GENERAL LIGHTING DEMAND LOAD
220-19	RANGES = 10KW x 100 UNITS	1,000,000	100,000	70,000	25KW + (3/4KW x 100); NEUTRAL = .7 x 100KW
	DRYERS = 5KW x 100 UNITS	500,000	125,000	87,500	.25 x 500,000W (NEUTRAL @ 70%)
	DISPOSALS = 120V x 6A x 100 UNITS	72,000	54,000	54,000	.75 x 72,000
	DISHWASHERS = 120 V x 12A x 100 UNITS	144,000	108,000	108,000	.75 x 144,000
	HEATERS (NEGLECT AIR CONDITIONING) = 9.6KW x 100 UNITS	960,000	960,000	—0—	
	TOTAL CONNECTED LOAD	3,426,000			
TABLE 220-32	APPLICATION OF DEMAND FACTORS		180	180	25% OF LARGEST MOTOR
	3,426,000 @ 23%	787,980			
	HOUSE LOADS	—0—			
	SERVICE LOAD	787,980 VA	1,548,630 VA	521,130 VA	SERVICE LOAD
220-22	LOAD BY OPTIONAL CALCULATION = 787,980 VA ÷ 240V = 3 283 AMPERES; LOAD BY STANDARD CALCULATION = 1,548,630 VA ÷ 240V = 6453 AMPERES				
	NEUTRAL LOAD = 521,130VA ÷ 240V = 2171 AMPERES				
	FURTHER DEMAND FACTOR FOR NEUTRAL				
	FIRST 200 AMPERES @ 100%	= 200 AMPERES			
	BALANCE (2171 - 200) @ 70%	= 1380 AMPERES			
	NET COMPUTED NEUTRAL LOAD	1580 AMPERES			

Figure 4-14. Main Service Calculation for 100-Unit Apartment Complex

heating, air-conditioning, and appliance load, the optional calculation in this case produced the smallest required service rating. The net computed service load is 3283 amperes, which is less than one-half the result obtained by the standard calculation method.

The wiring system for this dwelling would require parallel service-entrance conductors or bus bars. There would probably be an arrangement of main feeders supplying a number of meter banks with subfeeders to each individual dwelling unit.

QUIZ

(Closed-Book)   **1.** The connected load created by lighting, small appliance circuits, and laundry circuits in a 40-unit apartment complex is 400 kilovolt-amperes. Compare the service demand load by using both the standard calculation method and the optional calculation method for multifamily dwellings (demand factor = 28%).

(Open-Book)   **1.** Design the branch circuits, feeders, and the main 120/240-volt service for a 150-unit motel with cooking facilities for tenants. Each unit contains the following:
(a) 200 square feet of area
(b) 1/3-horsepower, 120-volt garbage disposal
(c) 10-ampere, 240-volt air conditioner
The cooking equipment is a gas oven. The service-entrance conductors need not be selected because a bus bar is used for this circuit. All other conductors are type THW copper. Use the standard method, and all permissible demand factors.

4-3 SPECIAL MULTIFAMILY DWELLING PROBLEMS

When a multifamily dwelling contains a large number of dwelling units, the main service from the utility company is usually a high-voltage, three-phase service. Transformers at the premises step down the voltage to supply the individual dwelling units and other loads. Many large dwelling complexes have central air-conditioning and heating systems which are three-phase loads while individual dwelling units are supplied by a single-phase feeder derived from two phases of the three-phase circuit.

Design Example. A 15-unit apartment building is supplied by a 12 470Y/7200-volt main service. A transformer on the premises provides a 208Y/120-volt circuit to supply the individual dwelling units and a central utility building. Each apartment contains the following loads:

a. 1640 square feet of living area
b. 12-kilowatt single-phase electric range
c. 1.3-kilovolt-amperes, 120-volt dishwasher

Each apartment is supplied by a separate feeder. The central utility room contains a 50-kilowatt, 208-volt, three-phase central heating system and a 20-horsepower, 208-volt, three-phase air conditioner compressor driven by a conventional motor.

Determine the ampacity of the apartment feeders, the kilovolt-ampere rating of the service transformer, and the load current of the main service at 12 470 volts.

A summary of the calculations is shown in Figure 4-15. The general lighting demand load on each dwelling unit feeder is 5247 volt-amperes. The dishwasher adds another 1300 volt-amperes. A 12-kilowatt range represents a demand load of

TABLE 220-19

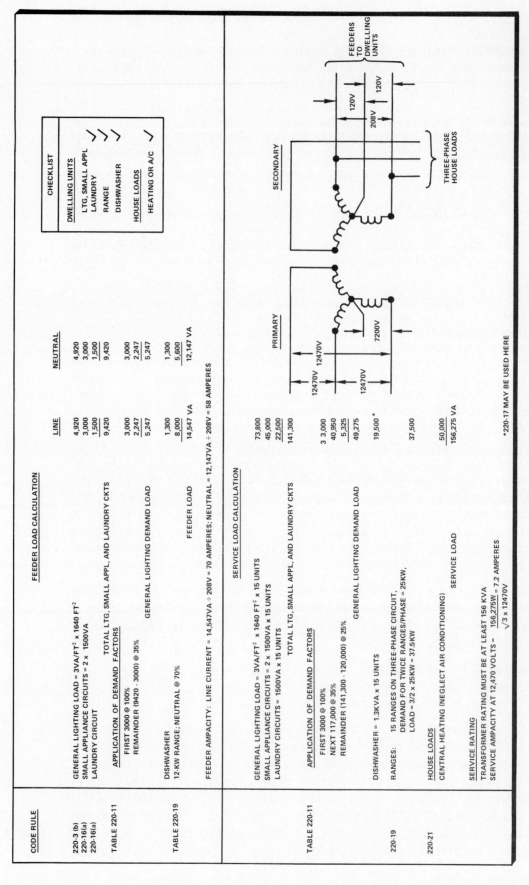

FEEDER LOAD CALCULATION

CODE RULE		LINE	NEUTRAL
220-3 (b)	GENERAL LIGHTING LOAD = 3VA/FT² x 1640 FT²	4,920	4,920
220-16(a)	SMALL APPLIANCE CIRCUITS = 2 x 1500VA	3,000	3,000
220-16(a)	LAUNDRY CIRCUIT	1,500	1,500
	TOTAL LTG, SMALL APPL, AND LAUNDRY CKTS	9,420	9,420
TABLE 220-11	APPLICATION OF DEMAND FACTORS		
	FIRST 3000 @ 100%	3,000	3,000
	REMAINDER (9420 - 3000) @ 35%	2,247	2,247
	GENERAL LIGHTING DEMAND LOAD	5,247	5,247
TABLE 220-19	DISHWASHER	1,300	1,300
	12-KW RANGE; NEUTRAL @ 70%	8,000	5,600
	FEEDER LOAD	14,547 VA	12,147 VA

FEEDER AMPACITY: LINE CURRENT = 14,547VA ÷ 208V = 70 AMPERES; NEUTRAL = 12,147VA ÷ 208V = 58 AMPERES

SERVICE LOAD CALCULATION

CODE RULE		
	GENERAL LIGHTING LOAD = 3VA/FT² x 1640 FT² x 15 UNITS	73,800
	SMALL APPLIANCE CIRCUITS = 2 x 1500VA x 15 UNITS	45,000
	LAUNDRY CIRCUITS = 1500VA x 15 UNITS	22,500
	TOTAL LTG, SMALL APPL, AND LAUNDRY CKTS	141,300
TABLE 220-11	APPLICATION OF DEMAND FACTORS	
	FIRST 3000 @ 100%	3 3,000
	NEXT 117,000 @ 35%	40,950
	REMAINDER (141,300 - 120,000) @ 25%	5,325
	GENERAL LIGHTING DEMAND LOAD	49,275
	DISHWASHER = 1.3KVA x 15 UNITS	19,500 *
220-19	RANGES: 15 RANGES ON THREE-PHASE CIRCUIT, DEMAND FOR TWICE RANGES/PHASE = 25KW, LOAD = 3/2 x 25KW = 37.5KW	37,500
220-21	HOUSE LOADS	
	CENTRAL HEATING (NEGLECT AIR CONDITIONING)	50,000
	SERVICE LOAD	156,275 VA

SERVICE RATING
TRANSFORMER RATING MUST BE AT LEAST 156 KVA
SERVICE AMPACITY AT 12,470 VOLTS = $\dfrac{156,275W}{\sqrt{3} \times 12470V}$ = 7.2 AMPERES

*220-17 MAY BE USED HERE

CHECKLIST

DWELLING UNITS
LTG, SMALL APPL
LAUNDRY
RANGE
DISHWASHER

HOUSE LOADS
HEATING OR A/C

Figure 4-15. Calculations for Multifamily Dwelling With High-Voltage Service

8 kilowatts (volt-amperes). The total load is 14 547 volt-amperes on the individual dwelling unit feeder. The fact that the feeders consist of two phases and the neutral of a three-phase circuit does not affect the calculation.

The 12 470Y/7200-volt to 208Y/120-volt transformer carries the load of the 15 units as well as the central utility "house" load. The 15 single-phase electric ranges are supplied by *two phases of a three-phase system;* therefore, the load must be computed on the basis of twice the maximum number connected between any two phases. The number of ranges for phase of the three-phase system is 15 ÷ 3 = 5 ranges. The demand given by the Code for twice that number, or 10 ranges, is 25 kilowatts. The load per phase then is 25 kilowatts ÷ 2 = 12.5 kilowatts and the three-phase load is 3 × 12.5 kilowatts = 37.5 kilowatts. **220-19**

The load of the central utilities must be added to the load from the dwelling units. The 20-horsepower, 208-volt compressor-motor has a full-load current of 54 amperes + (.1 × 54) amperes = 59.4 amperes after adjusting the 230-volt value from the Code table to a 208-volt value. The three-phase load is $\sqrt{3}$ × 208 volts × 59.4 amperes = 21 399.8 volt-amperes. Since the 50-kilowatt central heater represents a larger load, the air-conditioning load is neglected. **TABLE 430-150** **220-21**

The total 208-volt load on the transformer is 156.27 kilovolt-amperes and thus requires a transformer capable of supplying more than 156 kilovolt-amperes. At 12 470 volts, the line load in amperes is 7.2 amperes.

Unbalanced Loads in Multifamily Dwellings. If a multifamily dwelling is supplied by one phase of a three-phase transformer, the load is unbalanced. A typical case is the 120/240-volt high-leg delta service. The transformer may also serve 240-volt, single-phase loads and 240-volt, three-phase loads. In these problems the loads should be converted to amperes for each line. A circuit diagram would also help in analyzing the design problem.

QUIZ
(Open-Book)   1. Determine the conductor size and the main overcurrent device for a 120/240-volt high-leg delta service supplying a ten-unit apartment building. Each apartment has an area of 500 square feet and there are no permanently installed appliances and laundry facilities. The complex also has a 25-horsepower, 240-volt, three-phase motor whose branch circuit is protected by fuses. Type THW aluminum conductors are used throughout the wiring system.

TEST CHAPTER 4

I. True or False *(Closed-Book)*

	T	F
1. To measure the area of a dwelling to determine the lighting load, the outside dimensions (including the garage) are used.	[ ]	[ ]
2. Under no circumstances can the continuous load supplied by a branch circuit exceed 80% of the branch-circuit rating.	[ ]	[ ]
3. A 20-ampere laundry circuit is always required in each dwelling unit.	[ ]	[ ]
4. Branch-circuit conductors supplying fixed resistive type space heaters must be rated at 125% of the heater load.	[ ]	[ ]
5. The feeder load shall not be less than the sum of the loads on the branch circuits after any applicable demand factors have been applied.	[ ]	[ ]
6. When the standard calculation method is used, the minimum feeder load for a household electric clothes dryer is 5000 watts.	[ ]	[ ]
7. The optional calculation method for a dwelling unit may be used when the service is a two-wire, 120-volt service.	[ ]	[ ]
8. The optional calculation method for multifamily dwellings does not apply to dwellings whose units have neither electric space heating nor air conditioning.	[ ]	[ ]

II. Multiple Choice *(Closed-Book)*

1. The maximum rating of the overcurrent device for a branch circuit supplying a non-motor operated appliance that draws 20 amperes is:
    - (a) 20 amperes
    - (b) 25 amperes
    - (c) 30 amperes

    1. _____

2. If the standard calculation method is used, the feeder demand for a dwelling with a 150-kilovolt-ampere general lighting load is:
    - (a) 52 500 volt-amperes
    - (b) 51 450 volt-amperes
    - (c) 54 450 volt-amperes

    2. _____

3. If the optional calculation method for a dwelling unit is used, the load of less than four separately controlled electric space-heating units is subject to a demand factor of:
    - (a) 100%
    - (b) 65%
    - (c) 40%

    3. _____

4. The optional calculation for additional loads in an existing dwelling unit counts the load other than air conditioning or heating at 100% demand for the first
    - (a) 10 000 volt-amperes
    - (b) 3000 volt-amperes
    - (c) 8000 volt-amperes

    4. _____

5. If the optional calculation method is used, the load for a 4.5-kilowatt, 240-volt electric clothes dryer is:
    - (a) 4.5 kilowatts
    - (b) 5 kilowatts
    - (c) 5.625 kilowatts

    5. _____

6. If the optional calculation method for a one-family dwelling is used, all "other loads" above the initial 10 kilowatts are subject to a demand factor of:
    - (a) 40%
    - (b) 50%
    - (c) 65%

    6. _____

7. The ampacity of the neutral in a service circuit supplying a dwelling is calculated to be 40 amperes. The grounded electrode conductor is a No. 4 copper conductor. The minimum size neutral is:
   (a) No. 8
   (b) No. 6
   (c) No. 4

7. _____

8. A branch circuit supplies a household electric range with a demand of 8000 watts. The ungrounded conductors have an ampacity of 40 amperes. The neutral ampacity must be at least:
   (a) 28 amperes
   (b) 24 amperes
   (c) 35 amperes

8. _____

III. Problems *(Open-Book)*

1. Design the branch circuits and the service for a one-family dwelling supplied by a 120/240-volt service. The dwelling contains the following:
   (a) 1800 square feet of area
   (b) Two 6-kilowatt ranges on separate circuits
   (c) 4.8-kilowatt, 240-volt electric clothes dryer
   (d) 4.5-kilowatt, 240-volt hot water heater
   (e) 9-kilowatt, 240-volt central space heating
   (f) 17-ampere, 240-volt air conditioner
   (g) Two 1/4-horsepower, 120-volt motors on separate circuits.

   Use THW copper conductors. The service capacity should be calculated by using the optional calculation method.

2. Design the entire 120/240-volt service for a one-family dwelling. Use type THW aluminum conductors. The dwelling contains the following:
   (a) 2000 square feet of living area
   (b) 14-kilowatt electric range
   (c) 5-kilowatt, 240-volt electric clothes dryer
   (d) 16-kilowatt, 240-volt space heating
   (e) 1.2-kilovolt-ampere, 120-volt dishwasher

   Use the standard method of calculation. What error in the neutral ampacity results because the dishwasher causes an unbalanced load?

3. Design the service for a 12-unit apartment complex. The service is 208Y/120 volts with the load evenly distributed among the three phases. Each apartment unit contains the following:
   (a) 900 square feet of living area
   (b) 8.4-kilowatt, 208-volt electric range
   (c) 1.5-kilowatt, 120-volt bathroom heater
   (d) Laundry facilities in each apartment

   The apartment complex has a 19-kilowatt, three-phase central heater. The service-entrance conductors are THW copper conductors.

4. Design the feeders and the 120/240-volt service for a 25-unit apartment complex. Each unit contains the following:
   (a) 1000 square feet of living area
   (b) 12-kilowatt electric range
   (c) 1.5-kilovolt-ampere dishwasher
   (d) 4.5-kilowatt, 240-volt space heater
   (e) 30-ampere, 240-volt air conditioner

   The complex has the following as "house" loads:
   (a) 1000 square feet of corridors
   (b) Four 4-kilovolt-amperes, 120-volt washing machines
   (c) Four 6-kilowatt, 240-volt electric clothes dryers
   (d) Four 300-watt, 120-volt outside floodlights

   Calculate the feeder and service rating by using the optional methods. Type THW copper conductors are used for all circuits. Use 500 MCM conductors in parallel for the service.

5

Electrical Circuit Design for Commercial and Industrial Occupancies

ARTICLE 220* The design of the electrical circuits for commercial and industrial occupancies is based on specific Code rules that relate to the loads present in such occupancies. The design approach is to separate the loads into those for lighting, receptacles, motors, appliances, and other special loads and apply the applicable Code rules. In general, the loads are considered continuous unless specific information is given to the contrary.

If the electric load is large, the main service may be a three-phase supply which supplies transformers on the premises. It is not uncommon to have secondary feeders supplying panelboards which, in turn, supply branch circuits operating at different voltages. In this case, the design of the feeder and branch circuits for each voltage is considered separately. The rating of the main service is based on the total load **215-5** with the load values transformed according to the various circuit voltages if necessary. A feeder circuit diagram is essential when loads of different voltages are present.

*References in the margins are to the specific applicable rules and tables in the *National Electrical Code*.

Chapter 5

Electrical Circuit Design for Commercial and Industrial Occupancies **159**

In most commercial and industrial occupancies, no demand factors are applied to the loads served.[1] The lighting load in hospitals, hotels and motels, and warehouses, however, is subject to the application of demand factors. In restaurants and similar establishments the load of electric cooking equipment is subject to a demand factor if there are more than three units. Optional calculation methods to determine feeder or service loads for schools and farms are also provided in the Code. **TABLE 220-11, TABLE 220-20** **220-34** **220-40**

Special occupancies, such as mobile homes and recreational vehicles, require the feeder or service load to be calculated in accordance with specific Code rules. The service for mobile home parks and recreational vehicle *parks* is also designed based on specific Code rules that apply only to those locations. The feeder or service load for receptacles supplying shore power for boats in marinas and boatyards is also specified in the Code. **ARTICLE 550, ARTICLE 551** **ARTICLE 555**

Most situations for which specific Code rules are provided are covered in this chapter either by example or as a quiz problem. Other Code rules that pertain to wiring system design are summarized in Table 5-1 at the end of this chapter.

Practical Electrical Design. The design examples given in the Guide represent a design based on Code requirements. In many cases, the equipment ratings that result from strict application of Code rules are not adequate for a specific practical situation, and this is recognized by the Code. The conductor sizes that result from calculations performed in accordance with Code rules do not take into account voltage drop caused by the length of conductors. The Code recommends that the total voltage drop from the service to the farthest branch-circuit outlet should not be greater than 5%. This problem and considerations for future expansion are not covered by the Code or this Guide. **90-1** **210-19(a), 215-2**

The circuit voltages are taken to be nominal voltages based on the main service voltage or a stepped-down voltage at a transformer. These voltages may be higher than the actual voltage at an outlet (which may be lower as a result of voltage drop on the conductors). For example, a 480-volt service may actually supply 460 volts or even 440 volts to the farthest outlet. A conservative design approach would be to use the lower voltage in the calculations and thereby obtain a larger required ampacity for the conductors. The voltages to be used in the Guide are, of course, those given in the problem under discussion.

Sizes and Ratings. Transformers used in lighting and power circuits are normally considered capable of carrying their full-rated load continuously. Thus, a 10-kilowatt continuous load requires service conductors rated for 1.25×10 kilowatts = 12.5 kilowatts, but a transformer rated at only 10 kilovolt-amperes. In practice, transformers are manufactured with a limited number of standard ratings.

Fuses used to protect circuits are supplied in the standard sizes listed in the Code. If the current exceeds the fuse size for a nontime-delay fuse, the fuse element will open. Many circuit breakers, especially those for low current values, have a fixed rating or setting. Other breakers for larger loads (100 to 6000 amperes) are supplied in standard *size* frames such as 100, 225, 400, 600, 800, 1200 amperes, etc. The tripping mechanism that determines the rating or setting can be adjusted at the factory and the standard *ratings* given in the Code can be selected (although all ratings are **240-6**

[1]Demand factors are allowed for feeders supplying receptacle circuits in nondwelling occupancies. These demand factors may be applied to receptacles for general use (i.e., not continuous loads) and where allowed by local authorities. **220-13**

not readily available commercially). The Code rules that apply to the rating or *setting* of overcurrent devices refer to the capacity of the tripping mechanism in the breaker if adjustable breakers are used.

Service-entrance panelboards protected by main circuit breakers are also produced with standard capacities that correspond to the standard circuit breaker frame *sizes*. The circuit breaker *setting* is adjusted to protect the circuit as determined by Code rules.

5-1 TYPICAL COMMERCIAL OCCUPANCY CALCULATIONS

Typical commercial establishments range from small stores with single-phase services to large office buildings with three-phase services and significant motor loads. The design of the electrical wiring system for such occupancies is treated in this section. The examples given become increasingly complex although the approach to the design problem is the same for all examples. The branch circuits are designed first, then the feeders if used, and finally the main service.

When transformers are not involved, a relatively simple design problem with a single voltage results, as in the first example in this section. If step-down transformers are used, the transformer itself must be protected by an overcurrent device which may also protect the circuit conductors in most cases. In any design, the rules for the protection of the transformer must be considered to assure a design in complete conformance with Code rules.

Switchboards and panelboards used for the distribution of electricity within a building are also subject to Code rules. In particular, a lighting and appliance panelboard cannot have more than 42 overcurrent devices to protect the branch circuits originating at the panelboard. This rule could affect the number of feeders required when a large number of lighting or appliance circuits are needed.

margin refs: 450-3; ARTICLE 384; 384-14, 384-15

5-1.1 Design Example for a Store with Show Window

A small store has outside dimensions of 50 feet by 60 feet and a 30-foot length of show window. The store has 20 general-purpose receptacles that do not supply continuous loads. The branch circuits and the 120/240-volt service for the building are to be designed using THW copper conductors. Twenty-ampere circuits are used for all branch circuits.

The design summary is shown in Figure 5-1. The branch circuits for lighting and receptacles required a total of seven 20-ampere circuits. The load for the 30-foot show window is computed on the basis of 200 volt-amperes per linear foot which is considered a continuous load value. The 6000-volt-ampere load requires three 20-ampere circuits. One additional 20-ampere circuit is needed for the required outside outlet for sign or outline lighting if the store is on the ground floor.

The feeder or service load in this example is simply the sum of the branch-circuit loads. If it is assumed that the sign circuit is to be continuously loaded, its maximum load could be as high as .8 × 120 volts × 20 amperes = 1920 volt-amperes. Since the actual load is not given, a 1200 volt-ampere minimum load may be used in the calculation. The total service load is then 22 050 volt-amperes, or 91.9 amperes at 240 volts. The standard 100-ampere overcurrent protective device would be used. The minimum neutral ampacity based on the connected load is 19800 VA ÷ 240 V = 82.5 amperes. The service-entrance conductors would be enclosed in a 1¼-inch conduit and a No. 8 grounding electrode conductor would be used at the service.

margin refs: 220-3(c); EXCEPTION 3; CHAPTER 9 EXAMPLE 3; 600-6(b); 220-3(a); 600-6(c); TABLE 310-16; 230-90; 240-6; 250-94

CODE RULE	BRANCH CIRCUITS REQUIRED		SUMMARY
TABLE 220-3 (b)	LIGHTING = 50 FT x 60 FT x 3VA/FT^2 x 1.25 = 11,250 VOLT-AMPERES		
	$\dfrac{11,250VA}{120V \times 20A}$ = 4.7, OR 5 CIRCUITS		FIVE 20-AMPERE, 2-WIRE CIRCUITS
220-3 (c)	RECEPTACLES = 20 x 180VA/RECEPTACLE = 3600 VOLT-AMPERES		TWO 20-AMPERE, 2-WIRE CIRCUITS
	$\dfrac{3600VA}{120V \times 20A}$ = 1.5, OR 2 CIRCUITS		
220-3 (c) EXCEPTION 3	SHOW WINDOW = 30 FT x 200VA/FT = 6000 VOLT-AMPERES		THREE 20-AMPERE, 2-WIRE CIRCUITS
	$\dfrac{6000 VA}{120V \times 20A}$ = 2.5 OR 3 CIRCUITS		
600-6(b)	SIGN CIRCUIT: REQUIRED TO BE 20-AMPERE CIRCUIT		ONE 20-AMPERE, 2-WIRE CIRCUIT

	SERVICE LOAD CALCULATION	LOAD	CHECKLIST
220-10(a)			
	LIGHTING = 50 FT x 60 FT x 3VA/FT^2 x 1.25	11,250	LIGHTING ✓
	RECEPTACLES = 20 x 180VA/RECEPTACLE	3,600	RECEPTACLES ✓
	SHOW WINDOW = 30 FT x 200VA/FT	6,000	SHOW WINDOW ✓
	SIGN CIRCUIT = 1200 VA MINIMUM	1,200	SIGN CIRCUIT ✓
	SERVICE LOAD	22,050 VA	

SELECTION OF SERVICE EQUIPMENT

TABLE 310-16	1. SERVICE RATING = 22,050VA ÷ 240V = 91.9 AMPERES MINIMUM
	2. CONDUCTOR SIZES: TWO NO. 3 THW COPPER CONDUCTORS AND A NO. 4 THW COPPER NEUTRAL
230-90,	3. OVERCURRENT PROTECTIVE DEVICE AND DISCONNECT: USE 100-AMPERE STANDARD SIZE DEVICE
250-95	4. EQUIPMENT GROUNDING CONDUCTOR: USE NO. 8 COPPER IF REQUIRED
CHAPTER 9	5. CONDUIT SIZE: USE 1-1/4 INCH CONDUIT
250-94	6. GROUNDING ELECTRODE CONDUCTOR: REQUIRED TO BE NO. 8 COPPER

Figure 5-1. Design Summary for Store With Show Window

5-1.2 Design Example for an Office Building

A 14 000-square foot office building is served by a 460Y/265-volt, three-phase service. The building contains the following loads:

a. 10 000-volt-ampere, 208/120-volt, three-phase sign
b. 80 duplex receptacles supplying continuous loads[2]
c. 30-foot long show window
d. 12-kilowatt, 208/120-volt, three-phase electric range
e. 10-kilowatt, 208/120-volt, three-phase electric oven
f. 20-kilowatt, 460-volt, three-phase hot water heater
g. Fifty 150-watt, 120-volt lighting fixtures for outside lighting
h. One hundred fifty 200-volt-ampere, 265-volt fluorescent fixtures for interior office lighting
i. 7.5-horsepower, 460-volt, three-phase air handler motor protected by an approved instantaneous trip circuit breaker
j. 35-kilowatt, 460-volt, three-phase electric heating unit
k. 50-ampere, 460-volt, three-phase air-conditioning unit

The ratings of the service equipment, transformer, and the feeders or branch circuits supplying the equipment are to be determined. Circuit breakers are used to protect each circuit and THW copper conductors are used throughout the wiring system.

[2]If the receptacle load is not considered continuous, the portion of the receptacle load over 10 kVA may be subjected to a demand of 50% according to Section 220-13

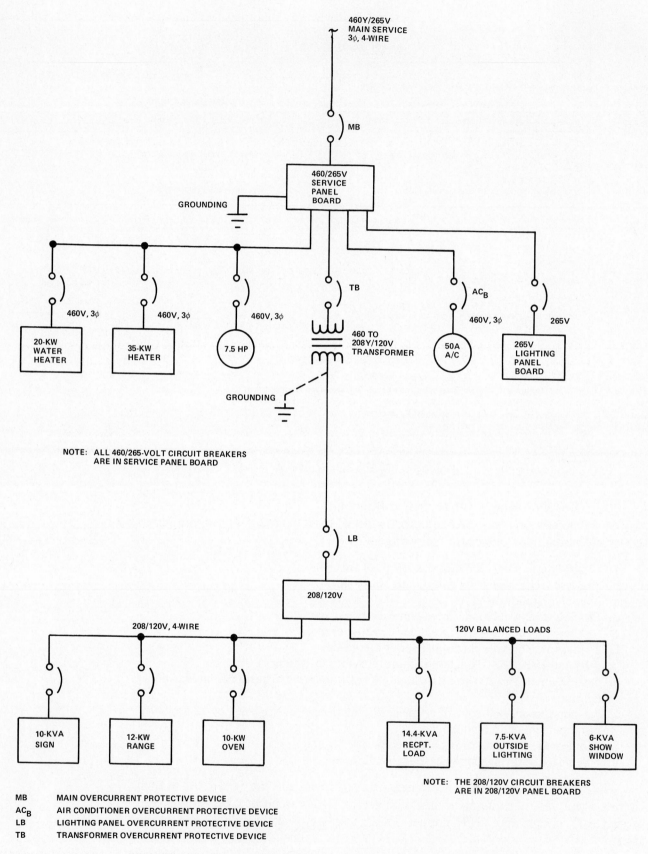

460Y/265V
MAIN SERVICE
3φ, 4-WIRE

MB

460/265V
SERVICE
PANEL
BOARD

GROUNDING

460V, 3φ   460V, 3φ   460V, 3φ   TB   AC_B   265V

20-KW
WATER
HEATER

35-KW
HEATER

7.5 HP

460 TO
208Y/120V
TRANSFORMER

50A
A/C

265V
LIGHTING
PANEL
BOARD

GROUNDING

NOTE: ALL 460/265-VOLT CIRCUIT BREAKERS
        ARE IN SERVICE PANEL BOARD

LB

208/120V

208/120V, 4-WIRE   120V BALANCED LOADS

10-KVA
SIGN

12-KW
RANGE

10-KW
OVEN

14.4-KVA
RECPT.
LOAD

7.5-KVA
OUTSIDE
LIGHTING

6-KVA
SHOW
WINDOW

NOTE: THE 208/120V CIRCUIT BREAKERS
        ARE IN 208/120V PANEL BOARD

MB   MAIN OVERCURRENT PROTECTIVE DEVICE
AC_B   AIR CONDITIONER OVERCURRENT PROTECTIVE DEVICE
LB   LIGHTING PANEL OVERCURRENT PROTECTIVE DEVICE
TB   TRANSFORMER OVERCURRENT PROTECTIVE DEVICE

Figure 5-2.   Simplified Feeder Diagram for 14,000-Square-Foot Office Building

A simplified feeder diagram is shown in Figure 5-2. The 208Y/120-volt lighting panel is designed first as shown in Figure 5-3(a). The total load on the panel is 73 375 volt-amperes, or 204 amperes at 208Y/120 volts. The 208Y/120-volt lighting **240-6** panel may be protected by a standard-size 225-ampere circuit breaker (L_B). The **TABLE 310-16** ungrounded conductors are No. 4/0 type THW copper.

Since the transformer supplying the lighting panel is rated for a continuous load, the kilovolt-ampere rating need only be 59.9 kilovolt-amperes. A commercially available 75-kilovolt-ampere transformer would be selected. The transformer requires **450-3(b)** individual overcurrent protection set at not more than 125% of the *rated* primary current. The rated primary current for the 75-kilovolt-ampere transformer is

$$\frac{75\ 000\ \text{VA}}{\sqrt{3}\ \times\ 460\ \text{V}} = 94.1\ \text{A}$$

The maximum setting of the transformer overcurrent protective device (T_B) is then **450-3(b)(1)** 1.25×94.1 amperes = 117.6 amperes. However, the primary conductors must be **EXCEPTION 1** protected at their ampacity if their ampacity is lower.

The 208Y/120-volt circuit is a separately derived system from the transformer **250-5(d)** and is grounded by means of a grounding electrode conductor which must be at least **250-26** a No. 2 copper conductor based on the No. 4/0 copper feeder conductors. **TABLE 250-94**

Figure 5-3(a). Calculations for 208Y/120-Volt Loads

CODE RULE	208Y/120-VOLT LOAD CALCULATIONS			
	NOTES: 1. ASSUME ALL LOADS ARE BALANCED ON 208/120-VOLT CIRCUIT. 2. ASSUME ALL LOADS ARE CONTINUOUS.	LOAD	TRANSFORMER	CHECKLIST
220-2(c)	RECEPTACLES = 1.25 x 180 VA x 80	18,000	14,400	RECEPTACLES ✓
220-12 EXAMPLE 3, CHAPTER 9	SHOW WINDOW = 200VA/FT x 30 FT (200VA/FT CONSIDERED CONTINUOUS LOAD VALUE)	6,000	6,000	SHOW WINDOW ✓ OUTSIDE LIGHTS ✓
	OUTSIDE LIGHTING = 1.25 x 150W x 50 10-KW SIGN = 1.25 x 10KW 12-KW RANGE = 1.25 x 12KW 10-KW OVEN = 1.25 x 10KW	9,375 12,500 15,000 12,500	7,500 10,000 12,000 10,000	SIGN ✓ RANGE ✓ OVEN ✓
	NET COMPUTED LOAD	73,375 VA	59,900 VA	
	SELECTION OF FEEDER EQUIPMENT			
220-22	1. FEEDER RATING = 73,375VA ÷ √3 X 208V = 204 AMPERES; NEUTRAL = 59900VA ÷ √3 X 208V = 166A			
TABLE 310-16 240-6, 384-16 TABLE 250-94	2. FEEDER OVERCURRENT PROTECTIVE DEVICE (L_B) AND PANELBOARD: USE 225-AMPERE STANDARD SIZE 3. FEED CONDUCTOR SIZES: TWO NO. 4/0 THW COPPER CONDUCTORS AND A NO. 2/0 NEUTRAL 4. GROUNDING ELECTRODE CONDUCTOR: USE NO. 2 COPPER CONDUCTOR 5. TRANSFORMER RATING: AT LEAST 59.9 KW; USE 75-KVA STANDARD SIZE (TRANSFORMER RATED FOR CONTINUOUS LOAD)			

CODE RULE	460/265-VOLT CIRCUIT CALCULATIONS			

NOTES: 1. ASSUME ALL LOADS ARE CONTINUOUS INCLUDING HOT WATER HEATER.
2. APPROPRIATE EQUIPMENT GROUNDING CONDUCTOR TO BE PROVIDED FOR EACH CIRCUIT.

CODE RULE	LOAD	CIRCUIT AMPACITY	THW COPPER CONDUCTORS	OVERCURRENT DEVICE
424-14(b) 250-95				

450-3(b)
EXCEPTION 1,
240-6
	FEEDER TO TRANSFORMER $= 73{,}375VA \div \sqrt{3} \times 460V$
	OVERCURRENT DEVICE (T_B) BASED ON RATED CURRENT
	   RATED CURRENT $= 75KVA \div \sqrt{3} \times 460V = 94.1A$
	   MAXIMUM RATING $= 1.25 \times 94.1A = 117.6A$,
	      T_B TO BE 100-AMPERE SIZE
		92.1	NO. 3	100-AMPERE

	265-VOLT LIGHTING PANEL $= 1.25 \times \dfrac{3.5VA/FT^2 \times 14000\ FT^2}{\sqrt{3} \times 460V}$ (ASSUME A BALANCED LOAD; NEGLECT SMALLER LOAD OF 150 FIXTURES)	76.9	NO. 4	90-AMPERE
422-14(b)	HOT WATER HEATER $= 1.25 \times \dfrac{20{,}000W}{\sqrt{3} \times 460V}$	31.4	NO. 8	35- OR 40-AMPERE
424-3(b)	ELECTRIC HEATING $= 1.25 \times \dfrac{35{,}000W}{\sqrt{3} \times 460V}$	54.9	NO. 6	60-AMPERE
TABLE 430-150 TABLE 430-152	7.5-HP MOTOR $= 1.25 \times 11A$ MAXIMUM BREAKER SIZE $= 7 \times 11A = 77A$ USE 80-AMPERE SETTING	13.75	NO. 14	80-AMPERE
440-32 440-22	AIR CONDITIONER $= 1.25 \times 50A$ MAXIMUM BREAKER SIZE $= 1.75 \times 50A = 87.5A$	62.5	NO. 6	90-AMPERE

Figure 5-3(b). Feeder or Branch Circuits for 460/265-Volt Loads

Figure 5-3(c). Main Service Load Calculation

CODE RULE	SERVICE LOAD CALCULATION	LINE	NEUTRAL
	208/120-VOLT LOAD	92.1	—0—
	265-VOLT LIGHTING PANEL	76.9	61.5
	HOT WATER HEATER	31.4	—0—
220-21	ELECTRIC HEATING (NEGLECTED)	—0—	—0—
	7.5-HP MOTOR	11.0	—0—
	AIR CONDITIONER	50.0	—0—
430-25	25% OF LARGEST MOTOR $= .25 \times 50A$	12.5	—0—
	SERVICE LOAD	273.9 AMPERES	61.5 AMPERES

SELECTION OF SERVICE EQUIPMENT

CODE RULE	
250-23(b)	1. SERVICE-ENTRANCE CONDUCTORS: THREE 300 MCM TYPE THW CONDUCTORS; NEUTRAL REQUIRED TO BE NO. 2 BASED ON SIZE OF GROUNDING ELECTRODE CONDUCTOR
430-63	2. MAIN PROTECTIVE DEVICE: MAXIMUM RATING $= 90A$ (LARGEST MOTOR DEVICE) $+ 11 + 31.4 + 76.9 + 92.1 = 301.4$ AMPERES USE 300-AMPERE STANDARD SIZE

CHAPTER 9	3. CONDUIT: AREA OF THREE 300 MCM THW $= 3 \times .5581\ IN.^2 = 1.6743\ IN.^2$
	   AREA OF ONE NO. 2 $= 0.1473\ IN.^2$
	   $1.8216\ IN.^2$ REQUIRES 2½-INCH CONDUIT

250-94	4. GROUNDING ELECTRODE CONDUCTOR REQUIRED TO BE NO. 2 BASED ON 300 MCM SERVICE-ENTRANCE CONDUCTORS

Design calculations for the primary feeder and the other 460/265-volt circuits are summarized in Figure 5-3(b).

The 460-volt feeder that supplies the transformer carries a load of

$$\frac{73\ 375\ \text{VA}}{\sqrt{3} \times 460\ \text{V}} = 92.1\ \text{A}$$

which requires a 100-ampere protective device and three No. 3 THW copper conductors.

The 265-volt lighting feeder load is the larger of the area load of 3.5 volt-amperes per square foot of building area or the connected load of the 150 fluorescent lighting fixtures. The area load is 1.25 × 3.5 volt-amperes/square foot × 14 000 square feet = 61 250 volt-amperes or a balanced load of 76.9 amperes. The connected load is 1.25 × 200 volt-amperes × 150 = 37 500 volt-amperes and is neglected. The feeder of No. 4 type THW copper conductors is protected by a 90-ampere circuit breaker.   **TABLE 310-16 / 240-6**

The continuous hot water heater requires a circuit ampacity of at least 31.4 amperes; therefore, No. 8 conductors are used for the circuit. The circuit supplying the electric heating requires a circuit ampacity of at least 54.9 amperes; therefore, No. 6 conductors are selected.   **TABLE 310-16**

An instantaneous trip circuit breaker was selected to protect the 7.5-horsepower motor circuit. This arrangement is allowed if the breaker is adjustable and is part of an approved controller. A full-load current of 11 amperes requires No. 14 conductors protected by a breaker set as high as 7 × 11 = 77 amperes. A standard 80-ampere setting would be selected.   **430-52 / TABLE 430-150, TABLE 430-152**

The air-conditioning circuit must have an ampacity of 1.25 × 50 amperes = 62.5 amperes for which No. 6 conductors are selected. An overcurrent protective device for this circuit cannot be set higher than 1.75 × 50 amperes = 87.5 amperes.   **440-32 / 440-22**

The design calculations for the main service are summarized in Figure 5-3(c). The loads are balanced and could be computed in terms of volt-amperes, but the loads are computed here in terms of amperes as is typically done with commercial and industrial systems in which unbalanced loads are usually present. Calculation of the load in amperes also simplifies the selection of the main overcurrent protective device which is based on the setting of the largest motor branch-circuit protective device plus the sum of the other loads in amperes.   **220-10 / 430-63**

The 73 375-volt-ampere, 208/120-volt lighting panel load represents a line load of 92.1 amperes at 460 volts. Even though no neutral load is present because the primary of the transformer is a 460-volt, three-phase delta winding, a neutral capacity at least equal to that of the grounding electrode conductor must be provided to the service.   **250-23(b)**

The 265-volt interior lighting load is 61 250 volt-amperes. As a balanced load, the line load is 76.9 amperes but a neutral to supply the connected load need carry only 61.5 amperes. The three-phase loads contribute a load in amperes according to their rating as shown in Figure 5-3(c). Twenty-five percent of the largest motor load is added, resulting in a total service load of 273.9 amperes.

The ungrounded conductors are selected to be 300 MCM type THW copper conductors. The neutral ampacity must be at least 61.5 amperes according to the calculation, but the neutral conductor cannot be smaller than the grounding electrode conductor. The neutral and the grounding electrode conductor at the service are therefore selected to be No. 2 copper based on the 300 MCM service-entrance conductors.   **TABLE 310-16 / 250-23(b) / TABLE 250-94**

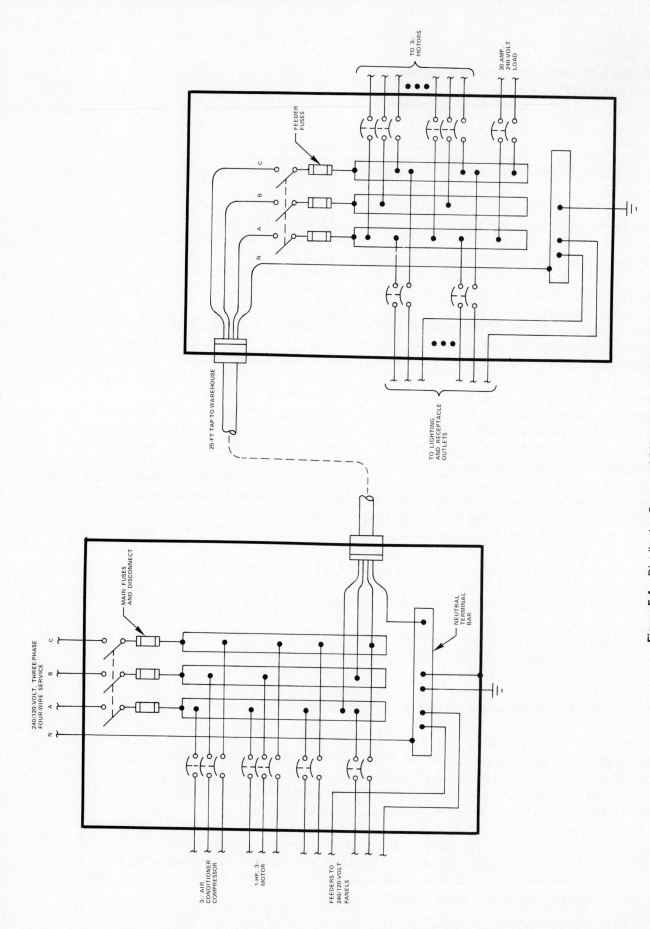

Figure 5-4. Distribution System of Office Building and Warehouse

The rating or setting of the overcurrent protective device is based on the rating of the largest motor branch-circuit protective device (90 amperes) plus the sum of the currents of the other loads. The maximum setting permitted is 301.4 amperes. A standard 300-ampere device is selected as the next lower standard size. The conduit must enclose three 300 MCM type THW conductors and one No. 2 THW conductor; therefore, a 2½-inch conduit is required.

430-63

CHAPTER 9

5-1.3 Design Example for an Office Building with Warehouse

An office building and a 40 000-square foot warehouse are supplied by a 240/120-volt, three-phase, four-wire service. The service-entrance equipment is located in the office building. The warehouse has its own feeder panel with overcurrent protection and a disconnecting means. The feeder panel is within 25 feet of the service-entrance equipment. Instantaneous trip breakers protect the motor circuits.

ARTICLE 230, PART J, 230-84

The office building contains the following loads:

a. 10 000 square feet of area lighted by 120-volt incandescent lamps
b. 240 duplex receptacles for noncontinuous loads
c. Six ⅙-horsepower, 240-volt, single-phase motors on separate circuits
d. 28-ampere, 240-volt, three-phase air conditioner motor-compressor
e. 1-horsepower, 240-volt, three-phase air handler motor
f. Outside sign circuit connected between phase A and neutral

The service and feeder overcurrent protective devices are fuses and the other circuits are protected by circuit breakers. The phase A and C conductors are parallel 350 MCM type THW copper conductors in two conduits to provide for future expansion.[3]

The loads supplied from the warehouse panel are:

a. 40 000 square feet of area lighted by 120-volt incandescent lamps in heavy-duty type lampholders
b. 40 duplex receptacles (considered continuous)
c. Three 1½-horsepower, 240-volt, three-phase motors
d. 30-ampere, 240-volt, single-phase special appliance

The feeder for the warehouse is tapped from the main service-entrance conductors and is enclosed in conduit. As long as the distance is within 25 feet and the feeder conductors have an ampacity of not less than one-third that of the service-entrance conductors or the overcurrent protective device, a set of fuses in the warehouse feeder panel may protect the feeder conductors. This so-called *25-foot tap rule* eliminates the need for feeder overcurrent protective devices within the service equipment.

240-21 EXCEPTION 3

A simplified distribution system diagram is shown in Figure 5-4. The conductor ampacities of the service-entrance conductors and the feeder or branch circuits are to be determined as well as the ratings of the required overcurrent protective devices.

Warehouse Loads. The design of the warehouse branch circuits is summarized in Figure 5-5(a). Each of the three 1½-horsepower motor circuits is protected by a 40-ampere overcurrent protective device. All of the loads are considered to be continuous loads.

The calculations for the warehouse protective devices are also shown in Figure 5-5(a). Although the total ampacity is calculated, the feeder conductor sizes cannot be determined until the service-entrance conductors are selected because of the re-

220-10
240-21

220-35
430-62(b)

---

[3]The Code provides for future expansion at existing installations.

quirements of the 25-foot tap rule. The total load on phases A and C is 143.9 amperes and only 16.9 amperes on phase B. The overcurrent protective devices must be rated or set at not more than 177.4 amperes for phases A and C and 50.4 amperes for phase B based on the largest branch-circuit protective device (40 amperes) plus the sum of the other loads. Two 175-ampere fuses and one 50-ampere fuse would be used.

Office Loads. Figure 5-5(b) presents the design calculations for the other feeders and branch circuits originating at the service equipment and the branch circuits originating from the 240/120-volt lighting and appliance panelboards supplied from the service equipment.

The branch circuits for the 120-volt lighting load and receptacle load are determined as shown in part A of Figure 5-5(b). Phases A and C and the neutral supply these loads. A minimum load of 1200 volt-amperes is assumed for the 20-ampere sign circuit which is supplied only by phase A and the neutral. The six ⅙-horsepower motors are supplied by six individual circuits in the example although several motors could be supplied by a single 15-ampere, 240-volt circuit.

The individual branch circuits for the three-phase motor loads are also shown in part B of Figure 5-5(b). The 1-horsepower motor and the 28-ampere motor-compressor are supplied by 25-ampere and 45-ampere circuits, respectively.

Two or more 240/120-volt lighting and appliance circuit panelboards are required since each panelboard may house 42 overcurrent devices with each pole of a circuit breaker counting as one device. The feeder or feeders supplying the panelboards must have a total overcurrent protection of 316.9 amperes for phase A as shown in part C of Figure 5-5(b). The conductor size would be based on the connected loads and the requirements for conductor overcurrent protection.

The main service load consists of the 240/120-volt loads, the three-phase motor loads, and the warehouse load as shown in Figure 5-5(c). The total loads are 497.5 amperes, 54.2 amperes, and 487.5 amperes for phases A, B, and C, respectively. The loads on phases A and C are carried by two 350 MCM type THW copper conductors in parallel per phase. If two conduits are used to enclose the conductors, then phase B and the neutral conductor must also be included in each conduit. Although the required ampacity of the phase B conductor is only 54.2 amperes, two No. 1/0 conductors must be used because that is the smallest size conductor permitted for use as parallel conductors. The neutral of each circuit must have an ampacity of at least ½ × 290 amperes = 145 amperes; thus, two No. 1/0 THW copper conductors in parallel would be used for the neutral. Two 2½-inch conduits would be used to enclose the service-entrance conductors.

The main fuses are selected based on the 45-ampere circuit breaker for the motor-compressor plus the sum of the other loads. The allowed fuse sizes are 600 amperes for phase A, 600 amperes for phase C, and 60 amperes for phase B. Since this is a motor feeder, strict adherence to Code rules requires the next smaller standard-size overcurrent protective device to be used. The Code, however, allows the rating or setting of the overcurrent protective device to be based on the rated ampacity of the conductors whenever higher capacity feeders are installed in order to provide for future additions or changes. In this case, since the 350 MCM conductors for phases A and C can carry 2 × 310 amperes = 620 amperes, 600-ampere fuses could be used for both phases and thereby provide capacity for future additions.

Margin references:
430-63
240-6
600-6(c)
430-53
384-15
220-10
240-3
300-20
310-4
TABLE 310-16
430-63
430-62(b)

WAREHOUSE BRANCH CIRCUITS REQUIRED

NOTES:
1. ASSUME ALL LOADS ARE CONTINUOUS
2. PROVIDE EQUIPMENT GROUNDING CONDUCTORS

CHECKLIST	
LIGHTS	✓
RECEPTACLES	✓
MOTORS (3)	✓
30-AMPERE LOAD	✓

LOADS

CODE RULE	Calculation	SUMMARY
TABLE 220-3 (b) 210-23(b)	LIGHTING = 1.25 x .25VA/FT² x 40,000 FT² = 12,500 VOLT-AMPERES USE 30-AMPERE CIRCUITS WITH HEAVY-DUTY LAMPHOLDERS $\frac{12,500VA}{120V} \times 30A$ = 3.5, OR 4 CIRCUITS	FOUR 30-AMPERE, 2-WIRE CIRCUITS
220-3 (c)	RECEPTACLES = 1.25 x 180VA x 40 = 9 000 WATTS $\frac{9,000VA}{120V} \times 20A$ = 3.8, OR 4 CIRCUITS	FOUR 20-AMPERE, 2-WIRE CIRCUITS
TABLE 430-150 TABLE 430-152	THREE 1-1/2-HP MOTORS = 1.25 x 5.2A = 6.5 AMPERES EACH MAXIMUM BREAKER SIZE = 7 x 5.2A = 36.4 AMPERES USE 40-AMPERE SETTING	THREE 40-AMPERE 240-VOLT, 3-PHASE CIRCUITS
220-3 (c)(1)	30-AMPERE, 240-VOLT LOAD = 1.25 x 30A = 37.5 AMPERES	ONE 40-AMPERE, 240-VOLT CIRCUIT

WAREHOUSE FEEDER LOAD CALCULATION

CODE RULE	Calculation	LINE A, C	NEUTRAL	LINE B
TABLE 220-3 (b) 220-10	LIGHTING = $\frac{1.25 \times .25VA/FT^2 \times 40,000FT^2}{240V}$ = 52 AMPERES	52	41.7	–0–
220-3 (c) 220-10	RECEPTACLES = $\frac{1.25 \times 180VA \times 40}{240V}$ = 37.5 AMPERES	37.5	30	–0–
TABLE 430-150	MOTORS = 3 x 5.2A = 15.6 AMPERES	15.6	–0–	15.6
	30-AMPERE LOAD = 1.25 x 30A = 37.5 AMPERES	37.5	–0–	–0–
430-25	25% OF LARGEST MOTOR	1.3	–0–	1.3
	FEEDER LOAD	143.9 AMPERES	71.7 AMPERES	16.9 AMPERES

SELECTION OF FEEDER EQUIPMENT

CODE RULE	
240-21	1. FEEDER CONDUCTORS: AMPACITY OF FEEDER CONDUCTORS REQUIRED TO BE NOT LESS THAN 1/3 THAT OF SERVICE-ENTRANCE CONDUCTORS OR OVERCURRENT PROTECTION. SEE SERVICE CALCULATIONS.
430-63	2. OVERCURRENT PROTECTIVE DEVICE: FEEDER PROTECTED BY FUSES WITH MAXIMUM RATINGS OF LINE A, C = 40A (LARGEST MOTOR PROTECTION) + 52A + 37.5A + 10.4 (OTHER MOTORS) + 37.5A = 177.4 LINE B = 40A (LARGEST MOTOR PROTECTION) + 10.4 (OTHER MOTORS) = 50.4 AMPERES
250-24(a) 250-95	3. A GROUNDING ELECTRODE CONDUCTOR IS REQUIRED AT THE WAREHOUSE. THE CONDUIT MAY SERVE AS THE EQUIPMENT GROUNDING CONDUCTOR

Figure 5-5(a). Calculations for Warehouse Branch Circuits and Feeder

169

CODE RULE		
	A. 240/120-VOLT PANEL BRANCH CIRCUITS REQUIRED	**SUMMARY**

CHECKLIST

LIGHTING ✓
RECEPTACLES ✓
MOTORS(6) ✓
SIGN ✓

A. 240/120-VOLT PANEL BRANCH CIRCUITS REQUIRED   **SUMMARY**

TABLE 220-3 (b)

LIGHTING = 1.25 × 3.5VA/FT² × 10,000 FT² = 43,750 VOLT-AMPERES

$$\frac{43,750VA}{120V \times 20A} = 18.23 \text{ OR } 19 \text{ CIRCUITS}$$

NINETEEN 20-AMPERE, 2-WIRE CIRCUITS, OR TEN 20-AMPERE, 3-WIRE CIRCUITS

RECEPTACLES = 180VA × 240 = 43,200 VOLT-AMPERES

$$\frac{43,200VA}{120V \times 20A} = 18 \text{ CIRCUITS}$$

TWENTY FOUR 20-AMPERE, 2-WIRE CIRCUITS, OR TWELVE 20-AMPERE, 3-WIRE CIRCUITS

600-6(b)

OUTSIDE SIGN: REQUIRED TO BE 20-AMPERE CIRCUIT

ONE 20-AMPERE, 2-WIRE CIRCUIT

SIX 1/6-HP MOTORS: USE SEPARATE 15-AMPERE CIRCUITS

SIX 15-AMPERE, 240-VOLT CIRCUITS

B. THREE-PHASE MOTOR CIRCUITS REQUIRED

430-22
TABLE 430-152

1-HP, 240-VOLT, 3-PHASE MOTOR:
CONDUCTOR AMPACITY = 1.25 × 3.6A = 4.5 AMPERES
MAXIMUM BREAKER SETTING = 7 × 3.6A = 25.2 AMPERES

USE 25-AMPERE, 3-PHASE CIRCUIT
NO. 14 THW COPPER CONDUCTORS

440-32
440-22

28-AMPERE, 240-VOLT, 3-PHASE MOTOR COMPRESSOR
CONDUCTOR AMPACITY = 1.25 × 28A = 35 AMPERES
MAXIMUM BREAKER SIZE = 1.75 × 28A = 49 AMPERES

USE 45-AMPERE, 3-PHASE CIRCUIT
NO. 8 THW COPPER CONDUCTORS

C. 240/120-VOLT PANEL FEEDER LOAD CALCULATION

CODE RULE		LINE A	LINE C	NEUTRAL
TABLE 220-3(b)	LIGHTING = $\dfrac{1.25 \times 3.5VA/FT^2 \times 10,000 FT^2}{240V}$	182.3	182.3	145.8
220-13	RECEPTACLES = $\dfrac{10,000 + .5(43,200 - 10,000)VA}{240V}$	110.8	110.8	110.8
600-6(c)	OUTSIDE SIGN CIRCUIT = 1200 VA(MIN) ÷ 120V	10	-0-	10
TABLE 430-148	SIX 1/6-HP MOTORS = 6 × 2.2A = 13.2 AMPERES	13.2	13.2	-0-
430-25	25% OF LARGEST MOTOR = .25 × 2.2 = .6	.6	.6	-0-
		316.9 AMPERES	306.9 AMPERES	266.6 AMPERES

D. SELECTION OF 240/120-VOLT PANEL FEEDER EQUIPMENT

384-15

1. NUMBER OF OVERCURRENT DEVICES EXCEEDS 42; THEREFORE, TWO OR MORE PANELBOARDS MUST BE USED.
2. CONDUCTOR AMPACITY AND OVERCURRENT PROTECTIVE DEVICE RATING BASED ON LOAD ON EACH PANELBOARD
3. NEUTRAL MAY BE REDUCED TO 200A + .7(266.6 - 200)A = 246.6A

Figure 5-5(b). Calculations for Branch Circuits and Feeders Originating From Main Service Equipment

SERVICE LOAD CALCULATIONS

CODE RULE		LINE A	LINE B	LINE C	NEUTRAL
	A. WAREHOUSE LOADS:				
220-3 (b)	LIGHTING = 1.25 × .25VA/FT² × 40,000 FT² ÷ 240V	52	-0-	52	41.6
220-3 (c)	RECEPTACLES = 1.25 × 180VA × 40 ÷ 240V	37.5	-0-	37.5	30
TABLE 430-150	THREE 1-1/2-HP MOTORS = 3 × 5.2A	15.6	15.6	15.6	-0-
	30-AMPERE LOAD = 1.25 × 30A	37.5	-0-	37.5	-0-
	B. 240/120-VOLT PANEL LOADS:				
220-3 (b)	LIGHTING = 1.25 × 3.5VA/FT² × 10,000 FT² ÷ 240V	182.3	-0-	182.3	145.8
220-13	RECEPTACLES = 10,000 + .5(43,200 - 10,000)VA ÷ 240V	110.8	-0-	110.8	110.8
220-3 (a)	OUTSIDE SIGN CIRCUIT = 1200VA ÷ 120V	10	-0-	-0-	10
TABLE 430-148	SIX 1/6-HP MOTORS = 6 × 2.2A	13.2	-0-	13.2	-0-
	C. THREE PHASE MOTOR LOADS:				
430-22	1-HP, 240-VOLT, 3-PHASE MOTOR	3.6	3.6	3.6	-0-
430-32	28-AMPERE MOTOR COMPRESSOR	28	28	28	-0-
430-25	**D.** 25% OF LARGEST MOTOR = .25 × 28	7	7	7	-0-
	SERVICE LOAD	497.5 AMPERES	54.2 AMPERES	487.5 AMPERES	328.2 AMPERES

SELECTION OF SERVICE EQUIPMENT

TABLE 310-16, NOTE 8 300-20

1. SERVICE-ENTRANCE CONDUCTORS:
   LOAD REQUIRES PARALLEL CONDUCTORS FOR PHASES A AND C. TWO CONDUITS ARE USED TO AVOID NECESSITY OF DERATING CONDUCTOR AMPACITIES. SINCE EACH CONDUIT MUST CONTAIN A COMPLETE CIRCUIT, PHASE B AND NEUTRAL MUST ALSO HAVE PARALLELED CONDUCTORS.

TABLE 310-16 310-4

   NUMBER OF 350 MCM THW COPPER CONDUCTORS : 2 CONDUCTORS PER PHASE
   (TOTAL AMPACITY = 620 AMPERES). USE TWO NO. 1/0 FOR PHASE B (TOTAL AMPACITY = 300 AMPERES)
   (NO. 1/0 IS SMALLEST CONDUCTOR PERMITTED FOR PARALLEL USE.)

220-22

   NEUTRAL LOAD REDUCED TO 200A ÷ .7(328.2- 200)A = 290A
   REQUIRED AMPACITY OF NEUTRAL CONDUCTORS IS 290A ÷ 2 = 145A. USE TWO NO. 1/0 CONDUCTORS

430-63

2. MAIN PROTECTIVE DEVICES: USE FUSES WITH RATINGS NOT TO EXCEED — (SEE TEXT)
   PHASE A = 45A (LARGEST MOTOR PROTECTION) + 52A + 37.5A + 15.6A + 37.5A + 182.3A + 110.8 + 10A + 13.2A + 3.6A = 497.5 AMPERES
   PHASE B = 45A + 15.6A + 3.6A = 64.2 AMPERES
   PHASE C = 45A + 52A + 37.5A + 15.6A + 37.5A + 182.3A + 110A + 13.2A + 3.6A = 487.5 AMPERES

430-62(b)

   USE 600-AMPERE, 60-AMPERE, AND 600-AMPERE STANDARD SIZE FUSES FOR PHASES A, B, AND C, RESPECTIVELY

CHAPTER 9

3. CONDUIT: TWO CONDUITS ARE USED WITH EACH CONTAINING THE FOLLOWING CONDUCTORS:

AREA OF TWO 350 MCM (PHASE A AND C) = 1.2582 IN.²	(2 × .6291)	
AREA OF ONE NO. 1/0 (PHASE B) = .2367 IN.²		
AREA OF ONE NO. 1/0 (NEUTRAL) = .2367 IN.²		
1.7316 IN.²	REQUIRES 2-1/2-INCH CONDUIT	

TABLE 250-94

4. GROUNDING ELECTRODE CONDUCTOR: BASED ON EQUIVALENT SERVICE-ENTRANCE CONDUCTOR OF 2 × 350 MCM = 700 MCM
   REQUIRES NO. 2/0 COPPER CONDUCTOR

Figure 5-5(c). Service Load Calculation for Office Building and Warehouse

240-21
EXCEPTION 3

Feeder Tap to Warehouse. The ampacity of the tap conductors must be at least one-third of the overcurrent protection for the paralleled service-entrance conductors. The minimum tap conductor ampacity then would be 200 amperes for phases A and C. Phase B requires a conductor with a 50-ampere ampacity based on the 50-ampere fuse in the warehouse panel. The neutral load is 71.7 amperes. Two No. 3/0, one No. 8, and one No. 4 conductors would be required.

250-24(a)
TABLE 250-94

A grounding electrode conductor is required at the warehouse and must be at least a No. 2/0 copper conductor based on the equivalent area of the two 350 MCM conductors.

QUIZ

(Closed-Book)

1. A main feeder uses No. 6 type THW copper conductors (65-ampere ampacity). A suitable tap circuit less than 25 feet long must have an ampacity of at least:
   (a) 65 amperes
   (b) 22 amperes
   (c) 32.5 amperes

2. A 240-volt, two-wire service is required for a building. The main utility circuit is a 120/240-volt circuit. It is a neutral conductor required for the 240-volt service? If so, how is it sized?

3. A secondary feeder is supplied by a 10-kilovolt-ampere, 480/240-volt, single-phase transformer. What is the largest primary feeder overcurrent device that may be used so that individual primary overcurrent protection for the transformer is not required? What is the maximum rating of the secondary side overcurrent protective device on the secondary side?

(Open-Book)

1. Design the circuits for an office building having three floors and a basement. The main service is a 460Y/265-volt circuit. Each floor contains the following:
   (a) 22 000 square feet of area
   (b) 265-volt fluorescent ceiling lighting supplied by a 460/265-volt panel with 30-ampere circuits
   (c) 120-volt duplex receptacles supplied by a 208Y/120-volt panel with 20-ampere circuits. Load is 1 VA per square foot.
   (d) A 208Y/120-volt transformer to supply the receptacles
   The basement has a 5000-square foot storage area containing the following equipment:
   (a) A 10 000 watt, 120-volt lighting and receptacle load for the machinery room on 20-ampere circuits (load is continuous)
   (b) Three 50-horsepower, 480-volt, three-phase synchronous motors each protected by time-delay fuses
   (c) Two 15-horsepower, 480-volt, three-phase squirrel-cage motors
   (d) Five 10-horsepower, 480-volt, three-phase squirrel-cage motors
   Draw a feeder diagram and select the circuits, overcurrent protective devices, and transformer ratings for the building.

5-2 FEEDER AND SERVICE DESIGN FOR OTHER COMMERCIAL OCCUPANCIES

This section discusses the design of feeder or service equipment for establishments with commercial cooking facilities, motels and hotels, schools, and farms. The design problem for each of these occupancies differs from previous examples because certain Code rules must be followed that do not apply generally.

5-2.1 Commercial Establishments with Electric Cooking Facilities

The load of three or more cooking appliances and other equipment for a commercial kitchen may be reduced in accordance with a Code table of demand factors. This provision would apply to restaurants, bakeries, and similar locations.

Design Example. A small restaurant is supplied by a 240/120-volt, four-wire, three-phase service. The restaurant has the following loads:

  a. 1000-square foot area lighted by 120-volt lamps
  b. Ten duplex receptacles
  c. 20-ampere, 240-volt, three-phase motor-compressor
  d. 5-horsepower, 240-volt, three-phase roof ventilation fan protected by an inverse-time circuit breaker
  e. More than six units of kitchen equipment with a total connected load of 80 kilovolt-amperes. All units are 240-volt, three-phase equipment
  f. Two 20-ampere sign circuits

The main service uses type THW copper conductors and is designed as shown in Figure 5-6 on the following page.

Lighting and receptacle loads contribute 27.9 amperes to phases A and C and 25.8 amperes to the neutral. The 80-kilovolt-ampere kitchen equipment load is subject to the application of a 65% demand factor which reduces it to a demand load of .65 × 80 kilovolt-amperes = 52 kilovolt-amperes. This load requires a minimum ampacity of 125 amperes per phase at 240 volts. The load of the three-phase motors and 25% of the largest motor load bring the service load total to 193.1 amperes for phases A and C and 165.2 amperes for phase B.

TABLE 220-20

If the phase conductors are two No. 3/0 type THW copper conductors and one No. 2/0 THW copper conductor, the grounding electrode conductor and the neutral conductor must each be at least a No. 4 copper conductor.

250-23(b)
TABLE 250-94

The fuses are selected in accordance with the Code rules for motor feeder protection. Phases A and C, therefore, are protected at 200 amperes each, and phase B is protected at 175 amperes.

430-63

5-2.2 Services for Hotels and Motels

The portion of the feeder or service load contributed by general lighting in hotels and motels without provisions for cooking by tenants is subject to the application of demand factors. In addition, the receptacle load in the guest rooms is included in the general lighting load at 2 watts per square foot. The demand factors, however, do not apply to any area where the entire lighting is likely to be used at one time, such as the dining room or a ballroom. All other loads for hotels or motels are calculated as shown previously.

TABLE 220-11

TABLE 220-3(b)

Simplified Design Example. It is required to determine the 120/240-volt feeder load contributed by general lighting in a 100-unit motel. Each guest room is 240 square feet in area. The general lighting load is

TABLE 220-3(b)

$$2 \text{ VA/ft}^2 \times 240 \text{ ft}^2/\text{unit} \times 100 \text{ units} = 48\ 000 \text{ VA}$$

but the reduced lighting load is

TABLE 220-11

First 20 000 at 50%	= 10 000
Remainder (48 000 − 20 000) at 40%	= 11 200
	21 200 volt-amperes

This load would be added to any other loads on the feeder or service to compute the total capacity required.

SERVICE LOAD CALCULATION

CODE RULE		LINE A, C	NEUTRAL	LINE B
	A. 240/120-VOLT LOADS			
TABLE 220-3 (b)	LIGHTING = $\dfrac{1.25 \times 2\text{VA/FT}^2 \times 1000 \text{ FT}^2}{240\text{V}}$ = 10.4 AMPERES	10.4	8.3	-0-
220-3 (c)	RECEPTACLES = $\dfrac{180\text{VA} \times 10}{240\text{V}}$ = 7.5 AMPERES	7.5	7.5	-0-
	B. THREE-PHASE LOADS			
TABLE 220-20	KITCHEN EQUIPMENT (6 OR MORE UNITS) =			
	$\dfrac{80,000\text{W} \times .65}{\sqrt{3} \times 240\text{V}}$ = 125 AMPERES	125	-0-	125
440-22	20-AMPERE, THREE-PHASE MOTOR-COMPRESSOR BREAKER SETTING = 1.75 x 20A = 35 AMPERES	20	-0-	20
TABLE 430-150	5-HP, THREE-PHASE MOTOR	15.2	-0-	15.2
TABLE 430-152 240-6	BREAKER SETTING = 2.5 x 15.2A = 38 AMPERES USE 40-AMPERE STANDARD SIZE			
430-25	C. 25% OF LARGEST MOTOR LOAD = .25 x 20A = 5 AMPERES	5	-0-	5
600-6(c)	D. SIGN CIRCUIT = 1200VA ÷ 120V EACH	10	10	-0-
	SERVICE LOAD	193.1 AMPERES	25.8 AMPERES	165.2 AMPERES

CHECKLIST

LIGHTS	✓
RECEPTACLES	✓
A/C	✓
FAN	✓
KITCHEN EQUIP.	✓

SELECTION OF SERVICE EQUIPMENT

TABLE 310-16, 250-23(b)
1. CONDUCTORS: USE NO. 3/0 THW COPPER FOR PHASES A AND C; USE NO. 2/0 FOR PHASE B. USE NO. 4 THW COPPER CONDUCTOR FOR NEUTRAL (NEUTRAL BASED ON SIZE OF GROUNDING ELECTRODE CONDUCTOR)

430-63
240-6
2. OVERCURRENT PROTECTIVE DEVICE:
PHASES A AND C = 40A (LARGEST MOTOR DEVICE) + 10.4A + 7.5A + 125A + 20A + 10A = 212.9 AMPERES
USE STANDARD SIZE 200-AMPERE FUSES.
PHASE B = 40A + 125A + 20A = 185 AMPERES; USE 175-AMPERE STANDARD SIZE FUSE

TABLE 250-94
3. GROUNDING ELECTRODE CONDUCTOR REQUIRED TO BE NO. 4 COPPER.

Figure 5-6. Calculation of Service Load for Small Restaurant

5-2.3 Optional Calculation for Schools

The Code provides an optional method for determining the feeder or service **220-34**
load of a school equipped with electric space heating or air conditioning, or both.
This optional method applies to the building load, not to feeders within the building.

The optional method for schools basically involves determining the total *connected* load in volt-amperes, converting the load to volt-amperes/square foot, and applying the demand factors from the Code table. If both air-conditioning and electric space-heating loads are present, only the larger of the loads is to be included in the calculation.

Simplified Design Example. A school building has 200 000 square feet of floor area. The electrical loads are as follows:

a. Interior lighting at 3 volt-amperes per square foot
b. 300-kilovolt-ampere power load
c. 100-kilovolt-ampere water heating load
d. 100-kilovolt-ampere cooking load
e. 100-kilovolt-ampere miscellaneous loads
f. 200-kilovolt-ampere air-conditioning load
g. 300-kilovolt-ampere heating load

The service load in volt-amperes is to be determined by the optional calculation method for schools.

As shown in Figure 5-7, the combined *connected* load is 1500 kilovolt-amperes. Based on the 200 000 square feet of floor area, the load per square foot is

$$\frac{1\ 500\ 000\ \text{VA}}{200\ 000\ \text{ft}^2} = 7.5\ \text{VA/ft}^2$$

The demand factor for the portion of the load up to and including 3 volt- **TABLE 220-34**
amperes/square foot is 100%. The remaining 4.5 volt-amperes/square foot in the example is added at a 75% demand factor for a total load of 1 275 000 volt-amperes.

Figure 5-7. Optional Calculation for School

CODE RULE	SERVICE LOAD CALCULATION	
		LOAD
220-34	LIGHTING = 3VA/FT² x 200,000 FT²	600,000
	POWER LOAD	300,000
	WATER HEATING LOAD	100,000
	COOKING LOAD	100,000
	MISCELLANEOUS LOAD	100,000
	HEATING LOAD (NEGLECT AIR CONDITIONING)	300,000
	CONNECTED LOAD	1,500,000 VA
	CONNECTED LOAD/SQ FT = 1,500,000VA ÷ 200,000 FT² = 7.5 VA/SQ. FT	
	APPLICATION OF DEMAND FACTOR	
TABLE 220-34	3VA/FT² x 200,000 FT² @ 100%	600,000
	4.5VA/FT² x 200,000 FT² @ 75%	675,000
	SERVICE LOAD	1,275,000 VA

TABLE 220-19,
NOTE 5

220-20

Household Electric Cooking Appliances in Instructional Programs. If the standard method is used to compute the feeder or service load for a school, the Code range table may be used to compute the load for household cooking appliances used for instructional programs. The kitchen equipment load in the school cafeteria may also be reduced if there are three or more units.

5-2.4 Feeder and Service Calculations for Farms

ARTICLE 220,
PART D

The Code provides a separate method for computing farm loads other than the dwelling. Tables of demand factors are provided for use in computing the feeder loads of individual buildings as well as the service load of the entire farm.

220-40

The demand factors may be applied to the 120/240-volt feeders for any building or load (other than the dwelling) that is supplied by two or more branch circuits. All loads that operate without diversity, that is, the entire load is on at one time, must be included in the calculation at 100% of connected load. All "other" loads may be included at reduced demands. The load to be included at 100% demand, however, cannot be less than 125% of the largest motor and not less than the first 60 amperes of the total load. In other words, if the nondiverse and largest motor load is less than 60 amperes, a portion of the "other" loads will have to be included at 100% in order to reach the 60-ampere minimum.

220-41

After the loads from individual buildings are computed, it may be possible to reduce the total farm load further by applying additional demand factors.

Design Example. A farm has a dwelling and two other buildings supplied by the same 120/240-volt service. The electrical loads are as follows:

a. Dwelling—100-ampere load as computed by the calculation method for dwellings
b. Building No. 1—
   1. 5-kilovolt-ampere continuous lighting load operated by a single switch
   2. 10-horsepower, 240-volt motor
   3. 21 kilovolt-amperes of other loads
c. Building No. 2—
   1. 2-kilovolt-ampere continuous load operated by a single switch
   2. 15 kilovolt-amperes of other loads

Determine the individual building loads and the total farm load as illustrated in Figure 5-8. The nondiverse load for building No. 1 consists of the 5-kilovolt-ampere lighting load and the 10-horsepower motor for a total of 83.3 amperes. This value is included in the calculation at the 100% demand factor. Since the requirement for adding at least the first 60 amperes of load at the 100% demand factor has been satisfied, the next 60 amperes of the 87.5 amperes from all other loads are added at a 50% demand factor and the remainder of $87.5 - 60 = 27.5$ amperes is added at a 25% demand factor.

In the case of building No. 2, the nondiverse load is only 8.3 amperes; therefore, 51.7 amperes of "other" loads must be added at the 100% demand factor in order to meet the 60-ampere minimum.

220-41

Using the method given for computing total farm load, we see that the service load is

Largest load at 100% =	120.2 A (building No. 1)
Second largest at 75% =	49.1 A (building No. 2)
	169.3 A
Dwelling	100.0 A
	269.3 A

CODE RULE	BUILDING NO. 1 FEEDER LOAD		
		LINE	NEUTRAL
220-40	LIGHTING (5-KVA NONDIVERSE LOAD) = 5000VA ÷ 240V	20.8	20.8
TABLE 430-148	10-HP MOTOR = 1.25 x 50A	62.5	—0—
	TOTAL MOTOR AND NONDIVERSE LOAD	83.3	20.8
	"OTHER LOADS" = 21000VA ÷ 240V	87.5	87.5
	APPLICATION OF DEMAND FACTORS		
TABLE 220-40	MOTOR AND NONDIVERSE LOADS @ 100%	83.3	20.8
	NEXT 60 OF "OTHER" LOADS @ 50%	30.0	30.0
	REMAINDER OF "OTHER" LOADS (87.5 - 60) @ 25%	6.9	6.9
	FEEDER LOAD	120.2 AMPERES	57.7 AMPERES

	BUILDING NO. 2 FEEDER LOAD		
		LINE	NEUTRAL
	LIGHTING (2-KVA NONDIVERSE LOAD) = 2000VA ÷ 240V	8.3	8.3
	"OTHER" LOADS = 15,000KVA ÷ 240V	62.5	62.5
	APPLICATION OF DEMAND FACTORS		
TABLE 220-40	NONDIVERSE LOAD @ 100%	8.3	8.3
	REMAINDER OF FIRST 60 (60 - 8.3) @ 100%	51.7	51.7
	REMAINDER OF "OTHER LOAD" (62.5 - 51.7) @ 50%	5.4	5.4
		65.4 AMPERES	65.4 AMPERES

	TOTAL FARM LOAD		
220-41		LINE	NEUTRAL
	APPLICATION OF DEMAND FACTORS		
TABLE 220-41	LARGEST LOAD (BLD. NO. 1) @ 100%	120.2	57.7
	SECOND LARGEST LOAD (BLD. NO. 2) @ 75%	49.1	49.1
	FARM LOAD (LESS DWELLING)	169.3	106.8
	FARM DWELLING LOAD	100.0	100.0
	TOTAL FARM LOAD	269.3 AMPERES	206.8 AMPERES

Figure 5-8. Example of Farm Load Calculation

The total service load of 269 amperes requires the ungrounded service-entrance conductors to be at least 300 MCM type THW copper conductors. The neutral load of the dwelling was assumed to be 100 amperes which brought the total farm neutral load to 207 amperes.

QUIZ
(Closed-Book)

1. A reduced demand is allowed for commercial kitchen equipment when the number of units is:
   (a) Two or more
   (b) Three or more
   (c) Four or more
2. The demand factor for diversified loads in a farm building for a load greater than 60 amperes is:
   (a) 75%
   (b) 25%
   (c) 50%
3. Select the occupancies in which a portion of the general lighting load may be reduced by applying demand factors:
   (a) Motels
   (b) Office buildings
   (c) Warehouses
   (d) Stores

(Open-Book) 1. Calculate the service load for a small restaurant that contains the following:
   (a) 1500 square feet of floor area
   (b) Eight duplex receptacles
   (c) Cooking equipment:
       (1) 20-kilowatt, 240-volt, three-phase range
       (2) 1½-horsepower, 240-volt, three-phase food chopper
       (3) 10-kilowatt, 240-volt, three-phase deep fat fryer
       (4) 4-kilowatt, 240-volt, single-phase food warmer
       (5) 5-kilowatt, 240-volt, single-phase dishwasher
       (6) ¼-horsepower, 120-volt, single-phase vegetable peeler
       (7) 2.4-kilowatt, 240/120-volt toaster
       (8) 10-kilowatt, 240/120-volt waffle iron
   The service is a 240/120-volt, three-phase, four-wire service. Type THW copper conductors are used. Balance the loads as closely as possible.
2. Calculate the service load for a farm with the following loads:
   (a) 60-ampere dwelling load
   (b) 80-ampere load in building No. 1
   (c) 40-ampere load in building No. 2
   (d) 70-ampere load in building No. 3
3. Design the service and feeders for a 100-unit motel. The service is a 480-volt, three-phase circuit with a 480 to 208Y/120-volt transformer for the guest room service. Each guest room contains the following:
   (a) 240 square feet of floor area
   (b) 2-ampere, 208-volt air handler
   The motel contains a utility room with the following equipment:
   (a) 130-ampere, 208-volt, three-phase motor-compressor load; the locked-rotor current is 624 amperes
   (b) Ten 120-volt outside duplex receptacles
   Determine the motor branch-circuit rating and the size of the disconnecting means, the transformer kilovolt-amperes and feeder sizes, and the service rating. Type THW copper conductors are used for all circuits.

5-3 SPECIAL OCCUPANCIES AND EQUIPMENT

The Code provides specific rules for designing feeders and branch circuits for certain special occupancies and equipment. These special rules apply to the design of wiring systems for hospitals, for mobile homes and mobile home parks, and for marinas and boatyards. These occupancies are discussed in some detail in this section. The wiring design rules for all other special occupancies and equipment are summarized in Table 5-1 at the end of this section for reference purposes.

5-3.1 Service Calculations for a Small Hospital

ARTICLE 517   The design of the electrical system for health care facilities is a complex task which is usually undertaken by designers who specialize in such systems. The calculation of the service load, however, is similar to that for any commercial occupancy.

ARTICLE 517,   An electrical system in a hospital consists of feeders and branch circuits supply-
PART E   ing (a) nonessential loads, (b) equipment systems, and (c) the emergency system. The emergency system includes the life safety branch and the critical branch. Only the general lighting load, which is a part of the nonessential loads, may be reduced by applying the demand factors listed in the Code, but these demand factors do not apply to areas where the entire lighting capacity is likely to be used at one time, such as the operating room.

```
┌─────────────┬──────────────────────────────────────────────────────────────────────┐
│  CODE RULE  │                        SERVICE LOAD CALCULATION                        │
│             │                                                                        │
│             │                                                          LOAD          │
│             │    A.   LIGHTING                                         100,000       │
│             │         APPLICATION OF DEMAND FACTORS                                  │
│ TABLE 220-11│             FIRST 50,000 @ 40%                            20,000       │
│             │             REMAINDER OVER 50,000 @ 20%                   10,000       │
│             │                               LIGHTING DEMAND LOAD        30,000       │
│             │                                                                        │
│             │    B.   EMERGENCY SYSTEMS                                              │
│             │         LIFE SUPPORT BRANCH                              100,000       │
│             │         CRITICAL BRANCH                                   50,000       │
│             │                        TOTAL LOAD (LESS MOTOR)          180,000VA      │
│             │                                                                        │
│             │                                             180,000VA                  │
│             │              TOTAL LOAD (AMPERES) = ─────────────        216.5         │
│             │                                             √3 x 480V                  │
│ TABLE 430-150│   C.  THREE-PHASE MOTOR = 1.25 x 34A                     42.5         │
│             │                             TOTAL SERVICE LOAD           259.0   AMPERES│
│             │                                                                        │
└─────────────┴──────────────────────────────────────────────────────────────────────┘
```

Figure 5-9. Service Calculation for Small Hospital

Simplified Design Example. The feeder load for a small hospital consists of the following continuous loads:

a. 100-kilovolt-ampere general lighting load in ward and administrative areas
b. 100-kilovolt-ampere load for life safety branch
c. 50-kilovolt-ampere load for critical branch
d. 25-horsepower, 480-volt, three-phase motor equipment system load

The loads are balanced on a 480-volt, three-phase service. It is required to determine the service ampacity as shown in Figure 5-9. Since each load is given as a continuous load value, an additional 25% need not be added. The 100-kilovolt- **220-11** ampere lighting load represents a demand load of only 30 000 volt-amperes after the application of demand factors. The total load except for the equipment system motor load is 180 kilovolt-amperes, or 216.5 amperes per phase.

The 25-horsepower motor has a full-load current of 34 amperes and its load is **TABLE 430-150** 1.25 × 34 amperes = 42.5 amperes. The required service ampacity is the sum of these loads, or 259 amperes.

5-3.2 Power Supply and Feeder Design
for Mobile Homes and Mobile Home Parks

A mobile home is a factory assembled structure that is capable of being moved **ARTICLE 550** on its own running gear. The mobile home is designed for use as a permanent dwelling unit although its location may be moved on occasion. The Code provides specific rules for the design of *both* the interior branch circuits and the feeder or service. Since the design of the branch circuits is usually the responsibility of the manufacturer, only the design of the feeder or service circuit to a mobile home or to a mobile home park will be discussed here.

The size of the service for a mobile home park depends on the maximum number of mobile homes that can be accommodated by the park.

550-5   ***Design of Power Supply for Single Mobile Home.*** The power supply for a mobile home may be a *single* approved 40- or 50-ampere power supply cord if the load does not exceed the rating of the cord. If a larger load is to be served, a permanently installed circuit must be used.

550-13   The Code provides a separate calculation method that must be used for computing the supply cord or distribution panel load for a mobile home. The lighting, small appliance, and laundry circuit loads are calculated in basically the same manner as before. The nameplate rating in amperes of all other appliances and motors and 25% of the load of the largest motor are added to determine the total load on the 120/240-volt service. The service rating is then based on the higher current calculated for either ungrounded conductor.

ARTICLE 220

550-13(b)(5)   If a free-standing range is installed, the load current for the range is calculated by using a Code table of reduced values. In addition, the load of four or more appliances, excluding motors, heater loads, or free-standing ranges, may be reduced to 75% of the connected load.

The calculations for the power supply to a mobile home are shown in Figure 5-10 for a mobile home containing the following loads:

(a) Outside dimensions of 60 feet by 12 feet
(b) 500-watt, 120-volt dishwasher
(c) 3-ampere, 120-volt garbage disposal
(d) 10-kilowatt electric range
(e) 1000-watt, 240-volt heater

550-13(a)   The lighting load at 3 volt-amperes/square foot, the two small appliance circuits, and the laundry circuit contribute a load of 6660 volt-amperes. This load is

Figure 5-10. Example of Power Supply Calculations for Mobile Home

CODE RULE	POWER SUPPLY CALCULATION		
		LINE	
550-13 (a)	A. LIGHTING AND APPLIANCE LOAD		
	LIGHTING LOAD = 3VA/FT^2 x 60 FT x 12 FT	2,160	
	SMALL APPLIANCE = 2 x 1500VA	3,000	
	LAUNDRY CIRCUIT	1,500	
	TOTAL LTG, SMALL APPL, AND LAUNDRY CKTS	6,660	
	APPLICATION OF DEMAND FACTORS		
	FIRST 3000 @ 100%	3,000	
	REMAINDER (6660 - 3000) @ 35%	1,281	
	LIGHTING AND APPLIANCE DEMAND LOAD	4,281 VA	
		LOAD PER LINE	
		A	B
550-13 (b)	B. TOTAL LOAD		
	LIGHTING AND APPLIANCE = 4281VA ÷ 240V	17.8	17.8
	DISHWASHER = 500W ÷ 120V	4.2	–0–
	GARBAGE DISPOSAL	–0–	3.0
	1-KW HEATER = 1000W ÷ 240V	4.2	4.2
	10-KW RANGE = (.8 x 10,000)W ÷ 240V	33.3	33.3
		59.5 AMPERES	58.3 AMPERES
550-5	A PERMANENTLY INSTALLED CIRCUIT WITH A MINIMUM AMPACITY OF 60 AMPERES IS REQUIRED.		

then reduced to a demand load of 3000 volt-amperes plus 35% of the remainder, or a total of 4281 volt-amperes. This load must be divided by 240 volts in order to find the load in amperes on each ungrounded conductor.

The loads in amperes for motors and appliances are added to the load obtained for the lighting and small appliance load. The 120-volt loads should balance as closely as possible on the circuit. In this case, the only 120-volt loads are the dishwasher and garbage disposal with loads of 4.2 amperes and 3 amperes, respectively.　**550-13(b)**

The Code method also requires that 25% of the largest motor load be added to the total load but lists specific types of motor loads for which the requirement applies. The garbage disposal is not included in the list.　**550-13(b)**

The hot water heater has a load current of 4.2 amperes. Since there are only three appliances of this group, that is, the water heater, the garbage disposal, and the dishwasher, no reduction in the total load is allowed.

The load for the 10-kilowatt free-standing range is based on 80% of its rating and is 33.3 amperes.　**550-13(b)(5)**

The total load is 59.5 amperes on line A and 58.3 amperes on line B. Since these values are higher than the 50-ampere maximum for an approved power supply cord, a permanently installed power supply circuit is required.　**550-5**

Mobile Home Park Service Equipment. The minimum rating for a service supplying a mobile home park is based on 16 000 volt-amperes (at 120/240 volts) per mobile home. A table of demand factors is provided that reduces this load according to the number of mobile home sites in the park.　**550-22(a)**　**TABLE 550-22**

As an example, if the park contains 21 mobile home sites, the load per site is

$$.25 \times 16\ 000\ \text{VA} = 4000\ \text{VA/site}$$

The total service load is then

$$21\ \text{sites} \times 4000\ \text{VA/site} = 84\ 000\ \text{VA}$$

In any case, the minimum ampacity of each mobile home feeder is required to be at least 100 amperes.　**550-22(b)**

5-3.3 Power Supply and Feeder Design for Recreational Vehicles and Recreational Vehicle Parks

A recreational vehicle is a unit designed for temporary living and includes travel and camping trailers, truck campers, and motor homes. Since these vehicles are usually factory wired, the branch-circuit design is not covered in this Guide. The main power supply assembly that supplies power to the vehicle may be rated from 15 amperes to 50 amperes, depending on the number of branch circuits and the total load of the vehicle.　**ARTICLE 551**　**551-12**

For recreational vehicles with a large electrical load, the calculation for computing the power supply assembly load is essentially the same as the calculation for computing the load of the power supply cord to a mobile home. Such vehicles would have a load consisting of lighting and receptacles as well as a significant appliance load.　**551-10(d)**

The rating of the 120/240-volt service for a recreational vehicle park is determined by the number of recreational vehicle sites in the park. The basic load is 3600 watts per site when equipped with both 20- and 30-ampere supply facilities and　**551-44**

2400 watts per site if there is only a 20-ampere supply. This total load may be reduced when four or more sites are served.

For example, the total load for 50 sites with both 20-ampere and 30-ampere supplies is

$$50 \times 3600 \text{ W} = 180\ 000 \text{ W}$$

TABLE 551-44   A demand factor of 39% may be applied, which results in a demand load of

$$.39 \times 180\ 000 \text{ W} = 70\ 200 \text{ W}$$

This is the required service capacity for the 50-site park.

Table 5-1. Miscellaneous Wiring System Design Rules

Load	Rule	Load	Rule
Neutral load for 5-wire 2-phase system	220-22	Manufactured Building	545-5
		Electric Signs	600-6
Rosettes	410-61	Cranes and Hoists	610-14, -31, -32, -33, -41, -42, -43, 53
Electrode-Type Boilers	424-82		
Fixed Outdoor Electric De-icing and Snow Melting Equip.	426-4		
		Elevators, Dumbwaiters, etc.	620-12, -13, -61
Fixed Electric Heating Equip. for Pipelines and Vessels	427-4	Electric Welders	630-11, -12, -13, -21, -22, -23, -31, -32, -33
Part-Winding Motors	430-3		
Torque Motors	430-6(b) 430-110(b)	Sound Recording Equip.	640-10
AC Adjustable Voltage Motors	430-6(c)	Data Processing Systems	645-2
		Organs	650-5, -7
Multispeed Motors	430-22(a)	X-ray Equip.	660-5, -6, -9
Other than continuous-duty motors, including wound rotor motors	430-22(a) Exception, 430-23	Induction and Dielectric Heating	665-41, -42, -43, -61, -62, -63
Generators	445-5	Electroplating	669-5
Transformers over 600 volts	450-3	Metalworking Machine Tools	670-4
Grounding Auto-transformer	450-4	Irrigation Machines	675-7, -8, -9, -10, -11, -22
Capacitors	460-8, 9	Swimming Pools and Fountains	680
		Solar Photovoltaic Systems	690
X-ray Equip. in Health Care Facilities	517	Electrode Type Boilers (Over 600 volts)	710-72
Theaters	520-25(a) 520-52 520-53(h)	Less than 50 volts	720-4, -5, -8, -10
		Remote Control Circuits	725
Motion Picture and Television Studios	530-18, -19	Fire Protective Signaling Systems	760
Motion Picture Projectors	540-13		

5-3.4 Shore Power Circuits for Marinas and Boatyards

The wiring system for marinas and boatyards is designed by using the same **ARTICLE 555** Code rules as for other commercial occupancies except for the application of several special rules dealing primarily with the design of circuits supplying power to boats.

The smallest sized receptacle that may be used to provide shore power for boats **555-3** is 20 amperes. Each single receptacle that supplies power to boats must be supplied **555-4** by an individual branch circuit with a rating corresponding to the rating of the receptacle.

The feeder or service ampacity required to supply the receptacles depends on **555-5** the number of receptacles and their rating, but demand factors may be applied that will reduce the load of five or more receptacles. For example, a feeder supplying ten 30-ampere shore power receptacles in a marina requires a minimum ampacity of

$$10 \times 30\,\text{A} \times .8 = 240\,\text{A}$$

Although this computed feeder ampacity might seem rather large, this is the minimum required by the Code.

5-3.5 Miscellaneous Wiring System Design Rules

Table 5-1 lists wiring system design rules for various types of equipment, occupancies, or systems that are not otherwise discussed in any detail in this Guide. The specific rules listed in the table pertain to ratings of circuits, conductors, and equipment and not to installation methods or other Code requirements. These rules should be read and understood fully in order to complete your study of wiring system design.

QUIZ

(Closed-Book)

1. The lighting load for mobile homes and recreational vehicles is _____volt-amperes/square foot.
2. The mobile home service load before any demand factors are applied is _____ volt-amperes per mobile home site.
3. For a mobile home or recreational vehicle, the lighting and small appliance load over 3000 volt-amperes may be reduced by a demand factor of _____percent.
4. The feeder or service load for circuits supplying power to boats in a marina is based on:
   (a) The rating of the receptacles
   (b) The total load to be served
   (c) The branch-circuit ratings
5. The branch-circuit conductors supplying data-processing equipment must have an ampacity based on the connected load of:
   (a) 100%
   (b) 125%
   (c) 150%
6. The feeder circuit in a mobile home may be a power supply cord if the load does not exceed:
   (a) 40 amperes
   (b) 60 amperes
   (c) 50 amperes

(Open-Book)

1. What is the feeder load for a mobile home with the following loads?
   (a) 50 feet by 10 feet outside dimensions
   (b) Laundry area
   (c) 1000-watt, 240-volt heater
   (d) 5-kilowatt electric range
   (e) 1-kilowatt, 120-volt water heater
   (f) 500-watt, 120-volt dishwasher
   (g) 500-watt, 120-volt garbage disposal
   (h) 7-ampere, 240-volt air conditioner
2. What service load is required for a mobile home park with 20 homes?
3. A recreational vehicle has a 20-ampere air-conditioning circuit and two 15-ampere circuits. What rating is required for the power supply assembly?
4. A recreational vehicle park consists of the following:
   (a) 30 sites with both 20-ampere and 30-ampere receptacles
   (b) 20 sites with 20-ampere power supply facilities
   What is the total service load for the park?
5. What service capacity is required to supply five 30-ampere receptacles in a marina?

TEST CHAPTER 5

I. True or False *(Closed-Book)*

	T	F
1. No separate receptacle load is required in guest rooms of hotels and motels.	[ ]	[ ]
2. A lighting and appliance panelboard is a panelboard in which *more* than 10% of its overcurrent devices rated 30 amperes or less have neutral connections.	[ ]	[ ]
3. A multiwire branch circuit protected by fuses may supply only one load or only line-to-neutral loads.	[ ]	[ ]
4. Branch circuits in industrial establishments protected by 15-ampere overcurrent devices may supply loads operating at 300 volts or less.	[ ]	[ ]
5. The *size* and the *setting* for a circuit breaker used as an overcurrent protective device are identical.	[ ]	[ ]
6. The maximum voltage drop for branch circuits and feeders combined should not exceed 3% of the circuit voltage.	[ ]	[ ]
7. The rating of receptacles providing shore power for boats at marinas shall not be less than 20 amperes.	[ ]	[ ]
8. The optional calculation for computing the feeder or service load for a school would not apply to a school with electric space heating or air conditioning.	[ ]	[ ]
9. The minimum service load for each mobile home site in a mobile home park is 12 000 volt-amperes.	[ ]	[ ]
10. The demand factor for three kitchen unit loads in a restaurant is 90%.	[ ]	[ ]

II. Multiple Choice *(Closed-Book)*

1. If a branch circuit supplies a single nonmotor operated appliance rated at 13.3 amperes or more, the overcurrent device rating shall not exceed what percentage of the appliance rating?
    - (a) 125%
    - (b) 150%
    - (c) 167%

    1. _____

2. A 50-foot long show window requires 30 lighting fixtures with heavy-duty sockets. The minimum continuous load for this window is:
    - (a) 10 000 volt-amperes
    - (b) 12 500 volt-amperes
    - (c) 22 500 volt-amperes

    2. _____

3. The unit lighting load for a store is
    - (a) 3 volt-amperes/square foot
    - (b) 5 volt-amperes/square foot
    - (c) 2 volt-amperes/square foot

    3. _____

4. In most cases, the next larger size overcurrent device may not be selected to protect a circuit when the device rating exceeds:
    - (a) 800 amperes
    - (b) 500 amperes
    - (c) 600 amperes

    6. _____

5. The ampacity of conductors connecting a power factor correcting capacitor to a motor circuit shall not be less than:
    - (a) 135% of the ampacity of the motor circuit conductors
    - (b) One-third the ampacity of the motor circuit conductors and not less than 135% of the rated current of the capacitor
    - (c) 135% of the rated current of the capacitor

    5. _____

6. The rating of a branch circuit supplying an electric sign with both lamps and a transformer shall not exceed:
    (a) 15 amperes
    (b) 30 amperes
    (c) 20 amperes

                                                        6. _____

7. If there is no overcurrent protection on the secondary side, the rating of the primary overcurrent protective device for a 480-volt transformer with a rated primary current of 100 amperes should not exceed:
    (a) 125 amperes
    (b) 250 amperes
    (c) 100 amperes

    7. _____

8. The general lighting load for warehouses based on the floor area is:
    (a) $1/4$ volt-ampere/square foot
    (b) $1/2$ volt-ampere/square foot
    (c) 1 volt-ampere/square foot

    8. _____

9. The second largest load for a farm service is allowed a demand factor of:
    (a) 50%
    (b) 65%
    (c) 75%

    9. _____

10. In a motel or hotel the demand factor for the first 20 000 volt-amperes of the general lighting load is:
    (a) 100%
    (b) 40%
    (c) 50%

    10. _____

III. Problems *(Open-Book)*

1. Determine the lighting load for a 100 000-square foot warehouse.

2. Design the 480-volt, three-phase service and the 120/240-volt, single-phase circuits for a bank (a 480 to 120/240-volt, single-phase transformer is connected to phases A and B of the 480-volt circuit). The bank contains the following:
    (a) 9000 square feet of floor area
    (b) 80 duplex receptacles
    (c) 60-foot long show window (120-volt lighting)
    (d) 6-kilowatt, 120-volt sign (outside)
    (e) 7½-horsepower, 480-volt, three-phase motor protected by a circuit breaker
    (f) 21-ampere, 480-volt, three-phase motor-compressor protected by a circuit breaker

Design the branch circuits, the transformer circuit including its protection, and the motor circuits with their disconnecting means. The conductors are type THW copper.

3. A community hospital is served by a 480Y/277-volt service. The hospital has the following loads:
    (a) Cafeteria supplied by a 480 to 208Y/120-volt transformer

        (1) One hundred and fifty 100-watt, 277-volt fixtures
        (2) Six 12-kilowatt, 208/120-volt ranges
        (3) 7.5-kilowatt, 120-volt dishwasher supplied only by phase A and the neutral
        (4) Three 4-kilowatt, 208/120-volt deep fat fryers
    (b) Equipment system (three-phase loads at 480 volts)
        (1) 10-horsepower water pump
        (2) 5-horsepower hot water pump
        (3) 100 kilovolt-ampere hot water generator
        (4) Two 77-ampere hermetic motor-compressors protected by inverse-time circuit breakers
    (c) 100-kilovolt-ampere, 460-volt, three-phase load for critical branch
    (d) 90-kilovolt-ampere, 460-volt, three-phase load for life safety branch
    (e) Ward area (supplied by a 480 to 208Y/120-volt transformer
        (1) 150 duplex receptacles
        (2) One thousand 75-watt, 120-volt lamps

Design the main service, the feeders, and the transformers for the hospital. Type THW copper conductors are used for all circuits.

General Rules
for Installation

*Part II presents general rules for electrical installations
including rules for clearances, wiring methods, materials, and equipment.
The rules deal primarily with construction methods
rather than ampacity ratings, sizes, etc.
The rules are clarified by illustrations wherever appropriate.*

Installation Rules
for Specific Circuits
or Systems

6

This chapter presents a discussion of the Code rules for the installation of electrical wiring considered as a complete circuit or system. Branch circuits, feeders, services, systems operating at over 600 volts, and miscellaneous circuits and systems are discussed. These particular circuits and systems are covered separately by the Code and the requirements for them supplement or modify Code rules that apply to any installation.

The installation rules in this chapter apply to the circuits or systems discussed regardless of the type of equipment supplied except as specifically noted. Additional rules apply when specific loads such as appliances or motors are supplied.

In addition to the installation rules presented in this chapter, the Code specifies a number of rules concerned with the design of circuits to supply various loads. These rules are discussed in Part I of the Guide and are not repeated here.

6-1 INSTALLATION OF BRANCH CIRCUITS

When a circuit is installed as a branch circuit, various Code rules must be followed to assure a safe installation in complete conformance with the Code. These branch-circuit installation rules can be separated into (a) general installation rules for any construction; (b) rules that apply only for dwellings; and, (c) rules that apply for other specific locations, such as hotels or motels.

This section discusses the installation of branch circuits in such locations. Other chapters in the Guide cover the additional requirements for branch circuits that supply particular loads including motors, electric space-heating equipment, and other **210-2*** utilization equipment. In fact, the Code includes a detailed list of other articles which contain rules for specific-purpose branch circuits.

6-1.1 General Installation Rules for Branch Circuits

ARTICLE 100 Branch circuits in dwellings, commercial occupancies, and industrial areas are **DEFINITIONS** distinguished by their use. A general-purpose branch circuit supplies two or more outlets for lighting and appliances. An individual branch circuit can supply only a single unit that utilizes electrical energy. Restrictions are placed on the maximum rating and use of general-purpose branch circuits as explained in this section. Either type of branch circuit may be installed as a multiwire branch circuit, but a general-purpose multiwire branch circuit must conform to Code rules not applicable to individual branch circuits.

The grounded and the grounding conductors for any branch circuit that requires such conductors must be identified by color or other means to distinguish them from ungrounded conductors. Other Code rules that apply to every branch circuit specify the maximum voltage allowed under certain conditions and also specify rules that apply to receptacle outlets installed on branch circuits.

To protect personnel at a construction site, there must be either ground-fault circuit interrupters on certain circuits or a program of maintenance and testing for grounding conductors.

General rules for the installation of branch circuits are listed in Table 6-1 and a summary of the voltage limitations is presented in Table 6-2. The discussion that follows explains some of the more important rules although not every item in the tables is covered in the text.

210-3 *General-Purpose Branch Circuits.* General-purpose branch circuits are classified according to the rating or setting of the overcurrent protective device for the circuit. When several outlets are supplied, the branch-circuit rating must be either 15, 20, 30, 40, or 50 amperes. If a load requires a supply current of more than 50 amperes, an individual branch circuit is required except in an industrial establishment.

210-4 A general-purpose multiwire branch circuit may supply only line-to-neutral loads except when all ungrounded conductors of the circuit are opened simultaneously by the branch-circuit protective device. For example, a 120/240-volt, three-wire circuit with grounded neutral may normally be used to supply only 115-volt loads, but it can supply combinations of 120- and 240-volt loads if all the ungrounded circuit conductors are opened simultaneously, for example, by a multiple-pole circuit breaker protecting the circuit.

210-5(a) *Color Coding of Grounded Conductors.* The grounded conductor in a two-wire circuit or the neutral in a three-wire or four-wire circuit must, in most cases, be identified by a white or natural gray color. In certain cases, the identification can be at the terminals rather than continuously along the conductor. The grounded conductors of Type MI cable and grounded conductors of multiconductor cable

*References in the margins are to the specific applicable rules and tables in the *National Electrical Code.*

maintained by qualified persons may be identified in that manner. If several circuits are installed in the same raceway or enclosure, one system neutral must be white or natural gray and the other neutrals must be identified by other means.

Unless it is bare, the grounding conductor is required to be identified by a **210-5(b)** continuous green color or green with yellow stripes. The Code allows exceptions for insulated conductors larger than No. 6 or for an insulated conductor in a multiconductor cable maintained by qualified persons. In these cases, identification is required only where the conductor is accessible.

Table 6-1. General Branch-Circuit Installation Rules

Application	Rule	Code Ref.
General-purpose Branch circuits	General-purpose branch circuits are classified in amperes according to setting or ratings of overcurrent device. Standard ratings are 15, 20, 30, 40, and 50 amperes for other than individual branch circuits. (See exception.) Note: Individual branch circuits have no such restrictions.	210-3
Multiwire Circuits	General-purpose multiwire branch circuits shall supply only line to neutral loads. Exceptions: Where all ungrounded conductors of multiwire branch circuits are opened simultaneously by branch-circuit overcurrent device.	210-4
Color Coding: a) Grounded Conductor b) Grounded Conductor of Different Systems	Continuous white or natural gray color. The color of one neutral conductor must be white or natural gray; the others must be white with a colored stripe (not green) or other and different means of identification. Exceptions: 1) Grounded conductor of type MI cable must be identified by distinctive markings at terminals. 2) Multiconductor cables under conditions specified by Code.	210-5(a)
c) Equipment Grounding Conductor	The identification must be a) Continuous green color or continuous green with one or more yellow stripes. b) Bare conductor. (See exceptions.)	210-5(b)
Receptacles:	Receptacles installed on 15- and 20-ampere branch circuits must be grounding type. (See exceptions.)	210-7(a)
	Grounding contact must be connected to equipment grounding conductor. (See exception for extension to existing branch circuits.)	210-7(c)
	Attachment plugs for different types of circuits used on same premises must not be interchangeable.	210-7(f)
Receptacle Outlets	Receptacle outlets are required: a) Where flexible cords are used. Exception: where flexible cords are permitted to be permanently connected. b) Every 12 feet above a show window c) Within 75 feet of rooftop heating, air conditioning, and refrigeration equipment. Exception: One- and two-family dwellings.	210-50(b) 400-7 210-62 210-63
Construction Sites (GFCI)	At construction sites, all 120-volt, single-phase, 15- and 20- ampere receptacles not part of permanent wiring of building or structure must have ground-fault circuit interrupters for personnel protection. Exceptions: 1) Portable generators rated not over 5 kW, where circuit conductors are insulated from generator frame. 2) Where written, approved procedure for maintenance and testing of grounding means is enforced.	305-4

Table 6-2. Voltage Limitations for Branch Circuits

Occupancy	Maximum Voltage	Limitations	Code Reference
Dwellings	120 volts between conductors	Voltage not to exceed 120 volts for light fixtures and small plug-in loads	210-6(a)
General	120 volts between conductors	Can supply medium-base screw-shell lampholders (or other types within ratings), auxiliary equipment of electric discharge lamps, and utilization equipment (Exceptions)	210-6(b)
General	277 volts to ground	Can supply mogul-based screw-shell lampholders (or lampholders other than screw-shell within ratings), auxiliary equipment of electric discharge lamps, and utilization equipment (Exceptions)	210-6(c)
General	600 volts between conductors	Can supply auxiliary equipment of electric discharge lamps at minimum height of 22 feet on poles or 18 feet in structures such as tunnels, and utilization equipment (Exceptions)	210-6(d)

210-6(a)
210-6(b)   *Maximum Voltage.* As shown in Figure 6-1(a), the voltage for a branch circuit supplying lighting fixtures or small appliances (1380 VA or less) in dwellings must not exceed 120 volts to ground. The conditions listed in Table 6-2 permit industrial and commercial lighting fixtures to be supplied by branch circuits with a voltage to ground of up to 277 volts when all specified conditions are met. This allows the use of the common 277-volt lighting system. Lighting on poles and in tunnels may operate at up to 600 volts between conductors if the height restrictions listed in Table 6-2 are met.

210-7
250-50
210-7(d)
EXCEPTION
210-50(b)
400-7(b)
210-7(f)

Receptacles. Receptacles installed on 15- or 20-ampere branch circuits must be of the grounding type and must have their grounding contact connected to the equipment grounding conductor of the circuit as shown in Figure 6-1(b). In existing installations in which an equipment grounding conductor is not present, the grounding contact may be connected to a grounded cold water pipe. If no grounding means exists, a nongrounding type receptacle or a GFCI receptacle must be used.

A receptacle outlet is required to supply equipment using a flexible supply cord except when flexible supply cords are specifically permitted to be permanently connected. When the receptacle is required, the cord must be equipped with an attachment plug, as shown in Figure 6-1(c), where portable lamps or appliances that must be moved occasionally are supplied.

If several circuits with different characteristics (voltage, current, frequency, etc.) are present on the same premises, the receptacle for each circuit must prevent the use of interchangeable attachment plugs. Figure 6-1(d) shows two standard receptacles for different voltages that meet this requirement.

305-6   *Ground-Fault Circuit Interrupters for Construction Sites.* Ground-fault circuit interrupters (GFCI) are required at construction sites for 120-volt, single-phase, 15- and 20-ampere outlets that are not part of the permanent wiring system. This rule does not apply to portable or vehicle-mounted generators rated not more than 5 kilo-

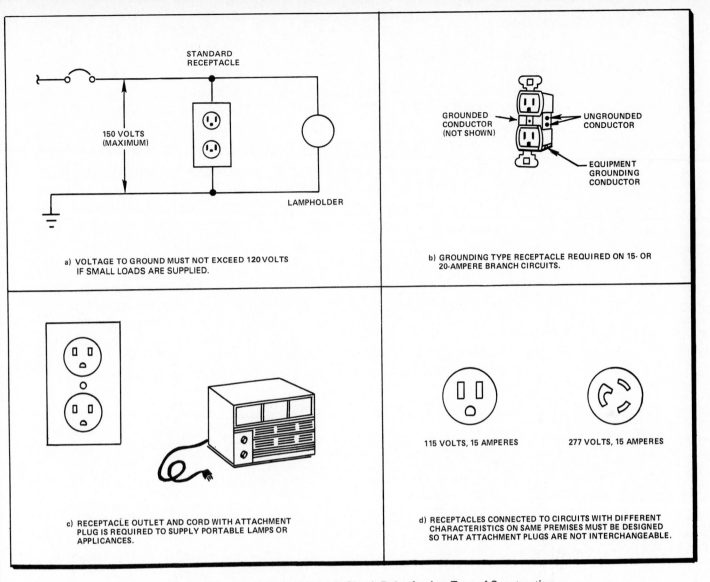

STANDARD RECEPTACLE

150 VOLTS (MAXIMUM)

LAMPHOLDER

a) VOLTAGE TO GROUND MUST NOT EXCEED 120 VOLTS IF SMALL LOADS ARE SUPPLIED.

GROUNDED CONDUCTOR (NOT SHOWN)

UNGROUNDED CONDUCTOR

EQUIPMENT GROUNDING CONDUCTOR

b) GROUNDING TYPE RECEPTACLE REQUIRED ON 15- OR 20-AMPERE BRANCH CIRCUITS.

c) RECEPTACLE OUTLET AND CORD WITH ATTACHMENT PLUG IS REQUIRED TO SUPPLY PORTABLE LAMPS OR APPLICANCES.

115 VOLTS, 15 AMPERES

277 VOLTS, 15 AMPERES

d) RECEPTACLES CONNECTED TO CIRCUITS WITH DIFFERENT CHARACTERISTICS ON SAME PREMISES MUST BE DESIGNED SO THAT ATTACHMENT PLUGS ARE NOT INTERCHANGEABLE.

Figure 6-1. Illustrations of Branch-Circuit Rules for Any Type of Construction

watts if the conductors are insulated from the generator frame. Ground-fault circuit protection may not be required when an approved procedure is enforced to assure that equipment grounding conductors are properly installed and maintained. This Code exception requires a designated individual at the construction site to test the grounding system, as specified in the Code, including all equipment connected by cord and plug if ground-fault circuit interrupters are not provided.

6-1.2 Branch-Circuit Installation Rules for Dwelling Units

Specific consideration is given to branch-circuit installations in dwellings since the safety of the public is involved. Since misuse of electrical equipment in homes has caused many unfortunate accidents, Code rules assure that an adequate number of circuits and outlets is provided to avoid overloaded circuits and the excessive use of extension cords.

To provide the safety required in a dwelling, the maximum voltage between conductors for certain branch circuits is limited to 120 volts. Additional protection for the occupants is provided by ground-fault circuit interrupters on circuits in areas where the shock hazard is considered greatest.

The Code specifies the location and placement of receptacle outlets and lighting outlets in great detail. A sufficient number of outlets placed conveniently in a dwelling unit is required to yield an installation that meets Code rules for such occupancies. Table 6-3 lists the pertinent installation rules for circuits and outlets in a dwelling.

210-6(a)   *Maximum Voltage.* On any branch circuit that supplies lighting fixtures or small appliances in dwelling units, the maximum voltage between conductors must not exceed 120 volts. Permanently connected appliances or large appliances rated 1380 watts or more connected by cord and plug may be supplied by a circuit with a higher voltage.

210-52(a)   *Receptacle Outlets Required.* As illustrated in Figure 6-2, receptacle outlets are required in habitable rooms and kitchens so that no point along the floor line is more than 6 feet from an outlet. On an unbroken wall, then, an outlet would be required at least every 12 feet. When a wall is broken by doorways, fireplaces, or similar openings, each wall space of 2 feet or more must be considered separately. The required receptacle outlets do not include those that are part of a lighting fixture or appliance or that are located in cabinets or cupboards. Also, those located over

Table 6-3. Branch-Circuit Installation Rules for Dwellings

Application	Rule	Code Ref.
Maximum Voltage	Voltage must not exceed 120 volts between branch circuit conductors supplying lighting fixtures or cord- and plug- connected loads of less than 1380 volt-amperes. Exceptions: 1) Permanently connected appliances 2) Cord- and plug-connected loads of 1380 volt-amperes or more	210-6(a)
Receptacle Outlets in Habitable Rooms	Receptacle outlets must be installed so that no point along floor line in any wall space is more than 6 feet horizontally from an outlet in that space. (A simple installation requires an outlet every 12 feet.) The wall space to be counted includes: a) Any wall space 2 feet or more in width b) Wall space occupied by sliding panels in exterior walls c) Wall space of fixed dividers (See definition of wall space in Code.) Note: Special circuits must serve the kitchen, pantry, dining room, breakfast room, and family room.	210-52(a)
Kitchen	A receptacle outlet must be installed at each counter space wider than 12 inches. Ground-fault protection required within 6 feet of sink.	210-52(b) 210-8
Bathroom	At least one wall receptacle must be installed in bathroom adjacent to basin location. In residential occupancies it must have ground-fault protection.	210-52(c) 210-8
Basement and Attached Garage	At least one receptacle outlet must be installed in basement and attached garage of one-family dwellings. Ground-fault protection is required for garage circuit (with exceptions) and for at least one receptacle in basement.	210-52(f) 210-8
Laundry	At least one receptacle outlet must be installed for laundry. Outlet must be placed within 6 feet of intended location of the appliance. Exceptions: 1) In unit of multifamily dwelling where central facilities are provided. 2) In other than one-family dwelling where laundry facilities are not installed or permitted.	210-52(e), 210-50(c)

5½ feet above the floor or located in the floor away from the wall do not count as required receptacles.

In the kitchen and dining areas a receptacle outlet must be installed at each **210-52(b)** countertop space wider than 12 inches. The countertop space separated by range tops, refrigerators, or sinks must be considered separately. These kitchen and dining **220-4(b)** area receptacle outlets must be supplied by two or more of the required 20-ampere small appliance branch circuits. Receptacles within 6 feet of the sink require GFCI **210-8** protection.

Additionally, at least one wall receptacle outlet must be installed in the bathroom **210-52(c)** adjacent to the basin and at least one receptacle outlet must be installed for the **210-52(e)** laundry in every dwelling unit where laundry facilities are permitted. A laundry outlet is not required in individual dwelling units of multifamily dwellings if central facilities are provided or if individual laundry facilities are not permitted. When required, the receptacle outlet(s) for the laundry equipment must be placed within **210-50(c)**

Table 6-3. Continued

Application	Rule	Code Ref.
Outdoors	For a one-family dwelling, at least one receptacle outlet must be installed outdoors; it must have ground-fault protection.	210-52(d) 210-8
Ground-Fault Circuit Interrupter Protection	All 120-volt, single-phase, 15- and 20-ampere receptacles must have ground-fault protection when installed in: a) bathrooms b) garages (with exceptions) c) outdoors d) basement (at least one receptacle) e) kitchens (within 6 ft of sink) The feeder may be protected by a ground-fault circuit-interrupter in lieu of branch circuits.	210-8(a) 215-9
Lighting Outlets Required	a) At least one wall-switch controlled lighting outlet must be installed: 1) In every habitable room 2) In bathrooms 3) In hallways 4) In stairways 5) In attached garages 6) At outdoor entrances b) At least one lighting outlet must be installed in: 1) Attics 2) Underfloor space, utility rooms, or basement used for storage or containing equipment requiring servicing. Exceptions: 1) In habitable rooms, other than kitchens, one or more receptacles controlled by a wall switch are permitted in lieu of lighting outlets. 2) In hallways, stairways, and at outdoor entrances, remote, central, or automatic control of lighting shall be permitted.	210-70(a)
Branch Circuits Required	a) Branch circuits for lighting and appliances must be provided to supply the load computed by Code rules for branch circuit loads. b) Two or more 20-ampere small appliance branch circuits must be provided for only receptacle outlets in kitchen, pantry, breakfast room, and dining room. The receptacle outlets in kitchen must be supplied by not less than two small-appliance branch circuits. c) At least one 20-ampere branch circuit must be provided to supply only the laundry outlet(s).	220-3

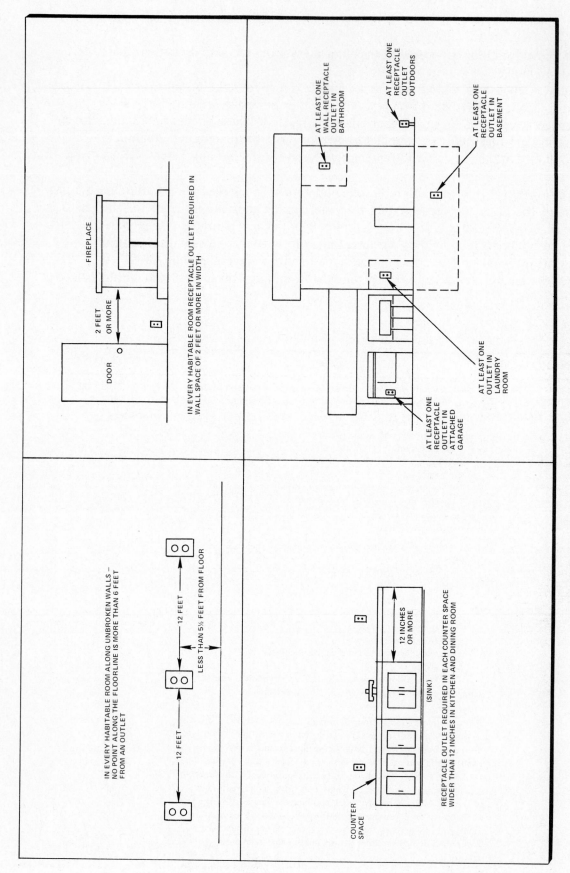

Figure 6-2. Placement of Receptacle Outlets in a Dwelling Unit

FIREPLACE

DOOR

2 FEET OR MORE

IN EVERY HABITABLE ROOM RECEPTACLE OUTLET REQUIRED IN WALL SPACE OF 2 FEET OR MORE IN WIDTH

AT LEAST ONE WALL RECEPTACLE OUTLET IN BATHROOM

AT LEAST ONE RECEPTACLE OUTLET OUTDOORS

AT LEAST ONE RECEPTACLE OUTLET IN BASEMENT

AT LEAST ONE OUTLET IN LAUNDRY ROOM

AT LEAST ONE RECEPTACLE OUTLET IN ATTACHED GARAGE

12 FEET

12 FEET

LESS THAN 5½ FEET FROM FLOOR

IN EVERY HABITABLE ROOM ALONG UNBROKEN WALLS — NO POINT ALONG THE FLOORLINE IS MORE THAN 6 FEET FROM AN OUTLET

COUNTER SPACE

(SINK)

12 INCHES OR MORE

RECEPTACLE OUTLET REQUIRED IN EACH COUNTER SPACE WIDER THAN 12 INCHES IN KITCHEN AND DINING ROOM

6 feet of the intended location of the appliance and the laundry receptacle outlet(s)   **220-4(c)**
must be supplied by a separate 20-ampere branch circuit.

In one-family dwellings other receptacle outlets are required in the basement, in   **210-52(d)**
each attached garage, and outdoors. Receptacle outlets in the garage (with excep-   **210-52(f)**
tions) and outdoors must have ground-fault circuit interrupter protection. One re-   **210-8(a)**
ceptacle installed in the basement must have GFCI protection.

Ground-Fault Circuit Protection. As shown in Figure 6-3(a), certain receptacle
outlets in dwelling units must have ground-fault circuit protection for personnel.
This protection may be provided by a ground-fault circuit interrupter built into the   **215-9**
receptacle itself or by a ground-fault circuit interrupter as part of either the branch-
circuit circuit breaker or the circuit breaker protecting the feeder or service conductors. The principle of operation of the GFCI is based on the device's sensing a
difference in current between the ungrounded conductor and the neutral caused by
a current flow to ground through a person's body in contact with the ungrounded
conductor. The GFCI opens the circuit when the detected difference exceeds a preset
value. GFCI devices to protect occupants are usually set to operate at approximately
5 milliamperes since tests have determined that a current flow through a person's
body in excess of that value may be harmful.

Lighting Outlets Required. In a dwelling as shown in Figure 6-3(b), wall-switch   **210-70(a)**
controlled lighting outlets are required in each habitable room, the bathroom, hallways, stairways, attached garages, and at outdoor entrances. A wall-switch controlled
receptacle may be used in place of the lighting outlet in habitable rooms other than

Figure 6-3. Rules for Ground-Fault Protection and Required Lighting Outlets

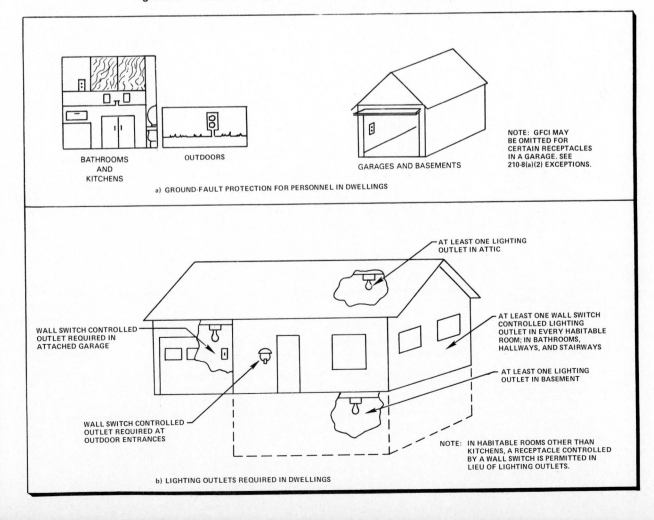

the kitchen. Providing a wall switch for room lighting is intended to prevent an occupant's groping in the dark for table lamps or pull chains. In certain locations, such as hallways, stairways, and outdoor entrances, the Code permits central or remote control of lighting.

Lighting outlets are also required in attics, underfloor space, utility rooms, and basements when these spaces are used for storage or contain equipment requiring servicing.

220-4   ***Branch Circuits Required.*** Branch circuits must be provided to supply the computed lighting load and appliance loads. In addition, two 20-ampere small appliance branch circuits and a 20-ampere laundry circuit must also be provided in dwelling units.

6-1.3 Branch-Circuit Installation Rules for Guest Rooms in Hotels and Motels

ARTICLE 100
DEFINITIONS

Table 6-4 summarizes the branch-circuit installation rules that apply to guest rooms of hotels and motels. These occupancies are not considered dwelling units if provisions for housekeeping, including eating, are not made available. Nevertheless, most rules for dwelling units apply to guest rooms, although receptacle outlets may be located where convenient for permanently installed furniture layouts.

Table 6-4. Branch-Circuit Installation Rules for Guest Rooms

Application	Rule	Code Ref.
Maximum Voltage	Same rule as for any dwelling unit.	210-6(c) (1)
Receptacle Outlets Required	Same rule as for any dwelling unit. Exception: Receptacle outlets may be located conveniently for permanent furniture layout. GFCI receptacles required in bathrooms	210-60 210-8(b)
Lighting Outlets Required	At least one wall-switch controlled lighting outlet or wall-switch-controlled receptacle outlet is required in guest rooms.	210-70(b)

6-2 INSTALLATION RULES FOR FEEDERS

ARTICLE 215

The Code provides specific installation requirements for feeders, but these requirements are not as extensive as those for branch circuits or services as shown by the

215-8 summary in Table 6-5. A four-wire delta connected feeder, such as a high-leg delta circuit, must have the higher voltage phase identified when the midpoint of one

215-9 phase is grounded. Also, the Code permits ground-fault circuit interrupter protection to be provided at the feeder rather than the branch circuit when GFCI is required.

Table 6-5. Installation Rules for Feeders

Application	Rule	Code Ref.
Identification of High-Leg	On 4-wire, delta-connected secondary where midpoint of one phase is grounded, the phase conductor with higher voltage to ground must be identified by orange outer finish or other effective means.	215-8
Ground-Fault Circuit Interrupter	Feeders supplying 15- and 20-ampere receptacle branch circuits may have ground-fault circuit-interrupters in lieu of required branch circuit device(s).	215-9, 210-8

6-3 INSTALLATION RULES FOR SERVICES

The Code article defining the design and installation of services contains an extensive coverage of installation requirements for service conductors and equipment. For convenience, this section of the Guide separates these installation rules into (a) general installation rules, (b) installation requirements for disconnecting means and over-current protection, and (c) grounding and bonding requirements for services.   **ARTICLE 230**

Many of the terms used to discuss services are defined in the Code and the reader should become familiar with them before proceeding. The service, which delivers energy from the utility distribution system to the premises served, is the control and distribution point for the electrical system within an occupancy. The service equipment consists of the main disconnecting means and main overcurrent devices as well as enclosures, watt-hour meters, and other required devices. When a system ground is required, the grounded and grounding conductors are connected at the service and then to earth through a grounding electrode system.   **ARTICLE 100 DEFINITIONS**

6-3.1 General Installation Rules for Services

In addition to installation rules that apply to any service, the Code provides specific installation rules for overhead services and service-entrance conductors. These rules and rules for the protection and support of service conductors are discussed in this section.

Number of Services. In general, a building may be served by only one set of service-drop or service lateral conductors, with the exceptions listed in Table 6-6. It is permissible to tap from two to six sets of service-entrance conductors to supply separate service-entrance enclosures if separate disconnecting means are supplied for each circuit.   **230-2**

Installation Rules for Overhead Services. When overhead service-drop conductors are installed, a minimum specified clearance over roofs, from ground, and from building openings is required. These clearances are listed in Table 6-7 and some examples are illustrated in Figure 6-4.   **230-24**

Installation Rules for Service-Entrance Conductors. The Code specifies that service-entrance conductors must not be bare, but the Code lists several conditions under which the grounded conductor may be bare. The Code also limits the methods that may be used to install service-entrance conductors. These wiring methods are listed in Table 6-8 and are discussed in further detail in Chapter 7 of this Guide.   **230-41**   **230-43**

With several exceptions, service-entrance conductors may not be spliced. If they are installed in a raceway, no other conductors other than grounding conductors or timing circuit conductors may share the same raceway. Unless the termination point is on a switchboard, any service raceway or service cable must terminate in an enclosure that encloses all live metal parts. Every service conductor must also be protected from physical damage whether installed underground or above ground.   **230-46, 230-7**   **230-55**   **230-49, 230-50**

Protection, Support, and Installation of Service Conductors. The Code specifies the method of protection and support of service conductors when run above ground. Connection of service conductors on a building to supply the service equipment must be by an approved method and must meet the requirements stated in the Code.

Service-entrance conductors may be individual conductors or part of a service-entrance cable. Type SE (service-entrance) cable is an approved wiring method used   **ARTICLE 338**

Table 6-6. Number of Services Allowed

Application	Rule	Code Ref.

Number of Services	A building or structure must be served by only one set of service drop or service lateral conductors except:
	  1) A separate service is allowed for fire pumps or emergency electrical systems.
	  2) By special permission in multiple occupancy buildings.
	  3) By permission when the capacity of a single service may not satisfy the load requirements.
	  4) In large area buildings by special permission.
	  5) When different characteristics (voltage, frequency, or number of phases) or different applications are required for a service.
		230-2

Sub-Sets of Service Conductors	Several sets of service-entrance conductors may be tapped from one service drop or lateral in certain cases. Service laterals of size 1/0 and larger may be connected in parallel at their supply end and serve up to six separate disconnecting means in separate enclosures.	230-2, 230-40

Table 6-7. Clearance for Overhead Services of 600 Volts or Less

Application	Rule	Code Ref.

Clearance over Roofs	Conductors must have clearance of not less than 8 feet from highest point of roofs over which they pass.
	Exceptions: Where voltage between conductors does not exceed 300 volts and
	  1) If roof has slope of not less than 4 inches in 12 inches, a clearance of 3 feet is permitted
	  2) If conductors do not pass over more than 4 feet of roof overhang and terminate in through-the-roof raceway or approved support, a minimum clearance of 18 inches is permitted.
		230-24(a)

Clearance from the Ground	The minimum clearances from ground are: 10 feet—above final grade or other accessible surface if voltage is limited to 150 volts to ground for cables. 12 feet—over residential property and driveways, and those commercial areas not subject to truck traffic if voltage is limited to 300 volts to ground. 15 feet—over residential property and driveways, and those commercial areas not subject to truck traffic. 18 feet—over public streets, alleys, roads, and driveways other than residential property.	230-24(b)
Clearance from any Building Opening	Clearance from windows, doors, and similar locations must be not less than 3 feet. Conductors run above top level of window are considered out of reach of window.	230-9(c)
Attachment	The point of attachment of conductors to building must provide minimum clearances specified and in no case be less than 10 feet above finished grade.	230-26

specifically for service-entrance conductors. In any case, if individual conductors or cable might be subject to physical damage, the Code rules listed in Table 6-9 must be followed.

230-51   When service-entrance cable is used to connect the service equipment to the service-drop conductors, the cable must be supported as shown in Figure 6-5(a). Other cables not approved for mounting in contact with a building must maintain a clearance of 2 inches and be supported at least every 15 feet.

Open individual conductors supported on buildings must meet the clearances and distances between supports summarized in Table 6-9. Figure 6-5(b) illustrates

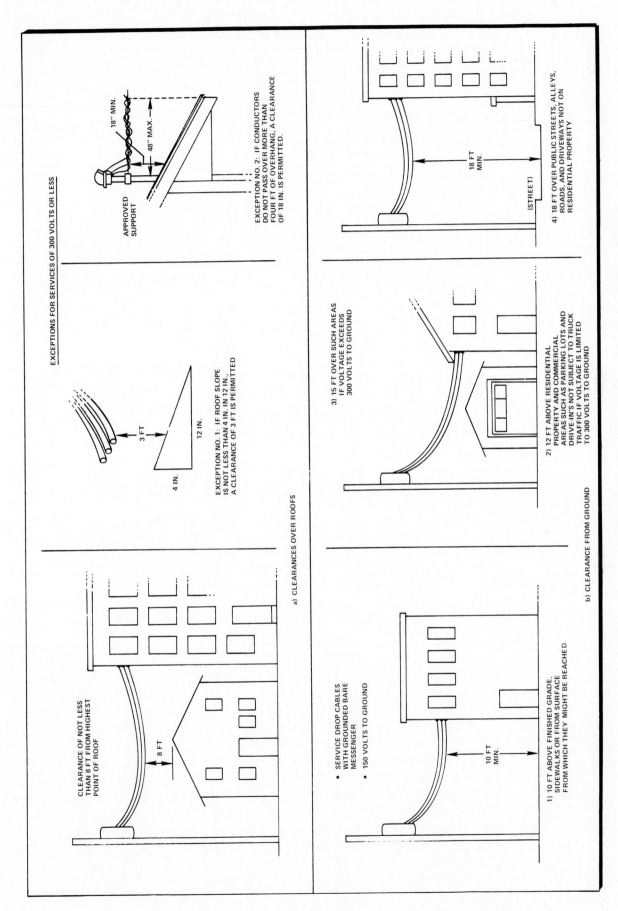

EXCEPTIONS FOR SERVICES OF 300 VOLTS OR LESS

18" MIN.

48" MAX.

APPROVED SUPPORT

EXCEPTION NO. 2: IF CONDUCTORS DO NOT PASS OVER MORE THAN FOUR FT OF OVERHANG, A CLEARANCE OF 18 IN. IS PERMITTED.

12 IN.

3 FT

4 IN.

EXCEPTION NO. 1: IF ROOF SLOPE IS NOT LESS THAN 4 IN. IN 12 IN., A CLEARANCE OF 3 FT IS PERMITTED

a) CLEARANCES OVER ROOFS

CLEARANCE OF NOT LESS THAN 8 FT FROM HIGHEST POINT OF ROOF

8 FT

18 FT MIN.

(STREET)

4) 18 FT OVER PUBLIC STREETS, ALLEYS, ROADS, AND DRIVEWAYS NOT ON RESIDENTIAL PROPERTY

3) 15 FT OVER SUCH AREAS IF VOLTAGE EXCEEDS 300 VOLTS TO GROUND

2) 12 FT ABOVE RESIDENTIAL PROPERTY AND COMMERCIAL AREAS SUCH AS PARKING LOTS AND DRIVE-IN'S NOT SUBJECT TO TRUCK TRAFFIC IF VOLTAGE IS LIMITED TO 300 VOLTS TO GROUND

b) CLEARANCE FROM GROUND

• SERVICE DROP CABLES WITH GROUNDED BARE MESSENGER

• 150 VOLTS TO GROUND

10 FT MIN.

1) 10 FT ABOVE FINISHED GRADE, SIDEWALKS OR FROM SURFACE FROM WHICH THEY MIGHT BE REACHED

Figure 6-4. Required Clearance of Service-Entrance Conductors Over Roofs and From Ground

Table 6-8. Installation Rules for Service-Entrance Conductors

Application	Rule	Code Ref.
Insulation Required	Service-entrance conductors entering or on exterior of building must be insulated. (See exceptions for grounded conductor.)	230-41
Wiring Methods	Service-entrance conductors must be installed using one of the following wiring methods: a) Open wiring on insulators b) Rigid metal conduit c) Intermediate metal conduit d) Electrical metal tubing e) Service-entrance cable f) Wireways g) Busways h) Auxiliary gutters i) Rigid nonmetal conduit j) Cablebus k) Type MC or MI cable l) Approved cable trays or flexible metal conduit	230-43
Splices	Service-entrance conductors must not be spliced. Exceptions: 1) In metering equipment. 2) Taps to supply several disconnecting means. 3) At a junction where an underground wiring method is changed to another method or for busways. 4) To connect an outside meter.	230-46
Service Raceways and Cables	a) Conductors other than service conductors must not be installed in the same service raceway or service-entrance cable. (See exception for grounding conductors or timing circuits.)	230-7
	b) A service raceway entering from an underground distribution system must be sealed. Spare or unused raceways must also be sealed.	230-8 300-5
	c) Where exposed to weather, service raceways must be raintight and arranged to drain. Raceways embedded in masonary must be arranged to drain.	230-53
	d) Any service raceway or cable must terminate at the inner end in an enclosure that encloses all live metal parts. Exception: The termination may be a bushing if the disconnecting means is mounted on a switchboard having exposed busbars on the back.	230-55
Underground Conductors	Underground service conductors must be protected against physical damage according to Code rules.	230-49 300-5
High-Leg Delta	The service-entrance conductor with higher phase voltage-to-ground must have orange finish or other effective marking.	230-56

the rules for installing open individual conductors that are not exposed to weather. If any open individual conductors enter a building, approved methods of passing conductors through a roof or wall must be used.

230-54 Figure 6-5(c) illustrates the most commonly used method of connecting service-entrance conductors to the service-drop conductors. When a service head and a service raceway are used, as shown, the service-entrance conductors are brought out of the service head and are spliced to the service drop. A raintight service head with individual bushings for each conductor must be used so that the whole arrangement does not allow water to enter the service raceway or equipment. Normally, the service head is located above the point of attachment of the service-drop conductors to a building; if this is impractical, the location must be within 24 inches of the point of attachment.

Table 6-9. Protection, Support, and Installation of Service Conductors

Application	Rule	Code Ref.

Protection of Conductors	a) Service-entrance cables or conductors installed above ground where subject to physical damage must be protected by
	    1) Rigid metal conduit
	    2) Intermediate metal conduit
	    3) Rigid nonmetallic conduit suitable for the location
	    4) Electrical metallic tubing
	    5) Other approved means
	
	b) Individual open conductors and cables other than service-entrance cable must not be installed
	    1) Within 10 feet of grade level
	    2) Where exposed to physical damage
		230-50

Mounting Supports	a) Service-entrance cables—Cables must be supported by straps or other approved means within
	    —12 inches of every service head, gooseneck, or connection to raceway or enclosure, and
	    —at intervals not exceeding 30 inches
	b) Other cables—Cables not approved for mounting in contact with binding must be mounted on insulating supports installed
	    —at intervals not exceeding 15 feet and
	    —to maintain clearance of not less than 2 inches from surface
	c) Individual Open Conductors
	    1) Conductors not exposed to weather must be mounted on glass or porcelain knobs as follows:
	    —maximum distance between supports of 4 1/2 feet and
	    —minimum clearance between conductors of 2 1/2 inches and
	    —minimum clearance from surface of 1 inch
	    2) Conductors exposed to weather must be installed on insulators, insulating supports, or other approved means and have a clearance from surface of at least 2 inches. Other supports and clearances are as listed herein.
		230-51

Conductors Entering Building	Open individual conductors entering building must enter through roof bushings or through wall in upward slant. Individual, noncombustible, nonabsorbent insulating tubes required for conductors passing through walls. Drip loops must be formed on conductors.	230-52

Connections at Service Head	a) Raceways—Service raceways must be equipped with raintight service head.
	b) Service cables—Unless cables are continuous from pole to service equipment, they must be installed either
	    1) With raintight service head, or
	    2) Formed in gooseneck.
	c) Location—Service heads or gooseneck must be located above point of attachment of service-drop conductors to building.
	Exception: Where not practicable to locate service head above point of attachment, it may be placed not further than 24 inches from point of attachment.
	d) Service Conductors
	    1) Conductors of opposite polarity must be brought out of service head through separately bushed holes.
	    2) Drip loops must be formed on individual conductors.
	    3) Service conductors must be arranged so that water will not enter service raceway or equipment.
		230-54

Maximum Voltage Between Conductors	Maximum Distance Between Conductor Support	Minimum Clearance Between Conductors
600 volts	9 feet	6 inches
600 volts	15 feet	12 inches
300 volts	4 1/2 feet	3 inches

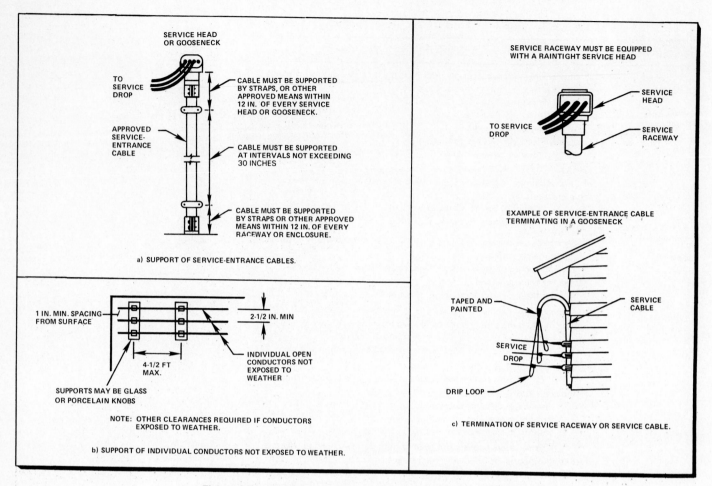

Figure 6-5. Installation Rules for Service-Entrance Conductors

6-3.2 Installation Rules for Service Disconnecting Means and Overcurrent Protection

Each service requires a disconnecting means to disconnect all conductors in a building from the service-entrance conductors. In addition, each service-entrance conductor must be protected from overcurrent created by overloads or short circuits. Specific installation rules from the Code for the disconnecting means and for the overcurrent protective device at the service are discussed in this section.

230-71 *Service Disconnecting Means.* The rules for the disconnecting means summarized in Table 6-10 restrict a disconnecting means to be not more than six switches or six circuit breakers. These devices may be mounted in a single enclosure, on a switch-

230-72(c) board, or in a group of separate enclosures. In multiple occupancy buildings, separate disconnecting means and metering equipment are commonly provided for each occupant. In multiple-occupancy buildings, each occupant must have access to his disconnecting means with certain exceptions.

230-84 When more than one building is on the same property and under single management, the Code requires that a disconnecting means be provided for each building served. This disconnecting means must be suitable for use as service equipment unless the second building is a garage or similar structure on residential property. In such locations, a snap switch or other switch suitable for use on branch circuits may be used as the disconnecting means for the second building.

204

Table 6-10. Installation Rules for Service Disconnecting Means

Application	Rule	Code Ref.
General	Means must be provided to disconnect all conductors in building from service-entrance conductors. Each disconnecting means must be identified.	230-70
Number of Disconnects	Service disconnecting means can consist of no more than six switches or circuit breakers for each service.	230-71
	Disconnecting means must be installed at readily accessible location near point of entrance of service-entrance conductors.	230-70(a)
	In multiple-occupancy building, each occupant must have access to his disconnecting means.	230-72(c)
Working Space	Rules for any electrical equipment apply.	110-16
Circuit Disconnecting	All ungrounded conductors must be disconnected simultaneously.	230-74

Type of Disconnect	Disconnecting means may be
	  a) Manually operable switch
	  b) Manually operable circuit breaker
	  c) Power-operated switch or circuit breaker
		230-76

Connections to Terminals	Service conductors must be connected to disconnecting means by pressure connectors, clamps, or other approved means. Soldered connections are forbidden.	230-81
More than One Building on Same Property	Each building must have an individual disconnecting means and each disconnecting means must be suitable for use as service equipment. Exception: In industrial establishments, the disconnecting means for several buildings may be conveniently located if conditions are met. Exception: A snap switch or set of 3-way or 4-way snap switches is permitted as disconnecting means for a garage or outbuilding on residential property.	230-84

Figure 6-6. Location of Disconnects in Multiple-Occupancy Building

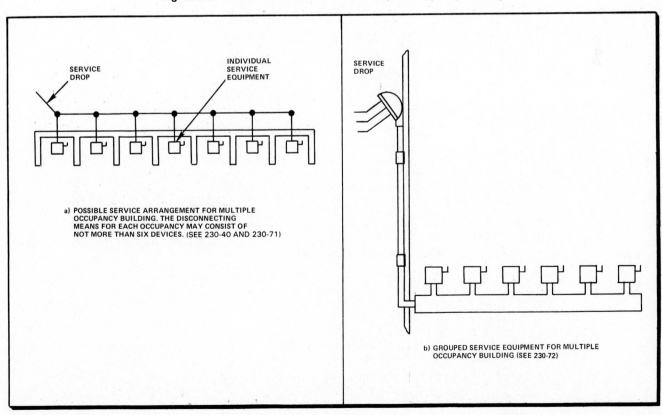

a) POSSIBLE SERVICE ARRANGEMENT FOR MULTIPLE OCCUPANCY BUILDING. THE DISCONNECTING MEANS FOR EACH OCCUPANCY MAY CONSIST OF NOT MORE THAN SIX DEVICES. (SEE 230-40 AND 230-71)

b) GROUPED SERVICE EQUIPMENT FOR MULTIPLE OCCUPANCY BUILDING (SEE 230-72)

230-90
230-91
240-21
EXCEPTION 4

Overcurrent Protection Device. The Code rules listed in Table 6-11 specify that the overcurrent protective device or devices for a service must be placed in series with each ungrounded service-entrance conductor. The overcurrent protection is normally provided in the service enclosure, which also contains the service disconnecting means. In this arrangement the service-entrance conductors are not protected at their source of supply from the service-drop or lateral; instead, they are protected at their load end, as permitted by the Code.

230-90(a)
230-91

Service-entrance conductors may be protected by not more than six circuit breakers or six sets of fuses. In multiple occupancy buildings, each occupant must have access to his overcurrent protective device(s).

Table 6-11. Installation Rules for Service Overcurrent Protection

Application	Rule	Code Ref.
Protection Required	Each ungrounded service-entrance conductor must have overcurrent protection. Device must be in series with each ungrounded conductor.	230-90(a)
Number of Devices	Up to six circuit breakers or sets of fuses may be considered as the overcurrent device.	230-90(a)
Location in Building	The overcurrent device must be part of the service disconnecting means or be located immediately adjacent to it.	230-91
Location for Accessibility	a) In multiple-occupancy buildings, each occupant must have access to overcurrent protective device protecting his occupancy. b) More than one building—the ungrounded conductors serving each building must be protected by overcurrent devices. The devices must be accessible to occupants of building served.	
Location in Circuit	The overcurrent device must protect all circuits and devices, except equipment which may be connected on the supply side including: 1) Service switch. 2) Special equipment, such as lightning arresters 3) Circuits for emergency supply and load management (where separately protected). 4) Circuits for fire alarms or fire pump equipment (where separately protected). 5) Meters, with all metal housings grounded, (600 volts or less). 6) Control circuits for automatic service equipment if suitable overcurrent protection and disconnecting means are provided.	230-94
Ground-Fault Protection of Equipment	Ground-fault protection of equipment must be provided for solidly grounded wye services of more than 150 volts to ground, but not exceeding 600 volts phase-to-phase if the disconnecting means is rated 1000 amperes or more. (The rating of highest fuse that can be installed or highest trip setting for the actual overcurrent device used determines the rating of the service in this case.) Exception: A continuous industrial process where shutdown would be hazardous or fire pumps.	230-95
Short-Circuit Rating	Service equipment must be suitable for the short-circuit current available at its supply terminals.	230-65

When more than one building is on the same property and under single manage- | **230-91**
ment, each building served requires protection for its supply conductors and the
protective devices must be accessible to the occupants of the building served.

The service overcurrent protective device is required to protect all circuits and | **230-94**
devices. This implies that all equipment must be connected on the load side of the
overcurrent protective device. Exceptions are provided that permit the service switch
and certain special devices (current-limiting devices, meters and other measuring
devices, lightning protection devices, emergency and fire alarm circuits, etc.) to be
connected on the supply side.

Ground-fault protection is required for solidly grounded wye services operating | **230-95**
at more than 150 volts to ground and less than 600 volts phase to phase if the rating
of the service disconnecting means is 1000 amperes or more. Thus, a 480Y/277-volt
system with a 1000-ampere disconnecting means would require such protection.
This provision is intended to protect equipment from a fault current resulting from
a line-to-ground fault on the load side of the overcurrent protective device. Without
such protection, the protective device may be set so high that it will not open the
circuit because of the current drawn by the fault; a fire or other damage could result
from the ensuing arc between the faulted conductor and ground.

6-3.3 Grounding and Bonding at Services

In a premises wiring system it is often required that a grounded conductor for | **ARTICLE 100**
the system be provided. The noncurrent-carrying metal parts of service equipment | **DEFINITIONS**
must also be connected together and grounded at the service. This grounding is
actually accomplished by electrically connecting the grounded conductor and the
equipment together and providing a conducting path to earth. As discussed in this
section, this path consists of a grounding electrode conductor connected to a
grounding electrode in intimate contact with the earth.

Bonding is the permanent joining of metal parts to form an electrically conductive path. Equipment and conductors required to be grounded at the service must be
bonded together as described in this section.

Grounding of Systems and Service Equipment. The Code requires a grounded | **250-5**
circuit conductor in many alternating-current systems to limit or stabilize the voltage
to ground. An ungrounded conductor may attain any voltage with respect to the
earth. Hence, unless one conductor is grounded, a system with a fixed potential
difference between conductors may not have a known voltage to ground. In most
cases, this is not desirable. As listed in Table 6-12, the Code specifies the circuits to
be grounded and the method of grounding.

Many alternating-current systems operating between 50 volts and 1000 volts
require grounding as shown in Figure 6-7(a). Grounding is required on any system
that can be grounded in such a manner that the maximum voltage to ground does not
exceed 150 volts, such as the common 120/240-volt system. Other systems that must
be grounded include 480Y/277-volt services and the 240/120-volt high-leg delta
service. Exceptions are made for special circuits such as control circuits and systems
requiring isolation.

Table 6-12. General Requirements for Service Grounding

Application	Rule	Code Ref.

Required Grounding	A system supplying a premises wiring system must be grounded if:
	   a) System can be grounded so that the maximum voltage to ground does not exceed 150 volts.
	   b) System is 480Y/277-volt system using the neutral as circuit conductor.
	   c) System is high-leg delta (240/120) using the midpoint of one phase as circuit conductor.
	   d) If an uninsulated service conductor (grounded conductor) is used.
	
	(See exceptions in Code for systems that do not require grounding or that must not be grounded.)
		250-5

Grounding Connection	a) AC systems grounded on a premises require:
	   1) A grounding electrode conductor connected to a grounding electrode at each service. The grounding connection must be on supply side of service disconnecting means.
	   2) For supply systems originating outside building, additional grounding connection to grounding electrode on secondary side of transformer supplying system.
	
	(See exception noted in Code.)
		250-23(a)

	b) AC system operating at less than 1000 volts and grounded at any point—the grounded conductor from supply system must be run to each service.	250-23(b)

Conductor to be Grounded	The conductor to be grounded is:
	   a) One conductor in single-phase, 2-wire circuit.
	   b) The neutral conductor in single-phase, 3-wire circuit.
	   c) The common conductor in multiphase systems with common conductor.
	   d) One phase conductor in multiphase systems with one phase grounded.
	   e) The neutral conductor in multiphase systems where one phase supplies single phase, 3-wire loads.
		250-25

Equipment Grounding	Metal enclosures for service conductors and equipment must be grounded.	250-32

250-23(a)   In addition to these specific cases, a system grounded on the premises requires a grounding electrode at each service. If the supply system originates outside the building, a second grounding electrode is required at the transformer. Figure 6-7(b) shows a 120/240-volt system with the required grounding electrode conductor connections to the grounding electrodes. A grounding electrode is required at the service connected to the grounding electrode conductor on the supply side of the service disconnecting means. The other grounding electrode conductor is connected to the transformer on the secondary side.

250-23(b)   Figure 6-7(c) shows an ungrounded premises system tapped from a grounded distribution system. The premises system requires grounding even though the loads do not use the neutral conductor. A grounded conductor must be brought to the service enclosure to connect the grounding system at the service to that of the distribution system.

250-25
250-32   The Code also specifies which conductor is to be grounded for various alternating-current systems, as the examples of Figure 6-7(a) show. Another general rule requires that metal enclosures for service conductors and equipment be grounded.

208

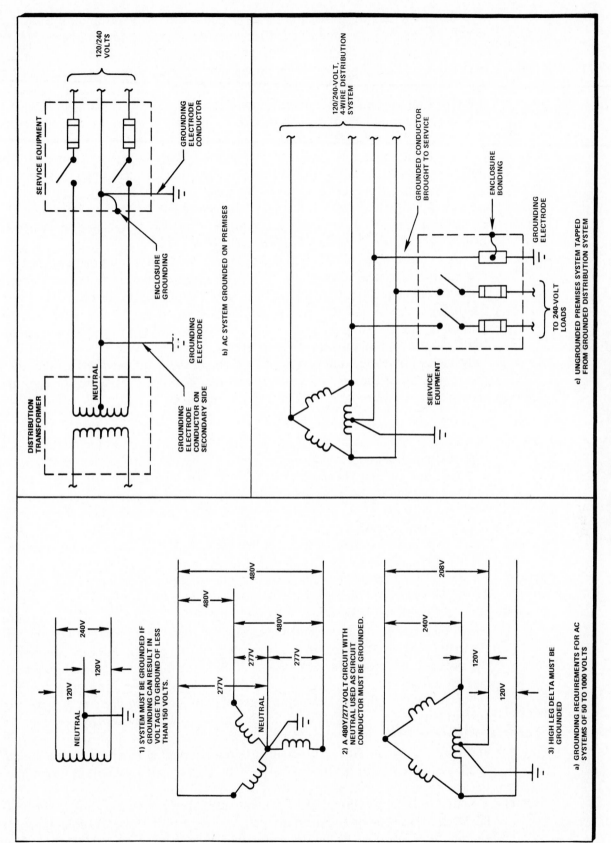

Figure 6-7. System Grounding Requirements

Rules for service grounding specify requirements for grounding both circuits and equipment. Grounding of the distribution system and grounding at the service as just described provide a path to earth outside the building if the outside circuits are struck by lightning. Thus, equipment and interior wiring are protected. Equipment grounding is required so that personnel are protected from contacting accessible metal equipment that is not at earth potential.

If an ungrounded conductor were to accidentally contact an ungrounded metal object, that object would assume the potential of the conductor with respect to the earth. A fault would then have occurred in the system. A person touching the metal object would be subjected to the voltage of the conductor, causing a fault current to flow through the person's body, with possible serious consequences. Grounding metal objects assures that any contact between the object and an ungrounded conductor would draw excessive current (since the resistance is low) and cause the circuit overcurrent protective devices to open the circuit.

250-50

A well-protected system is grounded so that all metal equipment required to be grounded is electrically connected to the grounding electrode system at the service, thus providing a low resistance path to earth if a fault were to occur. If the system has a grounded circuit conductor, this must also be electrically connected to the grounding electrode system at the service to assure one common grounding point for both circuits and equipment. The grounded circuit conductor must be grounded only at this point to assure current flow through the conductor rather than through the equipment or the earth. The Code rules for bonding specify how the grounding connections must be made using bonding jumpers.

250-53(a)

250-21

Bonding at the Service. The rules for service bonding specify the required bonding, the methods of bonding, and the sizes of the bonding conductors. Table 6-13 summarizes these rules for bonding.

Bonding jumpers at the service connect conductors and equipment to assure electrical continuity for adequate grounding. A bonding jumper is not necessarily a wire, but it must maintain the required electrical continuity between metal parts required to be grounded.

The grounded system in Figure 6-8(a) requires a *main bonding jumper* to connect the equipment grounding conductors and the service equipment enclosure to the grounded circuit conductor. The main bonding jumper may be a wire, bus (as shown), screw, or similar conductor within the service equipment enclosure.

250-53(b)

250-50

The grounding connection between the circuit, equipment, and the grounding electrode conductor is provided by bonding the equipment grounding conductors to the grounding electrode conductor and thus to the grounding electrode system. If the system is grounded, the grounded circuit conductor must also be bonded to this connection.

Equipment grounding requirements refer to service equipment as well as non-current-carrying metal parts of other equipment and raceways that are part of the system. The service equipment must be bonded together, as described in Table 6-13, and then grounded. Any equipment grounding conductors from utilization equipment supplied by the system must also be connected together and grounded at the service.

250-71

250-61

The method of grounding equipment at the service is by bonding to the grounding electrode conductor as previously described, except that specific equipment on the supply side of the service disconnecting means, such as meter enclosures and metal

Table 6-13. Rules for Bonding at the Service

Application	Rule	Code Ref.

Main Bonding Jumper	For a grounded system, a main bonding jumper is required to connect the following together:
	    a) equipment grounding conductors
	    b) service-equipment enclosure
	    c) grounded conductor of system
	within the service equipment or service conductor enclosure. The main bonding jumper must be unspliced and may be a wire, bus, screw, or similar conductor.
		250-53(b)
		
		250-79

Bonding to Ground Equipment and Systems	Equipment grounding conductor connections at the service must be made on supply side of service disconnecting means and
	    a) For grounded systems bonding must connect the following together:
	        1) equipment grounding conductor
	        2) grounded circuit conductor
	        3) grounding electrode conductor
	    b) For ungrounded systems, bonding must connect together the equipment grounding conductor and the grounding electrode conductor.
	Exception: A cold water pipe may be used to ground a grounding-type receptacle on extensions to existing branch circuits that have no equipment grounding conductor.
		250-50

Equipment Bonding	Noncurrent-carrying metal parts of equipment that must be bonded together includes:
	    a) Service raceways, cable trays, or service cable armor or sheath (See exception.)
	    b) All service-equipment enclosures containing service-entrance conductors
	    c) Any metal raceway or armor enclosing a grounding electrode conductor
		250-71(a)

Method of Bonding	Bonding of service equipment must be assured by
	    a) Bonding equipment to grounded service conductor by pressure connectors, clamps, or other approved means, or
	    b) For rigid or intermediate conduit, threaded couplings made up wrench-tight, or
	    c) Threadless couplings made up tight for rigid and intermediate metal conduit or electrical metal tubing, or
	    d) Bonding jumpers, or
	    e) Other devices approved for the purpose such as bonding type locknuts and bushings
		250-72
		250-113

Bonding of Water Piping System	The interior metal water piping system must always be bonded to the grounding system at the service. The bonding jumper size depends on the service-entrance conductor size.	250-80(a)
Bonding of Other Metal Piping	Interior metal piping which may become energized must be bonded to the grounding system at the service. The bonding jumper size is determined by the size of the equipment grounding conductor of the circuit that may energize the piping. The equipment grounding conductor may also serve as the bonding means.	250-80(b)
Material	Main and equipment bonding jumpers must be of copper or other corrosion-resistant material.	250-79
Size	The bonding jumper must not be smaller than the grounding electrode conductor for service conductors not larger than 1100 MCM copper or 1750 MCM aluminum.	250-79(c)

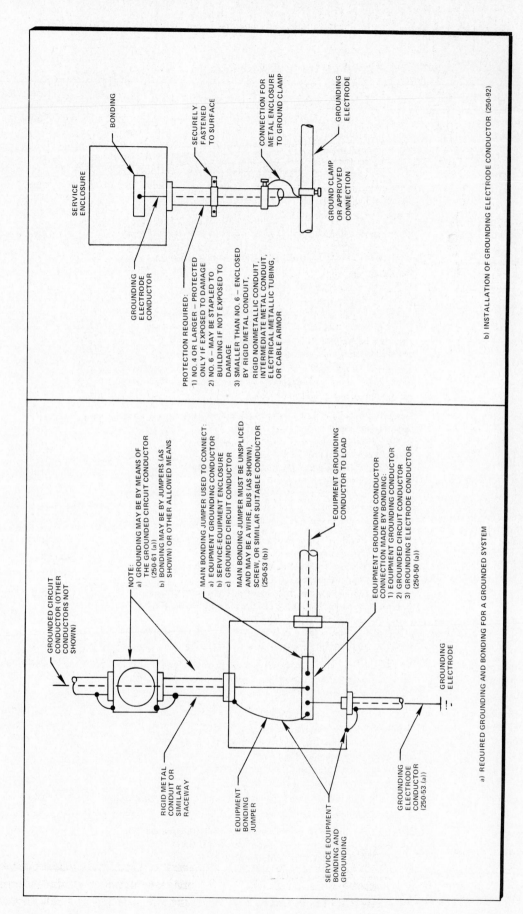

BONDING

SERVICE ENCLOSURE

SECURELY FASTENED TO SURFACE

CONNECTION FOR METAL ENCLOSURE TO GROUND CLAMP

GROUNDING ELECTRODE

GROUNDING ELECTRODE CONDUCTOR

GROUND CLAMP OR APPROVED CONNECTION

PROTECTION REQUIRED:
1) NO. 4 OR LARGER – PROTECTED ONLY IF EXPOSED TO DAMAGE
2) NO. 6 – MAY BE STAPLED TO BUILDING IF NOT EXPOSED TO DAMAGE
3) SMALLER THAN NO. 6 – ENCLOSED BY RIGID METAL CONDUIT, RIGID NONMETALLIC CONDUIT, INTERMEDIATE METAL CONDUIT, ELECTRICAL METALLIC TUBING, OR CABLE ARMOR

b) INSTALLATION OF GROUNDING ELECTRODE CONDUCTOR (250-92)

GROUNDED CIRCUIT CONDUCTOR (OTHER CONDUCTORS NOT SHOWN)

NOTE:
a) GROUNDING MAY BE BY MEANS OF THE GROUNDED CIRCUIT CONDUCTOR (250-61 (a))
b) BONDING MAY BE BY JUMPERS (AS SHOWN) OR OTHER ALLOWED MEANS

MAIN BONDING JUMPER USED TO CONNECT:
a) EQUIPMENT GROUNDING CONDUCTOR
b) SERVICE EQUIPMENT ENCLOSURE
c) GROUNDED CIRCUIT CONDUCTOR

MAIN BONDING JUMPER MUST BE UNSPLICED AND MAY BE A WIRE, BUS (AS SHOWN), SCREW, OR SIMILAR SUITABLE CONDUCTOR (250-53 (b))

EQUIPMENT GROUNDING CONDUCTOR TO LOAD

EQUIPMENT GROUNDING CONDUCTOR CONNECTION MADE BY BONDING:
1) EQUIPMENT GROUNDING CONDUCTOR
2) GROUNDED CIRCUIT CONDUCTOR
3) GROUNDING ELECTRODE CONDUCTOR (250-50 (a))

RIGID METAL CONDUIT OR SIMILAR RACEWAY

EQUIPMENT BONDING JUMPER

SERVICE EQUIPMENT BONDING AND GROUNDING

GROUNDING ELECTRODE CONDUCTOR (250-53 (a))

GROUNDING ELECTRODE

a) REQUIRED GROUNDING AND BONDING FOR A GROUNDED SYSTEM

Figure 6-8. Grounding and Bonding of Services

service raceways, may be grounded by connection to the grounded circuit conductor of the service conductors.

In addition to the conductors and equipment that must be bonded together and grounded, the Code specifies that interior metal piping that may become energized must be bonded to the service equipment enclosures and to the grounded conductor at the service. The size of the service-entrance conductors determines the size of the bonding jumper for a water piping system. However, the size of the jumper required for other metal pipes depends on the rating of the circuit that may energize the piping. **250-80(b)**

The Code further specifies that the main bonding jumper and equipment bonding jumpers on the supply side of the service must be of copper or other corrosion-resistant material. The size must not be less than that of the grounding electrode conductor for service-entrance conductors as large as 1100 MCM copper or 1750 MCM aluminum. **250-79(c)**

The Grounding Electrode Conductor. Any grounded premises wiring system requires a grounding electrode installed according to the Code rules listed in Table 6-14. This conductor connects equipment grounding conductors, service equipment enclosures, and the grounded circuit conductor to the grounding electrode. If a single service supplies several buildings, each building, in most cases, requires a grounding electrode. The Code also specifies rules for grounding separately derived systems supplied from a generator, transformer, or similar unit. **250-23** **250-24**

The grounding electrode conductor must be copper, aluminum, or copperclad aluminum and must be resistant to any corrosive condition at the installation. In fact, the Code restricts the use of aluminum or copperclad aluminum conductors. These conductors must not be used in direct contact with masonry or the earth or outside within 18 inches of the earth. **250-91** **250-92**

Figure 6-8(b) illustrates rules for installing the grounding electrode conductor. The conductor or its enclosure must be securely fastened to the surface on which it is carried. If the conductor is No. 4 or larger, it requires no protection unless it is exposed to severe physical damage. A No. 6 conductor not exposed to physical damage need not be protected either, and it may be stapled to the building. Conductors smaller than No. 6 or a No. 6 conductor subject to physical damage must be enclosed in a raceway of the type listed in the Code.

If the grounding electrode conductor is enclosed in a metal raceway, the raceway must be electrically continuous from the service equipment to the grounding electrode and it must be connected to the ground clamp or fitting at the electrode. The ground clamp or other means that connects to the grounding electrode itself must be an approved type without solder connections. **250-115**

The Grounding Electrode. The most commonly used grounding electrode for a building has been an underground metal water piping system at least 10 feet long. A recent Code rule change (1978) requires that this electrode be supplemented by an additional electrode, either available on the premises or specially constructed for the purpose (made electrode). An acceptable grounding electrode system including various supplemental electrodes for a building is illustrated in Figure 6-9(a). All the elements of the grounding system must be bonded together as listed in Table 6-15, including the water pipe, other available electrodes, and any made electrodes that are used. The supplemental electrodes at the building served could include the grounded metal frame of the building, steel reinforcing bars or rods at least 20 feet in length and ½ inch or more in diameter encased in 2 inches of concrete, or a length of **250-81(a)**

250-80

Table 6-14. Installation Rules for Grounding Electrode Conductor

Application	Rule	Code Ref.
Requirements	An AC system grounded on the premises must have a grounding electrode conductor connected to a grounding electrode at each service.	250-23

Use of Grounding Electrode Conductor	A grounding electrode conductor must be used to connect
	    a) equipment grounding conductors,
	    b) service-equipment enclosure, and
	    c) grounded conductor (grounded systems)
	to the grounding electrode.
		250-53(a)

Two or More Buildings	Where two or more buildings are supplied by a grounded system from a single service, each building must have a grounding electrode conductor connected to the grounded circuit conductor on supply side of building disconnecting means. (See Code rules for ungrounded systems and exceptions.)	250-24
Separately Derived Systems	A grounding electrode conductor must be used to connect the grounded conductor of the derived system to the grounding electrode. (See Code for detailed installation requirements.)	250-26(b)
Material and Splices	The grounding electrode conductor must be copper, aluminum, or copper-clad aluminum. The conductor must be installed without a splice. Exception: Splices are allowed for busbars or for a tapped conductor to several enclosures.	250-91(a)
Use of Aluminum Conductors	Aluminum or copper-clad aluminum conductors must not be used in direct contact with masonry or earth. Where used outside, these conductors must not be installed within 18 inches of the earth.	250-92(a)

Installation	The grounding electrode conductor installation rules require
	    a) The conductor or enclosure must be securely fastened to surface on which it is carried,
	    b) If No. 4 or larger, protection is required if subject to severe physical damage,
	    c) If No. 6 but free from exposure to physical damage it may be stapled to building without metal covering or protection; otherwise it must be protected as in (d) below,
	    d) If smaller than No. 6 the conductor must be enclosed in rigid metal conduit, intermediate metal conduit, rigid nonmetallic conduit, electrical metallic tubing or cable armor.
		250-92(a)

Connection to Grounding Electrode	The connection to the grounding electrode must be accessible unless the electrode is concrete-enclosed, driven, or buried. The connection must be by means of suitable lugs, pressure connectors, clamps, or other approved means. Solder connections are forbidden. (See Code for detailed connection methods.)	250-112

copper conductor. If none of these additional electrodes is available, a made electrode or other electrodes must be used to supplement the water piping system.

250-83   If none of the above mentioned types of electrodes is on the premises, other electrodes, such as a gas piping system (if approved) or underground metal structures may be used for the grounding electrode as shown in Figure 6-9(b). If a made electrode must be used, the various types of electrodes shown in Figure 6-9(b) are accept-

250-84   able. For these electrodes, the Code specifies a maximum resistance to ground of 25 ohms. If the measured resistance exceeds this value, one additional electrode must be used to augment the grounding electrode system.

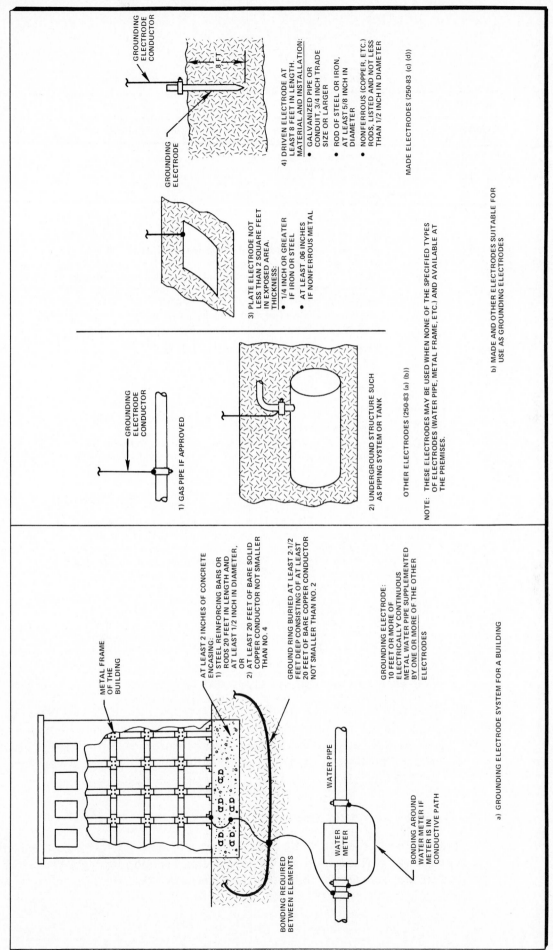

Figure 6-9. Grounding Electrode System

GROUNDING ELECTRODE CONDUCTOR

GROUNDING ELECTRODE

8 FT

4) DRIVEN ELECTRODE AT LEAST 8 FEET IN LENGTH. MATERIAL AND INSTALLATION:
- GALVANIZED PIPE OR CONDUIT, 3/4 INCH TRADE SIZE OR LARGER
- ROD OF STEEL OR IRON, AT LEAST 5/8 INCH IN DIAMETER
- NONFERROUS (COPPER, ETC.) RODS, LISTED AND NOT LESS THAN 1/2 INCH IN DIAMETER

MADE ELECTRODES (250-83 (c) (d))

3) PLATE ELECTRODE NOT LESS THAN 2 SQUARE FEET IN EXPOSED AREA. THICKNESS:
- 1/4 INCH OR GREATER IF IRON OR STEEL
- AT LEAST .06 INCHES IF NONFERROUS METAL

GROUNDING ELECTRODE CONDUCTOR

1) GAS PIPE IF APPROVED

2) UNDERGROUND STRUCTURE SUCH AS PIPING SYSTEM OR TANK

OTHER ELECTRODES (250-83 (a) (b))

NOTE: THESE ELECTRODES MAY BE USED WHEN NONE OF THE SPECIFIED TYPES OF ELECTRODES (WATER PIPE, METAL FRAME, ETC.) AND AVAILABLE AT THE PREMISES.

b) MADE AND OTHER ELECTRODES SUITABLE FOR USE AS GROUNDING ELECTRODES

METAL FRAME OF THE BUILDING

AT LEAST 2 INCHES OF CONCRETE ENCASING:
1) STEEL REINFORCING BARS OR RODS 20 FEET IN LENGTH AND AT LEAST 1/2 INCH IN DIAMETER, OR
2) AT LEAST 20 FEET OF BARE SOLID COPPER CONDUCTOR NOT SMALLER THAN NO. 4

GROUND RING BURIED AT LEAST 2-1/2 FEET DEEP CONSISTING OF AT LEAST 20 FEET OF BARE COPPER CONDUCTOR NOT SMALLER THAN NO. 2

GROUNDING ELECTRODE: 10 FEET OR MORE OF ELECTRICALLY CONTINUOUS METAL WATER PIPE SUPPLEMENTED BY ONE OR MORE OF THE OTHER ELECTRODES

BONDING REQUIRED BETWEEN ELEMENTS

WATER PIPE

WATER METER

BONDING AROUND WATER METER IF METER IS IN CONDUCTIVE PATH

a) GROUNDING ELECTRODE SYSTEM FOR A BUILDING

Table 6-15. Installation Rules for Grounding Electrode

Application	Rule	Code Ref.

Grounding Electrode System	The grounding electrode system must be formed by bonding together the following items where available on the premises:
	a) Metal underground water pipe in direct contact with the earth for 10 feet or more.
	b) Metal frame of building where effectively grounded.
	c) Electrode encased in at least 2 inches of concrete in foundation. The electrode must consist of either:
	    1) 20 feet or more of steel reinforcing bars on rods not less than 1/2 inch in diameter.
	    2) 20 feet of bare solid copper conductor not smaller than No. 4 AWG.
	d) Ground ring encircling building or structure, buried not less than 2 1/2 feet deep. The ring must consist of at least 20 feet of bare copper conductor not smaller than No. 2 AWG.
	The water pipe electrode must be supplemented by an additional electrode of a type specified, or by a made or other electrode.
		250-81

Made and Other Electrodes	Where none of the specified grounding electrodes is available, one or more of the following electrodes can be used.
	a) An electrically continuous metal underground gas piping system if approved by local authorities.
	b) Other metal systems or structures, such as underground tanks
	c) Rod and pipe electrodes—these electrodes must be not less than 8 feet in length and
	    1) Electrodes of pipe or conduit must have corrosion-resistant finish if iron or steel, and must not be smaller than 3/4 inch trade size.
	    2) Rods of steel or iron must be at least 5/8 inch in diameter
	    3) Nonferrous rods must be listed and must be not less than 1/2 inch in diameter (See definition of Listed.)
	    4) The electrode must be buried to a depth of 8 feet. If rock bottom is encountered, the electrode may be driven at an angle not exceeding 45 degrees or be buried in a trench at least 2½ feet deep.
	d) Plate electrodes—Each plate electrode must expose not less than 2 square feet of surface to exterior soil. The thickness must be
	    1) At least 1/4 inch if iron or steel.
	    2) At least 0.06 inch if nonferrous metal.
		250-83

Resistance of Made Electrodes	If a single made electrode does not have a resistance to ground of 25 ohms or less, it must be augmented by one additional electrode.	250-84

6-4 SYSTEMS OPERATING AT OVER 600 VOLTS

ARTICLE 710   The Code provides specific rules for systems operating at more than 600 volts. These rules supplement or modify the rules for general installations. Although a separate Code article is devoted to high-voltage systems, various rules for specific installations are set forth throughout the Code, as listed in Table 6-16. The reader should refer to these Code articles for installation rules applicable to such systems.

6-5 MISCELLANEOUS CIRCUITS AND SYSTEMS

Code references to installation rules for miscellaneous circuits and systems that are not otherwise covered in this Guide are listed on the opposite page. The reader should refer to the Code text in order to gain a complete understanding of the Code rules applicable to these specific circuits and systems.

Circuit or System	Code Article
Outside branch circuits and feeders	225
Temporary wiring	305
Floating buildings	553
Solar Photovoltaic System	690
Emergency systems	700
Standby power generation systems	701, 702
Interconnected electric power production systems	705
Circuits and equipment operating at less than 50 volts	720
Remote control, signaling, and power-limited circuits	725
Fire protective signaling systems	760
Optical Fiber Cables	770
Closed-Loop and programmed power distribution	780

Table 6-16. Installation Rules for Systems Operating at More Than 600 Volts

Coverage	Code Ref.	Coverage	Code Ref.
General rules	Article 710	Pull and Junction Boxes	Article 370 Part D
Definitions	Article 100 Part B	Portable Cables	Article 400 Part C
Installation and Clearance	Article 110 Part B	Electric-discharge Lighting	Article 410 Parts Q, R
Outside Branch Circuits and Feeders	Article 225	Motors	Article 430 Part J
Services	Article 230 Part H	Full-load Current Tables	Tables 430-149, 430,-150
Overcurrent Protection	Article 240 Part H	Transformers Overcurrent Protection	430-3(a)
Grounding (over 1000 volts)	Article 250 Part M	Capacitors	Article 460 Part B
Surge Arresters	Article 280	Resistors and Reactors	Article 470 Part B
Wiring Methods (General)	300-2, 300-3(b) Article 300 Part B	Motion Picture and Television Studios	530-61
Temporary Wiring	Article 305	Electric Signs	Article 600 Part C
		X-ray Equipment	660-4(c)
Conductors	Article 310	Induction and Dielectric Heating Equipment (Access)	665-3, 665-22
Wiring Methods	See 710-2	Swimming Pool (Clearances)	680-8

TEST CHAPTER 6

I. True or False (Closed-Book)

		T	F
1.	If a multiwire branch circuit supplies line-to-line loads, the branch-circuit protective device must open each ungrounded conductor simultaneously.	[ ]	[ ]
2.	Unless it is bare, an equipment grounding conductor must always be identified by a continuous green color or a continuous green color with one or more yellow stripes.	[ ]	[ ]
3.	Receptacle outlets in bathrooms of dwelling units require ground-fault circuit interrupter protection.	[ ]	[ ]
4.	The wall-switch controlled outlet required in dwelling unit kitchens may be a receptacle outlet in lieu of a lighting outlet.	[ ]	[ ]
5.	Service conductors passing over a flat roof must have a clearance of at least 8 feet from the roof.	[ ]	[ ]
6.	The clearance of service-drop conductors over a public street must be 15 feet or more.	[ ]	[ ]
7.	Service fuses may be placed on the supply side of the service disconnecting means.	[ ]	[ ]
8.	The service disconnecting means can consist of up to six switches or circuit breakers.	[ ]	[ ]
9.	All metal enclosures for service conductors and equipment must be grounded.	[ ]	[ ]
10.	A means must be provided for disconnecting the grounded conductor of a service from the premises wiring.	[ ]	[ ]
11.	A grounded circuit conductor may not be used to ground service equipment.	[ ]	[ ]
12.	A single metal underground water pipe may serve as a grounding electrode system.	[ ]	[ ]
13.	A single made electrode with a resistance to ground of 25 ohms must be augmented by an additional electrode.	[ ]	[ ]

II. Multiple Choice (Closed-Book)

1. Receptacle outlets in a dwelling must be installed in habitable rooms so that no point along the floor line is farther from an outlet than:
     (a) 12 feet
     (b) 6 feet
     (c) 5½ feet

1. _____

2. The wall space between two doors in a living room requires a receptacle outlet if the space is wider than:
     (a) 6 feet
     (b) 12 inches
     (c) 2 feet

2. _____

3. From the following list, select the locations that require receptacle outlets: 1. bathroom; 2. outdoors; 3. laundry; 4. basement; 5. attached garage.
     (a) All locations
     (b) 1, 2, 3, 4
     (c) 1, 3, 4, and 5

3. _____

4. Ground-fault circuit interrupter protection is required for all 120-volt, 15- or 20-ampere receptacles installed in the following dwelling unit areas:
     (a) Bathrooms
     (b) Garages (if receptacles are readily accessible)
     (c) Outdoors
     (d) All of the above

4. _____

5. A receptacle controlled by a wall switch is not permitted in lieu of the required wall-switch controlled lighting outlet in the following dwelling unit areas:
    (a) Kitchen
    (b) Bathroom
    (c) Stairway
    (d) All of the above
                                                                    5. _____

6. A receptacle outlet is required above a show window every:
    (a) 12 feet
    (b) 6 feet
    (c) 1½ feet
                                                                   6. _____

7. The minimum clearance required for service-drop conductors above commercial areas subject to truck traffic is:
    (a) 12 feet
    (b) 15 feet
    (c) 18 feet
                                                                  7. _____

8. The grounded (neutral) conductor for a service shall not be smaller than the size of the:
    (a) Grounding electrode conductor
    (b) Equipment grounding conductor
    (c) Ungrounded service conductors
                                                             8. _____

9. Unless the Code permits an exception, an alternating-current circuit must be grounded if such grounding can result in a voltage to ground on the ungrounded conductors that does not exceed:
    (a) 300 volts
    (b) 277 volts
    (c) 150 volts
                                                  9. _____

10. The main bonding jumper must not be smaller than the size of the:
    (a) Grounding electrode conductor
    (b) Equipment grounding conductor
    (c) Equipment bonding jumper connected on load side of service
                                       10. _____

III. Fill in the Blanks *(Closed-Book)*

1. In dwellings the voltage between conductors in a circuit supplying medium-base lampholders must not exceed _____ volts.

2. Required receptacle outlets in a dwelling unit must be located not more than _____ feet above the floor.

3. A receptacle outlet in a kitchen or dining area must be installed at each counter space wider than _____ inches.

4. The minimum height of service-drop conductors above a driveway is _____ feet.

5. Cables other than service-entrance cable may not be installed within _____ feet of grade level.

6. Service raceways must be equipped with a _____ _____ service head.

7. A service head must be located above the point of attachment of service-drop conductors to a building or within _____ inches of the point of attachment.

8. Rod and pipe electrodes must not be less than _____ feet in length.

9. If possible, a rod or pipe electrode must be driven to a depth of _____ feet.

10. A plate electrode must expose not less than _____ square feet of surface to exterior soil.

7

Installation
of General Circuits
and Equipment

**ARTICLE 100
DEFINITIONS***

This chapter deals with the installation of conductors and equipment for general use in electrical systems. Such equipment is normally used to conduct or control electricity or to perform some mechanical function rather than utilize electrical energy for useful purposes. The items considered include cabinets, conduit bodies, cutout boxes, devices, fittings, raceways, and switches. Distribution equipment and utilization equipment are covered in later chapters of the Guide.

For convenience, the Code rules for general installation can be separated as shown in Figure 7-1 into general rules, wiring rules, and rules for equipment and devices. These rules apply to all installations unless modified by other rules dealing with specific circuits, equipment, or occupancies.

The chapter divides the discussion of installation rules as follows: (a) use of conductors in circuits, (b) installation and protection of conductors, (c) wiring methods, (d) equipment grounding and bonding, and (e) equipment and devices. In each section summary tables and discussions concentrate on the basic installation rules. Information of primary concern to manufacturers is not covered except as it aids in understanding the intent of the Code. In many cases, an abbreviated discussion,

*References in the margins are to the specific applicable rules and tables in the *National Electrical Code*.

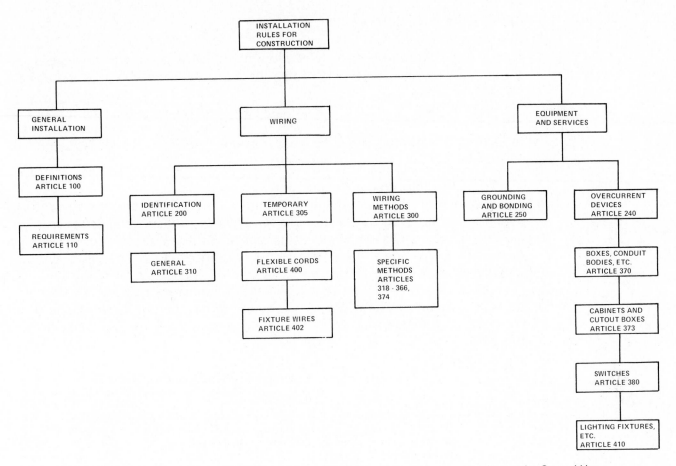

Figure 7-1. Organization of Code Rules for the Installation of Conductors and Equipment for General Use

or no mention at all, is given of certain Code rules that are considered less general than the ones covered but may nevertheless be of interest in a specific situation. The reader, therefore, is urged to study the Code text completely even though the Guide covers only a portion of the material presented there.

7-1 USE OF CONDUCTORS IN CIRCUITS

When current-carrying conductors are used in an electrical system, the Code requires that each conductor be protected from damage by an overcurrent protective device such as a circuit breaker or fuse. The conductors must also be identified so that the ungrounded conductors may be distinguished from the grounded and grounding conductors. Minimum sizes or required ampacity for any of these conductors used in a circuit are selected based on Code rules and tables. The Code also provides tables that list the physical and electrical properties of conductors to allow the designer to select the proper conductor for any application.

ARTICLE 240

7-1.1 Overcurrent Protection of Conductors

Table 7-1 summarizes the Code rules for protecting conductors from excess current caused by overloads, short circuits, or ground faults. The setting or sizes of the protective device are based on the ampacity of the conductors as listed in appropriate Code tables. Under certain conditions, the overcurrent device setting may be larger than the ampacity rating of the conductors as listed in the exceptions to the

240-3

basic rule. For convenience, a standard rating of a fuse or circuit breaker may be used even if this rating exceeds the ampacity of the conductor if the rating does not exceed 800 amperes.

240-21   In most cases, an overcurrent device must be connected at the point where the conductor to be protected receives its supply. The most common situations are shown in Figure 7-2 which illustrates the basic rule and several exceptions, including the 10-foot tap rule and the 25-foot tap rule.

7-1.2 Identification of Conductors

200-2   Premises wiring systems require a grounded conductor in most installations. A grounded conductor, such as the neutral, or a grounding conductor must be identified either by the color of its insulation, by markings at the terminals, or by other suitable means. Table 7-2 summarizes the rules for identification of both grounded and grounding conductors.

200-6,
310-12   Unless allowed by exception in the Code, a *grounded conductor* must have a white or natural gray finish. When this is not practical for conductors larger than No. 6, marking the terminals with white color is an acceptable method of identifying the conductor.

Table 7-1. Overcurrent Protection of Conductors

Application	Rule	Code Ref.
Protection of Conductors	Conductors, other than flexible cords and fixture wires, must be protected against overcurrent in accordance with their ampacities as specified in Code tables.	240-3 Chapter 3 Tables
	Exceptions: a) Remote-control circuits, transformer secondary conductors, capacitor circuits, circuits for welders, and circuits that cannot allow a power loss have special rules which apply. b) If the ampacity of a conductor does not correspond with the standard ampere rating of a fuse or a circuit breaker, the next higher standard device rating may be used only if this rating does not exceed 800 amperes and multioutlet branch circuits are not supplied. c) Tap conductors may be protected according to other Code rules. d) Motor and motor-control circuits may be protected according to other Code rules.	
Overcurrent Device Required	A fuse or overcurrent trip unit of a circuit breaker is required to be connected in series with each ungrounded conductor.	240-20(a)
	Note: Except as specifically allowed in the Code, no overcurrent device may be connected in series with any grounded conductor.	240-22
Location in Circuit	An overcurrent device must be connected at the point where the conductor receives its supply.	240-21

	Exceptions:
	a) Service-entrance conductors, branch-circuit taps, motor-circuit taps, busway taps, generator conductors, and taps in high-bay buildings may be protected as specified in the Code.
	b) Protection at the source of supply is not required for the following tap connections:
	     1) Where the overcurrent device protecting the larger conductor also protects the smaller conductor.
	     2) Feeder taps not over 10 feet long when all of the conditions listed in the Code are met.
	     3) Feeder taps not over 25 feet long when all of the conditions listed in the Code are met.
	     4) Transformer feeder taps with primary plus secondary not over 25 feet long when the conditions listed in the Code are met.	

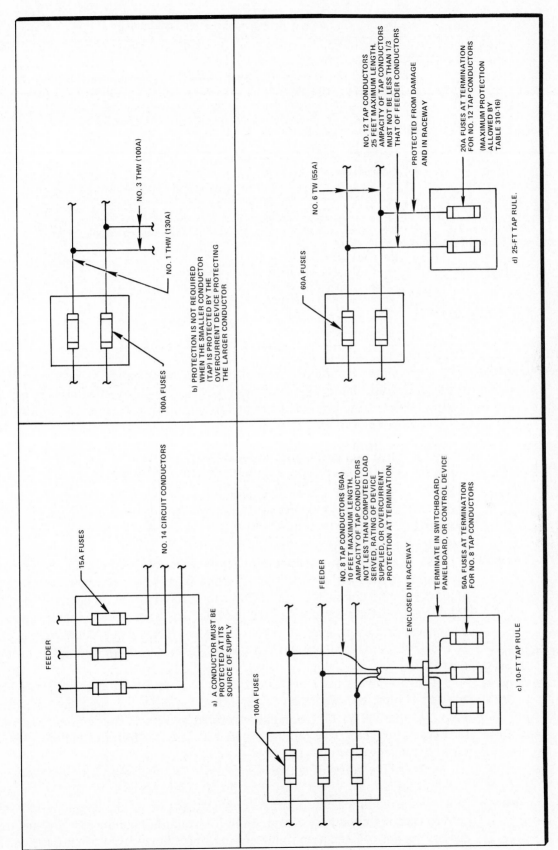

Figure 7-2. Location of Overcurrent Protection in Circuits

NO. 3 THW (100A)

NO. 1 THW (130A)

100A FUSES

b) PROTECTION IS NOT REQUIRED
WHEN THE SMALLER CONDUCTOR
(TAP) IS PROTECTED BY THE
OVERCURRENT DEVICE PROTECTING
THE LARGER CONDUCTOR

NO. 12 TAP CONDUCTORS
25 FEET MAXIMUM LENGTH.
AMPACITY OF TAP CONDUCTORS
MUST NOT BE LESS THAN 1/3
THAT OF FEEDER CONDUCTORS

PROTECTED FROM DAMAGE
AND IN RACEWAY

20A FUSES AT TERMINATION
FOR NO. 12 TAP CONDUCTORS
(MAXIMUM PROTECTION
ALLOWED BY
TABLE 310-16)

NO. 6 TW (55A)

60A FUSES

d) 25-FT TAP RULE.

15A FUSES

FEEDER

NO. 14 CIRCUIT CONDUCTORS

a) A CONDUCTOR MUST BE
PROTECTED AT ITS
SOURCE OF SUPPLY

FEEDER

NO. 8 TAP CONDUCTORS (50A)
10 FEET MAXIMUM LENGTH.
AMPACITY OF TAP CONDUCTORS
NOT LESS THAN COMPUTED LOAD
SERVED, RATING OF DEVICE
SUPPLIED, OR OVERCURRENT
PROTECTION AT TERMINATION.

ENCLOSED IN RACEWAY

TERMINATE IN SWITCHBOARD,
PANELBOARD, OR CONTROL DEVICE

50A FUSES AT TERMINATION
FOR NO. 8 TAP CONDUCTORS

100A FUSES

c) 10-FT TAP RULE

223

Table 7-2. Identification of Conductors

Application	Rule	Code Ref.

Identification of Grounded Conductor	a) No. 6 or smaller—An insulated grounded conductor must be identified by a continuous white or natural gray outer finish along its entire length.
	[Note: Exceptions are listed in the Code for both sections referenced.]
	b) Sizes larger than No. 6—Such an insulated grounded conductor must be identified by either
	    1) a continuous white or natural gray finish, or
	    2) distinctive white markings at its terminations
	(See exception in Code.)
	c) Identification for the grounded conductor of a branch circuit is required.
	d) In flexible cords, the intended grounded conductor must be identified.
		200-6(a)
		See also:
		310-12(a)
		200-7
		200-6(b)
		
		210-5(a)
		200-6(c)

Identification of Equipment Grounding Conductor	a) An equipment grounding conductor may be bare, covered, or insulated. If covered or insulated, identification is required by a continuous outer finish that is either
	    1) green or
	    2) green with one or more yellow stripes
	Exceptions are allowed for conductors larger than No. 6 and in other circumstances listed in the Code.
		310-12(b)
		See also:
		210-5(b)

Ungrounded Conductors	Ungrounded conductors must be clearly distinguishable from grounded and grounding conductors.	310-12(c)

310-12(b)
310-12(c)
A *grounding conductor,* unless it is bare, must be identified by a green outer finish or a finish that is green with yellow stripes. Ungrounded conductors must be distinguishable from grounded conductors and must not use white, natural gray, or green colors or markings.

7-1.3 Size and Ampacity of Conductors

310-5
The minimum size conductors allowed for general wiring are No. 14 copper or No. 12 aluminum or copperclad aluminum. Smaller conductors are allowed in special circumstances as listed in Table 7-3. In most cases, however, conductors are selected based on the required ampacity to supply a given load rather than to meet the minimum size requirement.

310-15
The maximum continuous ampacity of copper, aluminum, and copperclad aluminum conductors are specified in Code tables that relate the AWG size to the ampacity of the conductor. In addition to specifying the ampacity of a given size conductor, separate Code tables are provided for enclosed conductors and for conductors exposed in air. The tables for "Not More Than Three Conductors in Raceway in Free Air" are normally used in design problems to determine the proper conductor size.

TABLES 310-16, 310-18

TABLES 310-18, 310-19
If high-temperature (110°C to 250°C) conductors are used, separate tables of ampacities must be referenced to determine the allowable ampacity. For example, a No. 14, type THW, 75°C conductor is rated to carry 20 amperes in a raceway. The

TABLE 310-18
same size conductor with asbestos and varnished cambric insulation (AVA) may carry 29 amperes continuously.

Notes to the tables of ampacities modify the stated ampacities of conductors in certain installations. When more than three conductors are installed in a raceway or

NOTE 8
cable, the listed ampacities must be reduced because of the heating effect of many current-carrying conductors in proximity. Grounding conductors and neutral con-

NOTE 10
NOTE 11
ductors are not counted as current-carrying conductors except when electric discharge lighting is served, in which case the neutral is counted.

The ampacity stated in the tables must also be reduced if the ambient temperature for the conductor location exceeds 30°C (86°F). This reduction is required even if the reduction for more than three conductors in a raceway is also applied. Thus, if six No. 10 type TW conductors carry current in a raceway where the ambient temperature is 40°C, the ampacity of 30 amperes must be reduced to 80% because of conduit fill and then reduced again by a correction factor of .82 because of the ambient. The allowable ampacity is then only .8 × .82 = .656 or 65.5% of the original value. The ampacity for six No. 10 type TW conductors in a 40°C ambient becomes

TABLE 310-16

$$30 \ A \times .8 \times .82 = 19.68 \ A$$

In the case of residential services consisting of a three-wire, single-phase circuit, the ampacity listed in the Code tables may be increased for certain types of conductors. In practice, the local enforcing agency determines whether or not this increase in ampacity is allowed and in some cases prohibits it. For installations where raceway or cables are buried in earth, specific Code tables must be used to determine conductor ampacity according to appropriate conditions. If the circuits operate at over 2000 volts, appropriate Code tables of ampacities must be used.

NOTE 3

Table 7-3. Size and Ampacity of Conductors

Application	Rule	Code Ref.
Minimum Size	Whether solid or stranded, conductors must not be smaller than No. 14 copper or No. 12 aluminum or copper-clad aluminum. Exceptions: Flexible cords, fixture wires and other special conductors listed in the Code.	310-5
Ampacity from Code Tables	The maximum continuous ampacities for conductors are specified in the Code for the following cases: a) Cooper conductors in a raceway in free air (60°C to 90°C) b) Copper conductors in free air (60°C to 90°C) c) Aluminum or copper-clad aluminum conductors in a raceway in free air, or directly buried (60°C to 90°C) d) Aluminum or copper-clad aluminum conductors in free air e) Conductors with a rating of 110°C to 250°C in raceway f) High temperature conductors in free air Note: The tables for conductors in a raceway or cable apply for three or fewer current-carrying conductors so enclosed. The ambient temperature must be less than that specified in Code tables unless correction factors are applied.	Table 310-16 Table 310-17 Table 310-16 Table 310-17 Table 310-18 Table 310-19
Reduction in Ampacity	A reduction in ampacity is required in the following cases: a) When there are three or more conductors in a raceway or cable excluding: 1) a neutral in a balanced circuit not supplying electric discharge lighting, and 2) a grounding conductor (See exceptions in Code.) b) If the ambient temperature is above 30°C (86°F), correction factors given in the Code must be used.	Notes to Tables of Ampacities
Dwelling Services	An increase in ampacity is permitted for certain conductors used in residential occupancies when all of the conditions listed in the Code are met. Note: Other rules are presented in the Code by notes to the tables of ampacities.	Note 3

7-1.4 Properties of Conductors

Various Code tables listed in Table 7-4 define the physical and electrical properties of conductors. Electrical designers use the Code tables of physical properties to select the type of conductors and the size of conduit required to enclose the conductors in a specific application.

CHAPTER 9
TABLE 8

In other design problems the dc resistance and wire area are used to solve problems involving voltage drop. When large conductors are used in ac circuits, the *inductive reactance* increases the effective impedance and hence the voltage drop for a given length of wire. Table 7-4 also summarizes the properties of a No. 4/0 type THW copper conductor as an example and to show how these properties are used.

Table 7-4. Tables of Properties for Conductors

Application	Rule	Code Ref.
Properties of Conductors	Code tables tabulate properties of conductors as follows: a) Name, operating temperature, application, and insulation b) Dimensions of insulated conductors c) Physical properties and electrical resistance d) AC resistance and reactance	Table 310-13 Chapter 9 Tables 5, 6, 7 Table 8 Table 9

Example	Summary of properties for a No. 4/0 type THW copper conductor:
	a) Insulation
	  1) Moisture and heat resistant, thermoplastic
	  2) Maximum operating temperature 75° C
	  3) Allowed in dry and wet locations
	  4) Thickness 80 mils
	b) Dimensions
	  1) Diameter 0.705 in.
	  2) Area 0.3904 sq. in.
	c) Conductor Properties
	  1) Area 211600 cir. mils
	  2) No. of strands 19
	  3) DC resistance 0.0608 ohms per 1000 ft.
		Table 310-13
		
		Chapter 9
		Table 5
		
		Chapter 9
		Table 8

7-1.5 Miscellaneous Types of Conductors

ARTICLE 400,
ARTICLE 402

Flexible cords and cables and fixture wires are subject to special rules presented in the Code. The use of these conductors in circuits is restricted to the connections listed by the Code.

7-2 INSTALLATION AND PROTECTION OF CONDUCTORS

Conductors must be installed and protected from damage according to rules set forth in the Code. Additional rules specify the use of boxes or fittings for certain connections, specify how connections are made to terminals, and restrict the use of parallel conductors. When conductors are installed in enclosures or raceways, additional rules apply. Finally, if conductors are installed underground, the burial depth and other installation requirements are specified by the Code.

This section covers the rules that determine the method of installation and protection of conductors under various conditions. If the conductors form a specific circuit, such as a branch circuit, feeder, or service, additional rules may apply as discussed in Chapter 6 of the Guide.

7-2.1 General Rules for Conductor Installation

Table 7-5 lists the general rules that govern the installation of conductors. Insulated conductors that are identified for use in the intended location must be used as required by the Code. In wet locations, for example, only certain types of conductors specified in the Code such as type RHW may be used since the insulation of these conductors is able to resist water damage (as indicated by the letter W in the type designation).

110-11
310-8

Cables and conductors must also be adequately protected from physical damage and must have special protection when they pass through bored holes, notches in wood, or holes in metal frames. A steel plate at least $\frac{1}{16}$ inch in thickness must be installed to protect cables likely to be damaged from nails or screws that might accidently penetrate the cable.

300-4

Table 7-5. General Rules for Installation of Conductors

Application	Rule	Code Ref.
Conductors Subject to Deteriorating Effects	Conductors used in locations containing deteriorating agents or excessive temperatures must be indentified for use in such locations.	110-11, 310-8 310-9
Insulation	Conductors must be insulated unless specifically permitted to be bare by the Code.	310-2

Protection against Physical Damage	Cables and conductors must be installed and protected as specified in the Code when they pass through structural supports as follows:
	    a) Through bored holes in wood—hole to be bored in approximate center of joists and rafters. Holes in studs at least 1 1/4 in. from edge or protected by 1/16-in. steel plate
	        b) Through notches in wood—protected by 1/16-in. steel plate
	        c) Through holes in metal—protected by bushing or gromets.
	Protected from nails or screws by 1/16-in. steel plate
		300-4

Boxes or Fittings Required:	a) A box or fitting must be installed at:
	    1) each conductor splice point
	    2) each outlet, switch point, or junction point
	    3) each pull point for the connection of conduit and other raceways
	Exceptions: A box or fitting is not required for conductor splices in raceways or conduit bodies having removable covers accessible after installation.
	b) A box is required for wiring systems employing cables at:
	    1) each conductor splice point
	    2) each outlet, switch point, junction point, or pull-point
	    3) the connection point between cable system and raceway system
	(See exceptions in Code.)
		300-15

Change in Wiring Method	Boxes or other fittings are required when a change is made from conduit to open wiring.	300-16
Wiring in Ducts	Special rules apply for wiring in ducts, plenums, and other air-handling spaces.	300-22
Connection to Terminals	Connection of conductors to terminals must be made by means of pressure connectors, solder lugs, or splices to flexible leads. Exception: Connection by means of wire binding screws or studs and nuts having upturned lugs is permitted for No. 10 or smaller conductors.	110-14(a)
Conductors in Parallel	Conductors of size 1/0 and larger may be connected in parallel when the conductors are of the same length, same material, same area, same insulation, and the same termination. Exception: Traveling cables and some control circuits.	310-4

300-15   Whenever conductors are spliced, connected, or pulled, a box or fitting must be installed for the connection of conduit and other raceways. If a cable system is to be connected to a raceway system, a box is required at the point of connection.

110-14(a)   Connection of conductors to terminals must be made with pressure connectors, solder lugs, or splices if the conductors are larger than No. 10. Conductors of size No. 10 or smaller may be connected by wire binding screws or studs and nuts.

310-4   Conductors of size 1/0 and larger run in parallel must all have the same construction and electrical properties. Smaller size conductors are not permitted to be run in parallel in power circuits.

7-2.2 Conductors in Enclosures and Raceways

In many installations, conductors are routed in a raceway system for convenience and to provide physical protection. The circuits are connected in splice boxes and terminate at outlets or other enclosures such as panelboards. In such installations, the various Code rules listed in Table 7-6 apply.

300-3   Conductors of systems operating at different voltages up to 600 volts may occupy the same enclosure, cable, or raceway where all conductors are insulated for the maximum voltage.

300-13   When conductors are installed, they must be mechanically and electrically continuous between outlets, devices, etc.; splices are not permitted within a raceway itself except for wiring methods that specifically permit such splices. At each outlet or

300-14   switch point where splices are permitted or fixtures are to be connected there must be at least 6 inches of conductor left free for the connection.

Table 7-6. Installation of Conductors in Enclosures and Raceways

Application	Rule	Code Ref.
Conductors of Different Systems	a) Conductors of different systems operating at 600 volts or less may occupy the same enclosure, cable, or raceway only if all conductors are insulated for the maximum voltage of any conductor. b) Conductors of systems operating at over 600 volts cannot occupy the same enclosure, cable, or raceway with conductors operating at 600 volts or less except as specifically permitted.	300-3
Continuity	Conductors must be continuous between outlets, devices, etc. No splice or tap is allowed within a raceway except as specifically permitted.	300-13
Length of Free Conductor	At least 6 inches of free conductor must be left at each outlet and switch point for splices or connections.	300-14
Induced Currents	Where conductors carrying alternating current are installed in metal enclosures or raceways, all phase conductors, the neutral, and all equipment grounding conductors must be grouped together.	300-20
Inserting Conductors in Raceway	Raceways must be installed as a complete system before conductors or pull wires are inserted. Exception: Exposed raceways having a removable cover or capping.	300-18
Number and Size of Conductors	The Code limits the number and size of conductors in specific types of raceways.	300-17
Support of Conductors in Vertical Raceways	Conductors in vertical raceways must be supported at the top and at specified intervals.	300-19
Use of Stranded Conductors	Where installed in raceways, conductors of size No. 8 and larger must be stranded. See exceptions in Code.	310-3

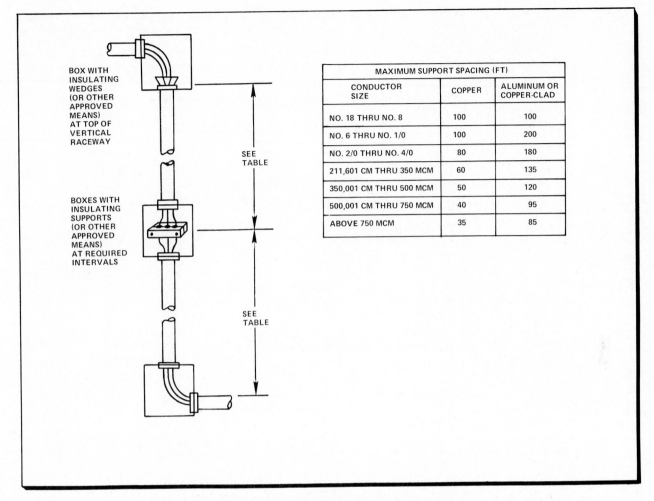

MAXIMUM SUPPORT SPACING (FT)		
CONDUCTOR SIZE	COPPER	ALUMINUM OR COPPER-CLAD
NO. 18 THRU NO. 8	100	100
NO. 6 THRU NO. 1/0	100	200
NO. 2/0 THRU NO. 4/0	80	180
211,601 CM THRU 350 MCM	60	135
350,001 CM THRU 500 MCM	50	120
500,001 CM THRU 750 MCM	40	95
ABOVE 750 MCM	35	85

Figure 7-3. Support of Conductors in Vertical Raceway

Unless a raceway is exposed with a removable cover, a raceway system must be **300-18** installed as a complete system without conductors. Pull wires may then be installed and the conductors pulled in as long as the number or size of conductors does not exceed that which allows dissipation of heat and safe installation or withdrawal of the conductors. The Code specifies the number and size of conductors that may be **300-17** used with a particular type of raceway of a given size. General conductors of size **310-3** No. 8 or larger must be stranded when installed in raceways.

As shown in Figure 7-3, conductors in vertical raceways must be supported at **300-19** the top and at fixed intervals. An approved method using clamps or other support is required except for short runs of lengths less than 25% of the distance between supports given in the figure.

7-2.3 Underground Installations

Conductors and cables installed underground must be listed for such use and be **310-8** protected from physical damage. The rules for underground installations and the minimum burial depths are listed in Table 7-7.

Table 7-7. Rules for Underground Installations

Application	Rule	Code Ref.
General	Conductors and cables used underground must be listed for such use.	310-8
Minimum Cover Requirements (600 V or less)	A minimum burial depth is required for the conditions indicated as follows:	300-5(a)

Conditions	Burial Depth	
Direct buried cable	24 inches	
Rigid metal conduit, Intermediate metal conduit	6 inches	
Rigid nonmetallic conduit and other approved raceways	18 inches	
Residential branch circuit rated 300 volts or less and 30 amperes or less	12 inches	EXCEPTION 4
Areas subject to heavy vehicular traffic	24 inches	EXCEPTION 3

Note: Other exceptions are listed in the Code. If a 2-inch thick concrete pad covers the installation, the burial depth may be reduced by 6 inches.

Application	Rule	Code Ref.
Minimum Cover Requirements (over 600 V)	The Code specifies minimum cover requirements for underground installations operating at over 600 volts.	710-3(b)
Under Buildings	Underground cable installed under a building must be in a raceway that is extended beyond the outside walls of the building.	300-5(c)
Protection of Conductors Emerging from Ground	a) Conductors emerging from the ground must be enclosed in enclosures or raceways approved for the purpose. The enclosures or raceways must extend from below the ground line to a point 8 feet above finished grade. b) Conductors subject to damage must be in rigid metal conduit, intermediate metal conduit, PVC schedule 80 (polyvinyl chloride, extra-heavy wall), or equivalent.	300-5(d)
Splices and Taps	Underground splices or taps without the use of splice boxes are permitted when made in an approved manner.	300-5(e)
Placement of Circuit Conductors	All conductors of the same circuit including any grounding conductors must be installed in the same raceway or in close proximity in the same trench.	300-5(i)
Miscellaneous	The Code also specifies rules for grounds, backfill, and seals for raceways of underground installations.	300-5

7-3 WIRING METHODS AND TECHNIQUES

110-8 The Code recognizes various wiring methods that may be properly installed in buildings or other occupancies. Table 7-8 lists the specific methods and techniques covered in the Code. For convenience, the wiring methods are separated into those using cables and conductors, those using raceways, and those using other methods. Selected wiring methods are discussed in this section of the Guide and summary tables are presented indicating the description, use, and installation rules for the more common methods.

7-3.1 Wiring Methods Using Conductors and Cables

In addition to the general rules for the installation of cables and conductors, specific wiring methods are treated in separate Code articles. Table 7-9 presents a brief summary of the Code rules for a few selected wiring methods in common use.

7-3.2 Wiring Methods Using Raceways

In industrial and commercial construction, raceways are used to contain conductors and cables for electrical distribution to panelboards, switchboards, and outlets. Each type of raceway is approved for certain uses with installation rules given in the Code that apply to each type. The most common raceways are of the conduit type consisting of circular metallic or nonmetallic tubes that enclose the conductors. Other types of raceways have very specific uses as outlined in the Code.

Conduit-Type Raceways. Table 7-10 summarizes the rules for raceways that normally come in standard lengths of circular tubing. The basic raceway is rigid   **ARTICLE 346** metal conduit that provides maximum protection for conductors in any installation. Because of its high cost and because it is difficult to install, other types of similar raceways have been developed that are less expensive, lighter in weight, and easier to install.

The similar rules of installation for intermediate metal conduit, rigid metal conduit, rigid nonmetallic conduit, and electrical metallic tubing are summarized in Table 7-11. The rules for flexible metal conduit differ significantly from those of the   **ARTICLE 350** other types and are not included in the table.

Table 7-8. Summary of Wiring Methods and Techniques

Conductors and Cables	Code Ref.	Raceways	Code Ref.
Open Wiring on Insulators	Article 320	Electrical Nonmetallic Tubing	Article 331
Messenger Supported Wiring	Article 321	Intermediate Metal Conduit	Article 345
Concealed Knob-and-Tube Wiring	Article 324	Rigid Metal Conduit	Article 346
Integrated Gas Spacer Cable	Article 325	Rigid Nonmetallic Conduit	Article 347
Medium Voltage Cable	Article 326	Electrical Metallic Tubing	Article 348
Flat Conductor Cable	Article 328	Flexible Metallic Tubing	Article 349
Mineral-Insulated Metal-Sheathed Cable	Article 330	Flexible Metal Conduit	Article 350
Armored Cable	Article 333	Liquidtight Flexible Metal Conduit	Article 351
Metal-Clad Cable	Article 334	Surface Raceways	Article 352
Nonmetallic-sheathed Cable	Article 336	Underfloor Raceways	Article 354
Shielded Nonmetallic-Sheathed Cable	Article 337	Cellular Metal Floor Raceways	Article 356
Service-Entrance Cable	Article 338	Cellular Concrete Floor Raceways	Article 358
Underground Feeder and Branch-Circuit Cable	Article 339	Wireways	Article 362
Power and Control Tray Cable	Article 340	Busways	Article 364

Other Methods or Techniques	Code Ref.
Cable Trays	Article 318
Nonmetallic Extensions	Article 342
Underplaster Extensions	Article 344
Multioutlet Assembly	Article 353
Flat Cable Assemblies	Article 363
Cablebus	Article 365
Auxiliary Gutters	Article 374

Table 7-9. Installation Rules for Selected Wiring Methods Using Cables or Conductors

Application	Open Wiring on Insulators	Mineral-Insulated Metal-Sheathed Cable (MI)	Armored Cable (AC)	Nonmetallic-Sheathed Cable (NM)	Service-Entrance Cable (SE)
Description	An exposed wiring method using insulators for support of single conductors run in or on buildings.	A factory assembly of one or more conductors with an insulation and sheath that is liquid tight and gas tight.	An assembly of insulated conductors in a flexible metallic enclosure. Example: BX cable	A factory assembly of two or more conductors having an outer sheath. Example: ROMEX	A single or multiconductor cable used primarily for services.
Uses Permitted	May be used for: a) 600 volt or less systems b) indoors or outdoors c) wet or dry locations d) where subject to corrosive vapors e) for services	In all locations except where subject to destructive corrosive conditions.	In dry locations where not subject to physical damage. Use is prohibited in certain locations.	In one- and two-family dwellings, multifamily dwellings and other structures. Use is restricted to normally dry locations and structures not over three floors.	As service-entrance cable or for branch circuits or feeders when grounded conductor is insulated.
Support	Rigidly supported: a) within 6 in. from a tap or splice b) within 12 in. of connection to outlet c) at intervals not exceeding 4-1/2 ft (with exceptions)	Unless in a raceway (fished-in) support is required at intervals not exceeding 6 ft.	Secured by approved means: a) at intervals not exceeding 4-1/2 ft b) within 12 ft in. of box or fitting (with exceptions)	Unless concealed, must be secured: a) at intervals not exceeding 4-1/2 ft b) within 12 in. of cabinet, box or fitting.	If cable is unarmored, the provisions for Type NM cable apply.
Installation	Protection is required for conductors within 7 ft of floor. Special rules apply for conductors in finished attics and roof spaces.	Bends must not damage cable; radius of at least 5 times cable diameter.	Bends must not injure cable; radius of at least 5 times cable diameter.	Special rules apply: a) for exposed work b) in unfinished basements c) in accessible attics	May be used to supply appliances if temperature limitations are observed.

Table 7-10. Rules for Installation of Conduit-Type Raceways

Application	Intermediate Metal Conduit	Rigid Metal Conduit	Rigid Non-Metallic Conduit	Electrical Metallic Tubing	Flexible Metal Conduit
Description	Lightweight steel conduit with thin walls as compared to rigid metal conduit.	Steel or aluminum conduit with heavy walls for maximum protection of conductors. The steel conduit may be galvanized and used where exposed to weather conditions.	Resistance to moisture and chemical atmospheres. A common type is heavy-wall polyvinyl chloride (PVC). Extra-heavy wall (schedule 80) must be used if PVC is subject to physical damage.	Lightweight conduit with thin walls as compared to rigid metal conduit. Special couplings are used since it is not threaded.	Consists of a strip of aluminum or steel wound in a spiral (along its length).
Uses Permitted	May be used under all atmospheric conditions and occupancies except as specifically prohibited. Permitted in or under cinder fill if approved for the purposes and encased in 2 in. of concrete or is below 18 in.	May be used under all atmospheric conditions and occupancies except as specifically prohibited. Rigid metal conduit protected from corrosion solely by enamel is permitted indoors only.	a) May be used to enclose conductors operating at 600 volts or less in locations except as specifically prohibited. b) For circuits operating at over 600 volts, the conduit must be encased in not less than 3 in. of concrete if underground.	Permitted for both exposed and concealed work. Permitted in or under cinder fill if approved for the purpose and encased in 2 in. of concrete or is below 18 in.	May be used in locations not prohibited by the Code. If used in wet locations, the enclosed conductors must be approved for the purpose. In hazardous locations, special flexible metal conduit approved for the purpose must be used.
Use Not Permitted	The conduit must not be used in concrete, in direct contact with the earth, or in corrosive atmospheres unless suitable for condition.	Rigid metal conduit must not be installed in concrete or in direct contact with the earth unless corrosion protected and suitable for condition.	Rigid nonmetallic conduit is not permitted in hazardous locations except for underground wiring in service stations and bulk-storage plants.	The tubing must not be used where subject to severe physical damage.	The conduit shall not be used as follows: a) In hoistways except as permitted b) In storage-battery rooms c) In hazardous locations except as permitted d) Underground or embedded in poured concrete
Trade Size	1/2 in. to 4 in.	1/2 in. to 6 in. (with exceptions)	1/2 in. to 6 in.	1/2 in. to 4 in. (with exceptions)	1/2 in. minimum (with exceptions)

Table 7-11. Rules for the Installation of Conduits and Tubing
(Except Flexible Types)

Application	Rule	Code Ref.
Number of Bends in One Run	Not more than four quarter bends (360° degrees total), including bends at the outlet or fitting.	345-11, 346-11 347-14, 348-10
Splices and Taps	Splices and taps must be made only in junction, outlet boxes or conduit bodies. Conductors, including splices and taps, must not fill a conduit body to more than 75 percent of its cross-sectional area.	345-14, 346-14 347-16, 348-14
Number of Conductors in Conduit or Tubing	The number of conductors as a percentage fill is limited by the Code Table 1, Chapter 9.	345-7, 346-6 347-11, 348-6

Every raceway must be supported at fixed intervals as specified in the Code. Figure 7-4 illustrates the support distances for the raceways listed in Table 7-10.

Other Raceways. In addition to conduit-type raceways, several other wiring methods using raceways are permitted by the Code. The rules for underfloor raceways, wireways, and busways are summarized in Table 7-12 to describe only three of the approved methods.

ARTICLE 354 The underfloor raceway system consists of ducts connected by special junction boxes. Floor outlets are easily attached by using special inserts set or screwed into the walls of the ducts. The raceways must be laid in straight lines and must be held firmly in place. When an outlet is removed, the circuit wiring must also be removed from the raceway.

ARTICLE 362 Wireways are installed exposed to allow access to conductors or cables at all points for tapping, splicing, etc., without disturbing existing wiring. The wireway must be supported at least every 5 feet unless approved for greater distances between supports or run vertically.

ARTICLE 364 Busways are used primarily to provide a flexible distribution system within a building (although approved outdoor types are available). A common use is in machine shops to supply power for power tools or machines that may be moved occasionally. A plug-in overcurrent device with an externally operable circuit breaker or switch is used to tap a branch circuit, with a power cord serving as the branch circuit conductors from the plug-in unit to the outlet for the machine.

7-3.3 Other Wiring Methods and Techniques

To meet particular installation requirements, the Code allows a number of special wiring methods and techniques in addition to the common methods previously **ARTICLE 374** discussed. A typical example of such special techniques is the use of auxiliary gutters. An auxiliary gutter is used to supplement wiring spaces at meter centers and similar locations. The gutters are sheet-metal troughs in which conductors may be routed after the gutter has been installed. In multifamily dwellings, for example, auxiliary gutters are used to enclose the wiring, including splices and taps, for individual metering of apartments.

The auxiliary gutter may extend not more than 30 feet beyond the equipment that it supplements (except for elevators). The trough may not contain more than 30 current-carrying conductors and may not be filled to more than 20% of the

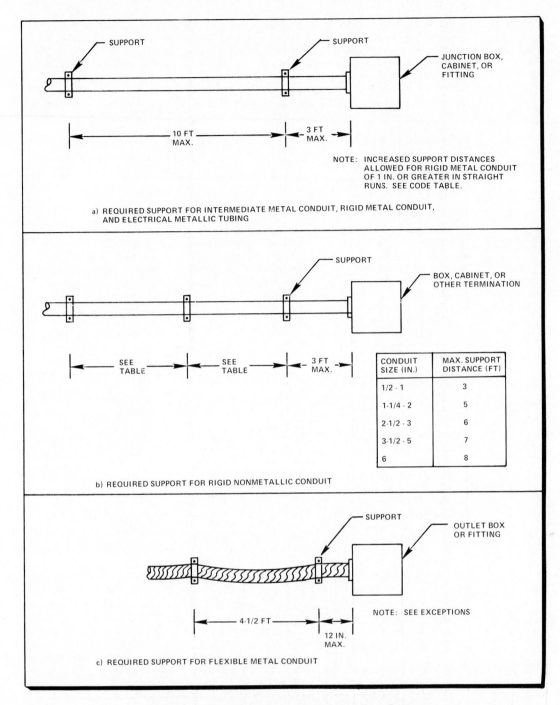

a) REQUIRED SUPPORT FOR INTERMEDIATE METAL CONDUIT, RIGID METAL CONDUIT, AND ELECTRICAL METALLIC TUBING

NOTE: INCREASED SUPPORT DISTANCES ALLOWED FOR RIGID METAL CONDUIT OF 1 IN. OR GREATER IN STRAIGHT RUNS. SEE CODE TABLE.

SUPPORT

SUPPORT

JUNCTION BOX, CABINET, OR FITTING

10 FT MAX.

3 FT MAX.

SUPPORT

BOX, CABINET, OR OTHER TERMINATION

SEE TABLE

SEE TABLE

3 FT MAX.

CONDUIT SIZE (IN.)	MAX. SUPPORT DISTANCE (FT)
1/2 - 1	3
1-1/4 - 2	5
2-1/2 - 3	6
3-1/2 - 5	7
6	8

b) REQUIRED SUPPORT FOR RIGID NONMETALLIC CONDUIT

SUPPORT

OUTLET BOX OR FITTING

NOTE: SEE EXCEPTIONS

4-1/2 FT

12 IN. MAX.

c) REQUIRED SUPPORT FOR FLEXIBLE METAL CONDUIT

Figure 7-4. Required Support for Conduit-Type Raceways

interior cross-sectional area (with certain exceptions). When splices or taps are made, the total area of conductors, including splices and taps, must not exceed 75% of the area of the gutter.

7-4 EQUIPMENT GROUNDING AND BONDING

The grounding and bonding of the noncurrent-carrying metal parts of equipment is treated extensively in the Code. Grounding is required for most equipment unless excepted or specifically prohibited in the Code. Bonding is normally only required to assure electrical continuity when the continuity may be broken by connections, loose joints, or the like.

ARTICLE 250

Table 7-12. Installation Rules for Underfloor Raceways, Wireways, and Busways

Application	Underfloor Raceways	Wireways	Busways
Description	Underfloor raceways are used beneath the surface of a floor. The covering required depends on size of raceway. Raceways not over 4 in. in width must be covered by 3/4 in. or thicker concrete or wood except in offices (see below). Larger raceways (4 in. to 8 in.) require one in. of concrete if more than 1 in. apart.	Wireways are sheet-metal troughs with hinged or removable covers. They are permitted only for exposed work. An unbroken length of wireway may be extended transversely through dry walls.	Busways may be used as services, feeders, or branch circuits. Taps from busways require over-current protection with an externally operable circuit breaker or externally operable fusible switch. Special rules apply for installation operating at over 600 volts.
Uses Permitted	Underfloor raceways are permitted as follows: a) beneath the surface of concrete or other flooring material b) in office occupancies, where laid flush with concrete floor and covered with floor covering (not over 4 in. wide)	Wireways are permitted as exposed raceways indoors and outdoors where not subject to severe physical damage or corrosive vapor.	Busways may be installed only where located in the open and are visible except if access is provided behind panels with conditions specified in the Code. Overcurrent protection is not required if a busway is reduced in size (ampacity) provided the smaller busway is 50 ft long or less and protected at 1/3 its ampacity.
Uses not Permitted	Underfloor raceways must not be used a) where subject to corrosive vapors b) in hazardous locations except Class I, Division 2 locations (as permitted by Code)	Wireways shall not be used in hazardous locations except as permitted in Class 1 or Class 2, Division 2	Busways may not be installed: a) where subject to damage or corrosive vapors b) in hoistways c) in hazardous locations (unless approved for purpose) d) outdoors or in wet or damp locations (unless identified for purpose)
Other Rules	Combined cross-sectional area of all conductors or cables must not exceed 40 percent of the inside area of raceways.	a) Combined cross-sectional area of all conductors shall not exceed 20 percent of inside area of raceway. b) Not more than 30 current-carrying conductors shall be installed except as noted in Code.	a) Busways shall be securely supported at intervals not exceeding 5 ft (with exceptions). b) Busways which extend through floors shall be totally enclosed up to a distance of 6 ft above the floor.

7-4.1 Equipment Grounding

Table 7-13 summarizes the equipment grounding rules for any type of equipment that is required to be grounded. These general Code rules apply to all installations 250-2 unless amended by other Code rules for specific equipment. The Code provides a reference list of additional rules that modify or extend the basic rules for grounding 250-43, equipment. The Code also lists specific equipment that is to be grounded regard-250-44 less of voltage.

250-45   In all occupancies, major appliances and many handheld tools are required to be grounded. The appliances include refrigerators, freezers, air conditioners, clothes washing and drying machines, dishwashing machines, sump pumps, and electrical aquarium equipment. Other tools likely to be used outdoors and in wet or damp locations must be grounded or have a system of double insulation.

Although most appliance circuits require an equipment grounding conductor, **250-57,** the frames of electric ranges and clothes dryers may be grounded by the grounded **250-60,** circuit conductor under certain conditions. This is allowed if the supply circuit is **250-61(b)** 120/240 volts or 120/208 volts, single-phase and if the grounded conductor is of size No. 10 or larger and is insulated (with exceptions). Also, the grounding contacts of receptacles on the equipment must be bonded to the equipment. If these specified conditions are met, it is not necessary to provide a separate equipment grounding conductor, either for the frames or any outlet or junction boxes which are part of the circuit for these appliances.

Table 7-13. Equipment Grounding Rules

Application	Rule	Code Ref.
Fixed Equipment	Metal parts of fixed equipment must be grounded: a) Where within 8 feet vertically or 5 feet horizontally of ground or metal objects. b) In a wet or damp location. c) Where in electrical contact with metal. d) In hazardous locations. e) Where supplied by a metal raceway wiring method (with exceptions). f) Where any terminal operates at over 150 volts to ground. See exceptions in Code.	250-42
Equipment Connected by Cord and Plug	Cord-and plug-connected equipment must be grounded: a) In hazardous locations. b) Where operated at over 150 volts to ground (with exceptions).	250-45

Methods of Grounding	a) Fixed equipment may be grounded by:
	    1) An allowed equipment grounding conductor (see below).
	    2) Other means by special permission.
		250-57

	b) The structural metal frame of a building cannot be used as the required equipment grounding conductor.	250-58

	c) Cord- and plug-connected equipment must be grounded:
	    1) By means of the metal enclosure of the conductors supplying the equipment if a grounding-type attachment plug is connected (with exceptions).
	    2) By means of a grounding conductor connected to a grounding-type attachment plug (with exceptions).
	    3) By means of a separate flexible wire or strap in certain cases.
		250-59

Equipment Grounding Conductor:	The allowable equipment grounding conductors are: a) A copper or other corrosion-resistant conductor b) Rigid metal conduit c) Intermediate metal conduit d) Electrical metallic tubing e) Flexible metal conduit if approved (See Exceptions in Code.) f) Armor of Type AC cable g) Sheath of Type MI cable h) Sheath of Type MC cable i) Cable trays if certain conditions are met	250-91(b)
Installation	a) An equipment grounding conductor must be installed by approved means using approved fittings. b) A separate grounding conductor is subject to the restrictions for the protection of the grounding electrode conductor.	250-92
Continuity	Continuity of the equipment grounding conductor must be assured by a first-make, last-break connection for separable connections.	250-99
Connection	Connections that depend on solder must not be used to connect the grounding conductor.	250-113
Size	The size of the equipment grounding conductor is based on the rating or setting of the overcurrent device protecting the circuit.	250-95

7-4.2 Equipment Bonding

ARTICLE 100
DEFINITIONS
A bonding jumper is sometimes used to assure electrical conductivity between metal parts. When the jumper is installed to connect two or more portions of the equipment grounding conductor, the jumper is referred to as an *equipment bonding jumper*. The rules for the equipment bonding jumper are summarized in Table 7-14. Some specific cases in which a bonding jumper is required are also listed in the table.

250-75
Metal raceways, cable armor, and other metal noncurrent-carrying parts that serve as grounding conductors must be bonded whenever necessary in order to assure
350-5
electrical continuity. When flexible metal conduit is used for equipment grounding, an equipment bonding jumper is required if the length of the ground return path
250-91(b)
exceeds 6 feet or the circuit enclosed is rated over 20 amperes. When the path exceeds 6 feet, the circuit must contain an equipment grounding conductor and the bonding may be accomplished by approved fittings.

A short length of flexible metal conduit that contains a circuit rated greater than 20 amperes may not serve as a grounding conductor itself, but a separate bonding jumper can be provided in lieu of a separate equipment grounding conductor for the
250-79(e)
circuit. This bonding jumper may be installed inside or outside the conduit, but an outside jumper cannot exceed 6 feet in length.

In many instances, a bonding jumper is not required to bond equipment together
250-75
to assure electrical continuity. For raceways, cable armor, etc., suitable fittings or
250-76
threaded couplings may serve to bond sections together. If a circuit operates at over 250 volts to ground, the methods used for bonding service equipment can be used or threadless fittings made up tight may serve to bond conduit or metalclad cable. For these circuits, a connection to a box or cabinet requires two locknuts, one inside and one outside, to bond the conduit or cable to the box.

Table 7-14. Installation Rules for Equipment Bonding Jumper

Application	Rule	Code Ref.
Material	Equipment bonding jumpers must be of copper or other corrosion-resistant material.	250-79(a)
Size	The size of the bonding jumper is determined by the rating or setting of the overcurrent device protecting the circuit.	250-79(d)
Installation	Connection devices or fittings that depend on solder must not be used.	250-113
	The jumper may be inside or outside of a raceway or enclosure. If outside, the maximum length must not exceed six feet.	250-79(e)
Required Bonding Jumper	A bonding jumper is required as follows:	
	a) To connect a receptacle grounding terminal to a grounded box (with exceptions).	250-74
	b) To bond expansion joints and telescoping sections of raceways unless other approved means are used.	250-77
	c) Outline lighting parts may be bonded by a No. 14 copper conductor protected from physical damage.	250-97

7-5 EQUIPMENT AND DEVICES

This section discusses Code rules for the installation of various types of electrical equipment and devices that are used to construct a complete electrical system. The purpose of these items is to conduct or control electricity or to perform a mechanical function in the electrical system. Figure 7-5 illustrates this type of equipment, which includes fuses, device boxes, conduit bodies, and pull boxes. Other devices that are

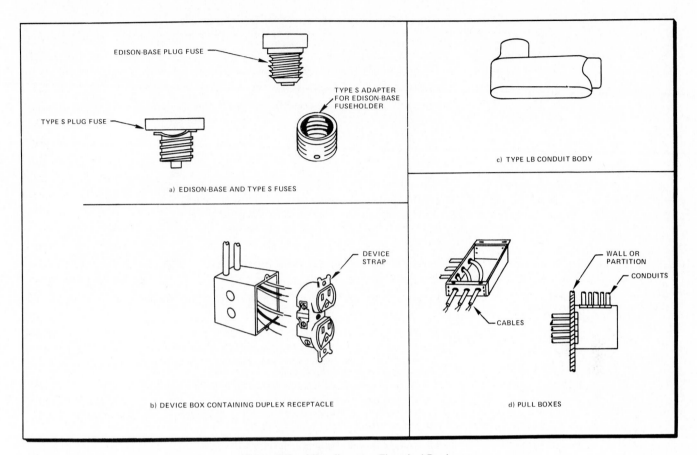

Figure 7-5. Miscellaneous Electrical Devices

used in systems include overcurrent devices, switches, receptacles, and certain types of lighting fixtures and associated apparatus.

Various boxes are used as junction boxes for splices or as device boxes to support switches and outlets. Small boxes come in standard sizes, and restrictions are placed on the number of conductors or devices they may contain. Larger boxes and conduit bodies are used as junction boxes and pull boxes for installations using conduit or other raceway methods.

Cabinets and cutout boxes normally enclose apparatus that must be accessible through a door or panel that can be opened easily. Circuit breakers and control equipment are placed in such enclosures.

7-5.1 Overcurrent Devices

This section deals with the installation of circuit breakers and fuses used for overcurrent protection of conductors. Table 7-15 summarizes the rules for these devices and the other rules that apply to all types of overcurrent devices. When fuses serve as the overcurrent protective device, the Code specifically considers plug fuses and cartridge fuses.

ARTICLE 240

Plug fuses have a screw-shell base and are commonly used in dwellings for circuits that supply lighting, heating, and appliances. Plug fuses are supplied with standard screw bases (Edison-base) or Type S bases. Edison-base fuses are permitted only as replacements in existing installations. These fuses are being replaced with Type S fuses that are constructed so that a larger ampere rated fuse will not fit the fuse holder or adapter for a lower rated size. In most industrial and commercial

ARTICLE 240, PART E

Table 7-15. Installation Rules for Overcurrent Devices

Application	Rule	Code Ref.
Enclosure Required	Overcurrent devices must be enclosed in cabinets or cutout boxes. See exceptions in Code.	240-30
Damp or Wet Locations	Enclosures in damp or wet locations must be identified for such use and be mounted with at least 1/4-inch air space between the enclosure and the supporting surface.	240-32
Mounting Enclosures	Where practical, enclosures must be mounted in the vertical position.	240-33
Disconnecting Means	Where accessible to other than qualified persons, disconnecting means are required on the supply side of fuses as follows: a) All fuses in circuits or over 150 volts to ground. b) Cartridge fuses in circuits of any voltage. See exceptions in Code.	240-40
Protection of Persons	Overcurrent devices should be located or shielded to avoid injury to persons from their operation.	240-41
Standard Sizes	Standard ampere ratings for fuses and inverse-time circuit breakers are listed in the Code.	240-6
Plug Fuses	The following rules apply to plug fuses: a) They are not to be used in circuits exceeding 125 volts between conductors unless the circuit has a grounded neutral and no conductor operates at over 150 volts to ground. b) Fuses, fuseholders and adapters must have no live part exposed after installation. c) The screw-shell of a plug-type fuseholder must be connected to the load side of the circuit.	ARTICLE 240, Part E
Edison-base fuses	Edison-base fuses are classified at not over 125 volts and 0 to 30 amperes. They may be used only as replacements in existing installations.	240-51
Type S Fuses	Type S fuses are classified at not over 125 volts and 0-15, 16-20, and 21-30 amperes.	240-53
Circuit Breakers Used as Switches	Circuit breakers used as switches in 120-volt and 277-volt, fluorescent lighting circuits must be approved and marked "SWD" (switching duty).	240-83(d)

applications, cartridge fuses are used because they have a wider range of types, sizes, and ratings than do plug fuses.

7-5.2 Boxes Containing Outlets, Receptacles, Switches, or Devices

Boxes are used to contain conductors as well as outlets, receptacles, switches, fittings, and devices. They differ from cabinets and cutout boxes in that the covers are normally screwed or bolted on rather than hinged for easy access. The most common smaller boxes are used for light fixtures, switches, and receptacles. Larger boxes of 100-cubic inch capacity or more are normally used as junction boxes or pull boxes as described in the next section of the Guide.

TABLE 370-6(a) In addition to the general installation rules listed in Table 7-16, the Code restricts the number of conductors allowed in a standard box of a given size. For a 4-inch by 1½-inch square box, for instance, ten No. 14 conductors may be enclosed but no No. 6 conductors.

Table 7-16. Installation Rules for Boxes Containing Outlets, Receptacles, Etc.

Application	Rule	Code Ref.
Round Boxes	Round boxes must not be used where conduits requiring the use of locknuts or bushings are to be connected to the side of box.	370-2
Nonmetallic Boxes	The use of nonmetallic boxes is restricted as defined in the Code.	370-3
Metal Boxes	Metal boxes used with Type NM cable and mounted in contact with metal must be grounded.	370-4
Number of Conductors	The maximum number of conductors of size No. 18 through No. 6, allowed in a standard box is specified in the Code.	370-6(a)
Volume per Conductor	For combinations of conductors, the minimum volume per conductor is as follows:	370-6(b)

Size of conductor	Free space required (cubic inches)
No. 14	2
No. 12	2.25
No. 10	2.5
No. 8	3
No. 6	5

Note: See Code to determine how conductors are counted.

Application	Rule	Code Ref.
Unused Openings	Unused openings in boxes must be effectively closed.	370-8
Installation in Walls or Ceiling	Boxes must be installed so that the front edge of the box will not set back more than 1/4 inch. If the wall or ceiling material is combustible, outlet boxes must be flush with the surface or project therefrom.	370-10
Support	Boxes must be securely and rigidly fastened to their support by an approved means. Threaded boxes may be supported by two or more conduits made up wrench tight as follows: a) If no devices or support fixtures are contained, support within 3 feet of the box on two or more sides is adequate. b) Other boxes may be supported by two or more conduits supported within 18 inches of the box	370-13
Depth of Outlet Boxes	The minimum depth of a box is 1/2 inch. Boxes enclosing flush devices must have an internal depth of not less than 15/16 inch.	370-14
Covers for Outlet Boxes	In completed installations, each outlet box must have a cover, faceplate or fixture canopy.	370-15

The opposite problem, that of determining the size of box needed to enclose a given number of conductors, is solved by calculating the total volume of conductors based on a Code table of volume required per conductor. In addition to the actual circuit conductors, other devices or conductors in the box are counted as follows: **TABLE 370-6(b)**

  a. Fixture studs, cable clamps, or hickeys count as one conductor each
  b. Each device strap counts as one conductor (for switches or receptacles)
  c. One or more grounding conductors count as only one conductor
  d. A conductor running through the box is counted as one conductor
  e. Each conductor originating outside the box and terminating within the box is counted as one conductor

By adding up the number of conductors and their volumes, the proper size box may be selected. The volume required per conductor is listed in Table 7-16.

If a box contains a duplex receptacle supplied by two No. 14, two-wire cables (plus one grounding conductor per cable), the number of conductors to be counted is:

Circuit conductors	4
Grounding conductors	1
Receptacle strap	1
Total	6

Each No. 14 conductor requires 2 cubic inches. Thus, a device box of over 12 cubic inches is required. A standard 3-inch by 2-inch by 2½-inch device box would be adequate.

7-5.3 Conduit Bodies, Pull Boxes, and Junction Boxes

300-15

ARTICLE 100 DEFINITIONS

At each splice point, junction point, or pull point for the connection of conduit or other raceways, a box or fitting must be installed. The Code specifically considers conduit bodies, pull boxes, and junction boxes and specifies the installation rules as listed in Table 7-17.

370-6(c)

Conduit bodies provide access to the wiring through removable covers. Typical examples are Type T, Type E, and Type LB. Conduit bodies enclosing No. 6 or smaller conductors must have an area twice that of the largest conduit to which they are attached, but the number of conductors within the body must not exceed that allowed in the conduit. If a conduit body has entry for three or more conduits such as Type T, splices may be made within the conduit body. Splices may not be made in conduit bodies having one or two entries unless the volume is sufficient to qualify the conduit body as a junction box or device box.

370-18

When conduit bodies or boxes are used as junction boxes or as pull boxes, a minimum size box is required to allow conductors to be installed without undue bending. The calculated dimensions of the box depend on the type of conduit arrangement and on the size of the conduits involved.

For straight pulls or junctions in straight runs of conduit, the length of the box must be at least *eight* times the trade diameter of the largest conduit. If the conduits are smaller than ¾ inch or if the conductors are smaller than No. 4, the length restriction does not apply. Figure 7-6(a) shows the minimum length for a box connected to one 3-inch conduit and two 2-inch conduits all containing No. 1/0 conductors. The minimum length is 8×3 inches $= 24$ inches.

In angle or U-pulls, two conditions must be met in order to determine the length and width of the required box. First, the minimum distance to the opposite side of the box from any conduit entry must be at least six times the trade diameter of the largest raceway plus the sum of the diameters of the raceways on the same wall. Figure 7-6(b) shows the minimum length of a box with two 3-inch conduits, two 2-inch conduits, and two 1½-inch conduits in a right-angle pull. The minimum length based on this configuration is:

6×3	inches	=	18 inches
1×3	inches	=	3 inches
2×2	inches	=	4 inches
$2 \times 1½$	inches	=	3 inches
			28 inches

Since the number and size of conduits on the two sides of the box are equal, the box is square and has a minimum side dimension of 28 inches. This size box is *not* sufficient, however, because it does not meet the complete Code requirements.

The second condition given in the Code must also be checked to assure that the box is large enough. In addition to the minimum length for the box, the spacing between raceways enclosing the same conductor must not be less than six times the raceway diameter. Figure 7-6(b) shows that the diagonal spacing must be

$$6 \times 1\tfrac{1}{2} \text{ inches} = 9 \text{ inches for } 1\tfrac{1}{2}\text{-inch conduits}$$
$$6 \times 2 \phantom{\tfrac{1}{2}}\text{ inches} = 12 \text{ inches for 2-inch conduits}$$
$$6 \times 3 \phantom{\tfrac{1}{2}}\text{ inches} = 18 \text{ inches for 3-inch conduits}$$

The distance from the corner of the box to the center of the conduits (x) is then the diagonal distance (d) divided by the square root of 2 (1.414). For the 1½-inch conduits, this distance is 9 inches ÷ 1.414 = 6.4 inches. In order to determine the required length, this distance must be added to the spacing of the other conduits, including locknuts or bushings and clearance for a wrench. As shown in Figure 7-6(b), the minimum length calculated this way is 29.4 inches; therefore, this larger length must be used. The spacing for the conduits was based on a measurement of the locknuts with ½-inch clearance between locknuts.

Table 7-17. Installation Rules for Conduit Bodies, Pull and Junction Boxes

Application	Rule	Code Ref.
Conduit Bodies	Conduit bodies must be installed as follows: a) A conduit body enclosing No. 6 conductors or smaller must have a cross-sectional area not less than twice the cross-sectional area of the largest conduit to which it is attached. b) The maximum number of conductors in the body must not exceed that of the conduit to which it is attached. c) Bodies with less than three conduit entries must not contain splices, taps, or devices unless they meet Code requirements for junction or device boxes.	370-6(c) Chapter 9, Table 1

Minimum Sizes	Boxes and conduit bodies used as pull or junction boxes containing conductors or cables No. 4 or larger or connected to 3/4-inch conduit or larger must have the following minimum sizes:
	a) Straight pulls: The length of the box must not be less than 8 times the diameter of the largest raceway.
	b) Angle or U-pulls: Two conditions must be met:
	    1) The minimum distance between the raceway entries and the opposite wall of the box must be 6 times the diameter of the largest raceway plus the sum of the diameters of all of the other raceways on the same wall of the box.
	    2) The distance between raceway entries enclosing the same conductor must not be less than 6 times the diameter of the larger raceway.
		370-18(a)

Boxes over 6 feet	If a box has any dimension over 6 feet, all conductors must be cabled or racked up.	370-18(b)
Covers	All pull boxes, junction boxes, and fittings must have covers.	370-18(c)
Accessibility	Junction, pull, and outlet boxes must be accessible without removing part of a building or digging. Exception: Listed boxes may be covered by gravel, etc. if the location is identified and accessible for excavation.	370-19
Over 600 V	Special requirements apply to boxes used on systems of over 600 volts.	ARTICLE 370, Part D

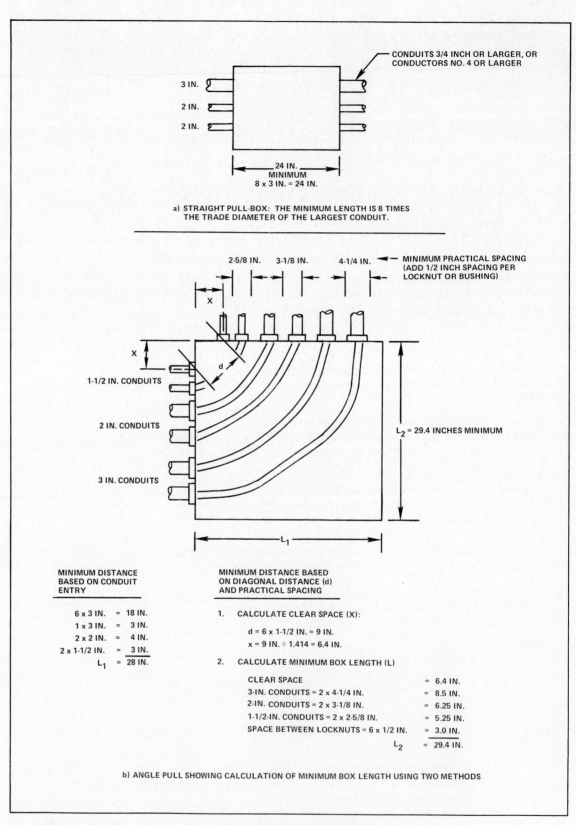

Figure 7-6. Minimum Size Pull Boxes

7-5.4 Cabinets and Cutout Boxes, Switches, and Lighting Fixtures

The Code provides rules for miscellaneous devices and equipment such as cabinets and cutout boxes, switches, lighting fixtures, and similar devices. Although many of the rules deal with the manufacture or construction of the units, a summary of the important references for installation is listed below.

Device or Equipment	Code Reference
Cabinets and cutout boxes	Article 373, Part A
Switches	
Installation	Article 380, Part A
Specifications and ratings:	
Knife switches	380-13
ac general use snap switches	380-14(a)
ac-dc general use snap switches	380-14(b)
Lighting equipment	
Fixture locations	Article 410, Part B
Fixture outlet boxes	Article 410, Part C
Fixture supports	Article 410, Part D
Grounding	Article 410, Part E
Wiring of fixtures	Article 410, Part F
Lampholders	Article 410, Part H
Receptacles, etc.	Article 410, Part L
Rosettes	Article 410, Part M
Flush and recessed fixtures	Article 410, Part N
Electric-discharge lighting	Article 410, Part Q
Electric-discharge lighting	Article 410, Part R
Track lighting	Article 410, Part S

TEST CHAPTER 7

I. True or False *(Closed-Book)*

		T	F
1.	A conductor in a general circuit with an ampacity of 735 amperes may be protected by an 800-ampere fuse.	[ ]	[ ]
2.	A feeder tap less than 25 feet long does not require overcurrent protection if the ampacity of the tap conductors is at least one-half of the feeder conductors.	[ ]	[ ]
3.	A No. 14 grounded conductor in a circuit must have a white or natural gray color.	[ ]	[ ]
4.	When six current-carrying conductors are installed in a raceway, the maximum load current of each must be reduced to 70%.	[ ]	[ ]
5.	As the area of a wire increases, the dc resistance decreases, but the ac resistance increases as a percentage of the dc resistance.	[ ]	[ ]
6.	Flexible cords may not be used if they are concealed behind building walls, ceilings, or floors.	[ ]	[ ]
7.	The smallest fixture wire permitted by the Code is size No. 18.	[ ]	[ ]
8.	Holes in studs for cables must be in the center and not less than 1 inch from the nearest edge of the stud.	[ ]	[ ]
9.	At least 4 inches of free conductor must be left at each outlet to connect fixtures.	[ ]	[ ]
10.	A connection of a No. 8 conductor to a terminal must be by means of a pressure connector or solder lug.	[ ]	[ ]
11.	Conductors of size No. 1 for general use may not be run in parallel.	[ ]	[ ]
12.	A No. 6 copper conductor in a vertical raceway must be supported at least every 80 feet.	[ ]	[ ]
13.	Under normal conditions, direct buried cables must be buried at least 24 inches.	[ ]	[ ]
14.	Flexible metal conduit may be used to protect conductors emerging from the ground and run on a pole.	[ ]	[ ]
15.	Cable tray systems must not be used in hoistways.	[ ]	[ ]
16.	Open wiring on insulators above 6 feet from the floor does not require protection.	[ ]	[ ]
17.	Type MV cable is rated 2001 volts or higher.	[ ]	[ ]
18.	The largest size NM cable has No. 2 copper conductors or No. 2 aluminum conductors.	[ ]	[ ]
19.	EMT installed in the ground in cinder concrete must have a protective layer of non-cinder concrete at least 2 inches thick.	[ ]	[ ]
20.	Flexible metallic tubing may not be used in lengths over 5 feet.	[ ]	[ ]
21.	Splices and taps from bare conductors in auxiliary gutters must leave the gutter opposite their connections.	[ ]	[ ]
22.	Bonding is required for equipment used in a 480Y/277-volt circuit.	[ ]	[ ]
23.	Edison-base fuses are acceptable for standard installations.	[ ]	[ ]
24.	A duplex receptacle in a box reduces the number of allowed conductors by two.	[ ]	[ ]
25.	Junction boxes must be accessible.	[ ]	[ ]
26.	An ac general-use snap switch must not be used to control a tungsten-filament lamp load.	[ ]	[ ]
27.	A lighting fixture is permitted in a clothes closet if an 18-inch clearance is maintained from the storage area.	[ ]	[ ]

II. Multiple Choice *(Closed-Book)*

1. A grounded conductor may be identified by a white or natural gray outer finish or by white markings at its terminations for conductors of size:
   - (a) Larger than No. 4
   - (b) Larger than No. 6
   - (c) Larger than No. 8

   1. _____

2. Which of the following is not a standard ampere rating for a fuse?
   (a) 110-ampere
   (b) 225-ampere
   (c) 450-ampere
   (d) 1500-ampere

2. _____

3. The highest voltage ungrounded circuit for which plug fuses may be used is:
   (a) 125 volts
   (b) 150 volts
   (c) 277 volts

3. _____

4. The highest current rating of a cartridge fuse is:
   (a) 30-ampere
   (b) 6000-ampere
   (c) 3000-ampere

4. _____

5. Which of the following is not required to be grounded in a residence?
   (a) Refrigerators
   (b) Clothes washer
   (c) Hot water heater

5. _____

6. If the frame of an electric range or clothes dryer is grounded by using the grounded circuit conductor, the minimum size conductor is:
   (a) No. 10
   (b) No. 8
   (c) No. 12

6. _____

7. The minimum size copper equipment grounding conductor for a 200-ampere circuit is:
   (a) No. 8
   (b) No. 6
   (c) No. 3

7. _____

8. Rigid metal conduit must be buried at least:
   (a) 12 inches
   (b) 18 inches
   (c) 6 inches

8. _____

9. A 120-volt, 20-ampere residential branch circuit may be buried:
   (a) 12 inches
   (b) 18 inches
   (c) 24 inches

9. _____

10. The listed ampacities of 75° conductors in a raceway must be reduced when the ambient air temperature exceeds:
    (a) 60°C
    (b) 30°C
    (c) 75°C

10. _____

11. Nonmetallic extensions must be run only from circuits with a rating of:
    (a) 15 or 20 amperes
    (b) 125 volts or less
    (c) 30 amperes

11. _____

12. The minimum field bend of 2-inch rigid metal conduit containing THW conductors has a radius of at least:
    (a) 10 inches
    (b) 12 inches
    (c) 15 inches

12. _____

13. In a straight run, 2-inch conduit with threaded couplings may be supported at intervals not exceeding:
    (a) 10 feet
    (b) 12 feet
    (c) 16 feet

13. _____

14. The maximum size of electrical metallic tubing is:
    (a) 6-inch
    (b) 4-inch
    (c) 3-inch

14. _____

15. Auxiliary gutters must not be filled to greater than:
    (a) 20%
    (b) 30%
    (c) 75%
16. The volume per No. 14 conductor required in a box is:
    (a) 2.25 cubic inches
    (b) 2 cubic inches
    (c) 3 cubic inches
17. The largest conductor allowed in a standard box without other restrictions is:
    (a) No. 4
    (b) No. 8
    (c) No. 6
18. An ac-dc general-use snap switch may control what inductive load as a percentage of its ampere rating?
    (a) 80%
    (b) 50%
    (c) 75%
19. The ampacity of No. 14 fixture wire is:
    (a) 8 amperes
    (b) 15 amperes
    (c) 17 amperes
20. The screw shell of a lampholder must not support a fixture that weighs more than:
    (a) 6 pounds
    (b) 12 pounds
    (c) 10 pounds

15. _____

16. _____

17. _____

18. _____

19. _____

20. _____

III. Fill in the Blanks *(Closed-Book)*

1. A cabinet is an enclosure either for _____ _____ or flush mounting.
2. Conductor sizes are expressed in _____ _____ or in circular mils.
3. The cap or window of plug fuses of 15-ampere and lower rating have a _____ configuration.
4. The screw shell of a plug-type fuse holder must be connected to the _____ side of the circuit.
5. Type S fuses of different ampere classifications are not _____ .
6. An equipment bonding jumper outside a raceway must not exceed _____ feet in length.
7. A separate equipment grounding conductor smaller than _____ must be enclosed in rigid metal conduit, intermediate metal conduit, rigid nonmetallic conduit, EMT, or cable armor.
8. Rigid nonmetallic conduit protected by a 2-inch thick concrete pad may be buried at a minimum depth of _____ inches.

9. Conductors in raceways of size _____ _____ or larger must be stranded.
10. The ampacity of a No. 4 THW copper conductor used in a residential service is _____ amperes.
11. In most cases, open wiring on insulators must be supported every _____ feet.
12. Type AC cables run in attics need protection within _____ feet of the nearest edge of a scuttle hole if the attic is not accessible by permanent stairs or a ladder.
13. The standard taper of rigid metal conduit threads is _____ inch per foot.
14. The total number of bends in a run of conduit must not exceed _____ quarter bends or _____ degrees.
15. The largest size conductor allowed in cellular metal floor raceways is _____ .
16. Boxes intended to enclose flush devices must have a depth of not less than _____ inch(es).
17. The length of a box for a straight pull must be not less than _____ times the trade diameter of the largest raceway.

248

18. The current-carrying capacity of bare copper bars in auxiliary gutters is _____ ampere(s) per square inch.

19. Single-throw knife switches must be connected so that the _____ are dead when the switch is open.

20. The _____ conductor must be connected to the screw shell of a lampholder.

Installation Rules
for Distribution Equipment

Electrical circuits on a premises normally originate at a distribution center where the main overcurrent devices and other equipment are grouped together. A residential occupancy has only service-entrance equipment containing the disconnecting means and overcurrent devices. A large industrial complex may have several distribution centers as well as additional equipment to control and distribute electrical energy for various locations within the complex. This chapter deals with the installation of equipment associated with the distribution system on a premises.

Working space and clearances are required around electrical equipment that must be accessible for examination or service. These clearances apply to all electrical equipment, although service equipment, switchboards, panelboards, and motor control centers require careful installation to satisfy Code requirements. The Code rules for working space near such equipment is discussed in detail in the first section below.

The specific equipment discussed in this chapter includes switchboards and panelboards, transformers, and capacitors. Miscellaneous equipment such as generators, resistors and reactors, storage batteries, and surge arresters are also covered.

8-1 WORKING CLEARANCES

Access and working space must be provided around all electrical equipment that requires examination or service. Table 8-1 summarizes the required clearances for equipment operating at 600 volts or less and, in a separate column, the clearances for over 600 volts. Not only front working space, headroom, and working clearance must be provided, but also sufficient access space must be allowed for equipment located in confined areas.

110-16, 110-34*

Table 8-1. Working Space and Guarding of Electrical Equipment

Clearance	600 Volts or Less	Over 600 Volts	Code Ref.
In Front—Minimum	2½ ft	3 ft	110-16(a) 110-32
Headroom	6¼ ft Exception: Panel rated 200 ampere or less in dwelling	6½ ft	110-16(f) 110-32
Working Clearance as specified by conditions:			Table 110-16(a) Table 110-34(a)
Condition 1			
a) Exposed live parts on one side and no live or grounded parts on the other side	0-150V: 3 ft 151-600V: 3 ft	601-2500V: 3 ft 2501-9000V: 4 ft 9001-25000V: 5 ft 25001-75KV: 6 ft Above 75KV: 8 ft	
b) Exposed live parts on both sides effectively guarded			
Condition 2			
Exposed live parts on one side and grounded parts on the other side	0-150V: 3 ft 151-600V: 3½ ft	601-2500V: 4 ft 2501-9000V: 5 ft 9001-25000V: 6 ft 25001-75KV: 8 ft Above 75KV: 10 ft	
Condition 3			
Exposed live parts on both sides with the operator between	0-150V: 3 ft 151-600V: 4 ft	601-2500V: 5 ft 2501-9000V: 6 ft 9001-25000V: 9 ft 25001-75KV: 10 ft Above 75KV: 12 ft	
Access to working space (see exceptions)	Entrance of sufficient area 2 ft × 6½ ft at each end for switchboard rated 1200 ampere or more and over 6 ft wide	Access 2 ft × 6½ ft	110-16(c) 110-33
Guarding (Placement of equipment to guard live parts)	• Approved cabinets • Vault • Partitions or screens • Gallery • Elevation over 8 ft	• Vaults or locked area • Elevations not less than: 601-7500V: 8 ft 6 in. 7501-35KV: 9 ft over 35KV: 9 ft +0.37 in. per KV above 35	110-17 110-31 110-34

*References in the margins are to the specific applicable rules and tables in the *National Electrical Code.*

For equipment operating at 600 volts or less, a work space 30 inches wide or more is required in front of equipment. The additional clearance required in the horizontal direction from live parts is illustrated in Figure 8-1. This clearance depends on the operating voltage range, either from 0 to 150 volts or from 151 to 600 volts, and the nature of the closest surface or equipment. The working clearance from live parts operating at up to 150 volts must be a minimum of 3 feet under all conditions. The distance increases for equipment operating at 151 to 600 volts when grounded parts or other live parts are on the other side of the work space.

110-16(c) When equipment is placed in a room, sufficient access and exit space must be provided. Large switchboards and control panels rated 1200 amperes or more and over 6 feet wide generally require space at least 24 inches wide and 6½ feet high at both ends, thus providing an exit route from either side of the equipment.

Figure 8-1. Clearance From Live Parts Required Around Electrical Equipment

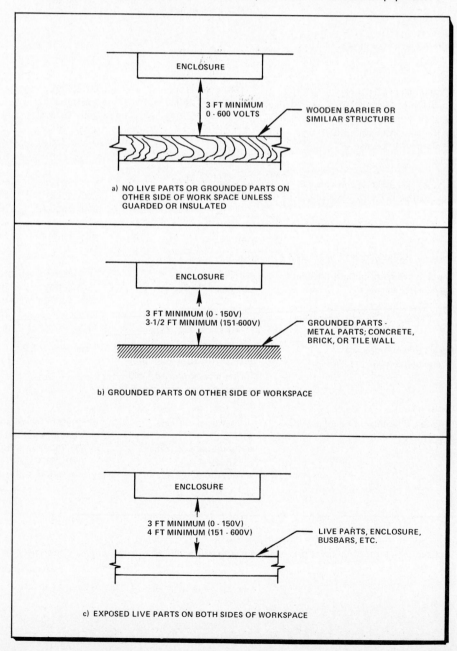

Except for service equipment or panelboards rated 200 amperes or less in dwell-   **110-16(e)**
ing units, illumination and sufficient headroom must be provided around service
equipment, switchboards, panelboards, or motor control centers.

Equipment with live parts must also be guarded to prevent accidental contact   **110-17**
by persons working near the live parts. Any live parts over 50 volts must be guarded
as described in Table 8-1. Stricter rules apply when equipment operates at over 600-   **110-34**
volts. Access or contact must be restricted by locked rooms or enclosures or by
elevation of the equipment.

8-2 SWITCHBOARDS AND PANELBOARDS

Switchboards and panelboards are used to control and distribute electricity to circuits   **ARTICLE 100**
within an occupancy. Switchboards are normally free-standing and are accessible   **DEFINITIONS**
from both the front and back. Panelboards are placed in cabinets or in cutout boxes
that are accessible from the front and contain fuses or circuit breakers. They are
used for controlling lighting circuits and similar circuits in a building or an area of
a building.

Switchboards and panelboards must be constructed and wired as specified by   **ARTICLE 384**
the Code rules listed in Table 8-2. The phase arrangement for three-phase bus bars
must be A, B, C, from front-to-back, top-to-bottom, or left-to-right, as viewed from
the front. If one leg has a higher voltage to ground, such as in a high-leg delta circuit,
phase B must have the higher voltage to ground. In this case, the phase having the
higher voltage must be marked.

Space must be provided in gutters or at terminals to route and terminate con-   **ARTICLE 384**
ductors. The spacing and width of wiring gutters must follow the rules for cabinets   **PART A**
and cutout boxes.

8-2.1 Switchboards

Switchboards are usually supplied by a raceway wiring method with the conductors connected to bus bars at the back of the switchboard. Commercially available
switchboards may be enclosed so that no live parts are normally exposed, but they
may allow access to live parts through a door or pull-out drawer. Since in certain
industrial plants critical circuits may not be shut down, it is not an uncommon
practice in construction to connect circuits to one portion of a switchboard while
other areas of the switchboard are already in use or "live." Therefore, the Code
rules listed in Table 8-3 must be observed when switchboards are installed.

Table 8-2. General Installation Rules for Switchboards and Panelboards

Application	Rule	Code Ref.
Support	Conductors and busbars must be located so as to be free from physical damage and be held firmly in place.	384-3(a)
Bonding	A main bonding jumper is required to connect the grounded service conductor to the frame of switchboards or panelboards when they are used as service equipment.	384-3(c)
	a) The phase arrangement for three-phase buses must be A, B, C, from front-to-back, top-to-bottom, or left-to-right, as viewed from the front. b) When required, Phase B must be the phase having the higher voltage to ground.	384-3(f)
	Minimum wire bending space and wiring gutter space must be provided as required for cabinets and cutout boxes.	384-3(g)

Table 8-3. Installation Rules for Switchboards

Application	Rule	Code Ref.
Clearances	Unless in a weatherproof enclosure, switchboards must be located in a permanently dry location and accessible only to qualified persons.	384-5, 384-6
	a) Standard working clearances must be provided around switchboards	384-8, 110-16
	b) A space of 3 feet or more must be provided between the top and any nonfireproof ceiling	
	Exceptions: Unless a fireproof shield is provided or the switchboard is totally enclosed.	
Clearance of Conductors and Raceways	A conduit or raceway including end fittings must not rise more than 3 inches above the bottom of the enclosure. The minimum spacing between the bottom of the enclosure and the busbars is 8 inches if busbars are insulated and 10 inches otherwise.	384-10
Grounding	Switchboard frames and supporting structures must be grounded. Instruments, relays, etc., located on switchboards must be grounded.	384-11,
Bonding	All sections of a switchboard must be bonded together using an equipment grounding conductor.	384-3(c)

Any switchboard that has exposed live parts must be in a permanently dry location and be accessible only to qualified personnel. The clearances from live parts for working space must conform to the general working clearances listed in Table 8-1. Figure 8-2 illustrates the specific clearances required for switchboards in certain cases.

384-8 A clearance of 3 feet from a nonfireproof ceiling is required unless the switchboard is totally enclosed or a fireproof barrier is provided between it and the ceiling.

Space must be provided to install conductors at the bottom of the switchboard enclosure when bus bars are supplied. A minimum clearance of 8 inches is needed when the bus bars are insulated and 10 inches otherwise. No part of a raceway may extend more than 3 inches above the bottom of the enclosure.

8-2.2 Panelboards

ARTICLE 384, PART B Panelboards normally are manufactured with standard sizes and ratings and include bus bars and space for the insertion of circuit breakers or fuses. Standard ratings are 100 amperes, 225-amperes, 600 amperes, etc., although smaller panelboards called *load centers* are available that have lower ratings. The panelboard fits into a cabinet that is placed against a wall and that has no access from the back.

384-13 *General Installation Rules for Panelboards.* The rating of the panelboard must not be less than that of the minimum feeder capacity for the load to be served. Individual circuit breakers or fuses are inserted in the panelboard in order to protect the branch circuits it supplies. The number of poles for circuit breakers depends on the construction, but normally panelboards may be purchased that have room for 12, 16, 20, 24, 30, or 42 poles. Each three-phrase breaker requires three poles in a panelboard designed for three-phase circuits.

Installation rules for panelboards and their devices are listed in Table 8-4. These rules apply in addition to the ones for both switchboards and panelboards given in Table 8-2. If snap switches rated 30 amperes or less are installed in a panelboard, the main protection must not exceed 200 amperes. In other cases, main protection is not required as long as the panelboard has a sufficient rating to carry the load.

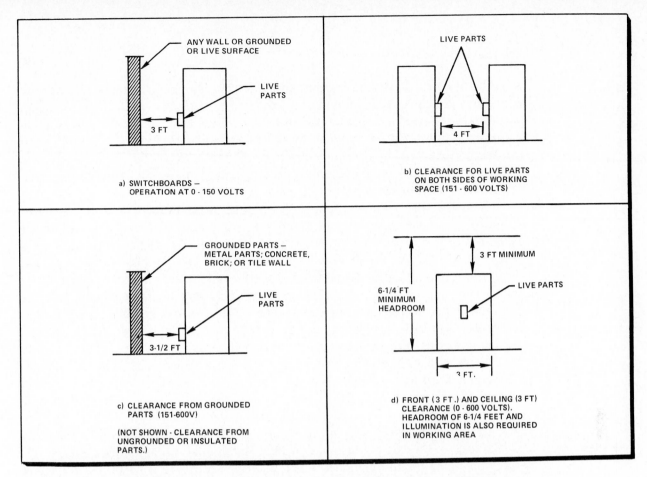

Figure 8-2. Required Clearance Around Switchboards Operating at 600 Volts or Less

Table 8-4. Installation Rules for Panelboards

Application	Rule	Code Ref.
Rating	Panelboards must have a rating not less than the minimum feeder capacity required for the load.	384-13
Snap Switches	Panelboards equipped with snap switches rated 30 amperes or less must have overcurrent protection of 200 amperes or less.	384-16(c)
Continuous Load	The continuous load on any overcurrent device must not exceed 80% of its rating. Exception: Where approved for continuous duty at 100% rating.	384-16(c)
Supplied by Transformer	When a panelboard is supplied by a transformer, overcurrent protection for the panelboard must be on the secondary side. Exception: For single-phase, 2-wire secondaries, protection may be on the primary side if the protection conforms to Code rules for transformers.	384-16(d)

Lighting and Appliance Branch Circuit Panelboards. A lighting and appliance **384-14** branch-circuit panelboard is one having *more* than 10% of its overcurrent devices rated 30 amperes or less for which neutral connections are provided. This definition normally applies to panelboards used in residential occupancies or those used to supply lights and receptacles in industrial or commercial buildings. However, no actual restriction is made on the use of such a panelboard if the stated conditions are met.

The basic rules for lighting and appliance branch-circuit panelboards are presented in Table 8-5. A maximum of 42 overcurrent devices (or 42 poles of circuit breakers) are allowed in a single panelboard in addition to the main protective devices.

384-16(a) Main protection must be provided by not more than two main circuit breakers or two sets of fuses having a combined rating not greater than that of the panelboard. This protection may be provided in the panelboard or by the feeder overcurrent device if it is rated or set to protect the panelboard.

As an example, assume that a feeder supplies a 200-ampere lighting and motor load through a panelboard with a 200-ampere rating, but because of the requirements of the motor loads, the feeder overcurrent protective device must be rated or set at 300 amperes. If there are 20 two-wire, single-phase branch circuits with neutrals, 40 poles are required and the panelboard therefore requires main protection since it must be considered to be a lighting and appliance branch-circuit panelboard. The panel may be protected by a single 200-ampere device or by two 100-ampere devices. The feeder protective device may not be used to protect the panelboard in this case because its setting exceeds the rating of the panelboard.

If a panelboard has 90% or more of its overcurrent devices rated over 30 amperes, it is not a lighting and appliance branch-circuit panelboard. Thus, no main protection is required and there is no restriction on the number of overcurrent devices the panel may contain.

384-16(a), EXCEPTION 2

230-71 An important exception to the requirement for main protection is made for panelboards in an existing installation in an individual residential occupancy. The exception states conditions under which main protection is not required for panelboards used as a distribution center for residential branch circuits. However, the rule for service equipment that limits the number of disconnects to not more than six switches or circuit breakers must still be observed.

8-3 TRANSFORMERS

ARTICLE 450 The Code rules for the installation and protection of transformers operating at voltages up to 600 volts are summarized in Table 8-6. The Code rules listed cover all transformers except special transformers used in specific applications covered by separate Code articles.

Table 8-5. Installation Rules for Lighting and Appliance Branch-Circuit Panelboards

Application	Rule	Code Ref.
Definition	A lighting and appliance branch-circuit panelboard is one having more than 10 percent of its overcurrent devices rated 30 amperes or less, for which neutral connections are provided.	384-14
Maximum Number of Devices	Not more than 42 overcurrent devices may be installed in a single panelboard not including any main protective devices. [Note: Each pole of a circuit breaker counts as one device.]	384-15
Main Protection	A lighting and appliance branch circuit panelboard must be individually protected by not more than two main circuit breakers or two sets of fuses. The combined rating must not exceed that of the panelboard. Exceptions: 1) The feeder protective device may protect the panelboard if its rating does not exceed that of the panelboard. 2) Protection is not required under certain conditions when the panelboard is used as service equipment in an individual residential occupancy.	384-16(a)

Table 8-6. Rules for the Installation and Protection of Transformers (0 to 600 volts)

Application	Rule	Code Ref.
Location	Transformers and transformer vaults must be readily accessible to qualified personnel. Exceptions: 1) Dry-type transformers located in the open. 2) Dry-type transformers not exceeding 600 volts and 50 kVA are permitted in fire-resistant hollow spaces of buildings under certain conditions.	450-13
Rules for Specific Types	Dry-type transformers and transformers with liquid, askarel, or oil fill must conform to the specific installation rules given in the Code.	Article 450, Part B
Vaults	The rules for location and construction of transformer vaults are specified in the Code.	Article 450, Part C
Overcurrent Protection	a) The primary protection must be rated or set as follows:	450-3(b)

Primary current	Percent of the rated primary current
9 amperes or more	125%
less than 9 amperes	167%
less than 2 amperes	300%

Exception:
1) If the primary current is 9 amperes or more, the next higher standard size protective device greater than 125% of the primary current may be used.
2) Individual primary protection is not required if the secondary and the feeder is protected as specified below.
b) Individual primary protection is not required when
     1) An overcurrent device on the secondary side is rated or set at not more than 125% of the rated secondary current.
     2) The primary feeder overcurrent device is rated or set at not more than 250% of the rated primary current.

Application	Rule	Code Ref.
Over 600V	Special rules apply to transformers operating at over 600 volts.	450-3(a)
Other Transformers	The Code specifies rules for potential transformers, autotransformers, and tie circuits.	450-3(c), 450-4, 450-5
Protection of Secondary Conductors	Conductors on the secondary side of a single-phase transformer with a two-wire secondary may be protected by the primary overcurrent device under certain conditions.	240-3, Exception 5

Transformers must normally be accessible for inspection except for dry-type **450-13** transformers under certain specified conditions. Certain types of transformers with a high voltage or kilovolt-ampere rating are required to be enclosed in transformer rooms or vaults when installed indoors. The construction of these vaults is covered in the Code.

The overcurrent protection for transformers is based on their *rated current*, **450-3(b)** not on the load to be served. The primary circuit may be protected by a device rated or set at not more than 125% of the rated primary current of the transformer for transformers with a rated primary current of 9 amperes or more.

In lieu of individual protection on the primary side, the transformer may be protected only on the secondary side if (a) the overcurrent device on the secondary side is rated or set at not more than 125% of the rated secondary current and (b) the primary feeder overcurrent device is rated or set at not more than 250% of the rated primary current.

As an example, the 480/120-volt, 10-kilovolt-ampere transformer shown in Figure 8-3 has a rated primary current of

$$\frac{10000 \text{ VA}}{480 \text{ V}} = 20.83 \text{ A}$$

and a rated secondary current of

$$\frac{10000 \text{ VA}}{120 \text{ V}} = 83.33 \text{ A}$$

450-3(b)(1) If individual primary protection is used, the device must be set at

$$1.25 \times 20.83 \text{ A} = 26.04 \text{ A}$$

A standard 30-ampere fuse or circuit breaker could be used.

450-3(b)(2) Individual primary protection for the transformer is not necessary in this case if the feeder overcurrent protective device is rated at not more than

$$2.5 \times 20.83 \text{ A} = 52.08 \text{ A}$$

and the protection on the secondary side is set at not more than

$$1.25 \times 83.33 \text{ A} = 104.2 \text{ A}$$

A standard 110-ampere device could be used.

It is important to note that the protection just considered is for the transformer, not for the conductors, since the conductors must be protected as specified in other **240-3** Code rules. In one case, however, the conductors supplied by the secondary of a **EXCEPTION 5** transformer having a two-wire secondary may be protected by the overcurrent device on the primary side. For the example shown in Figure 8-3(c), a 30-ampere device in the primary circuit may protect secondary conductors with an ampacity not less than

$$30 \text{ A} \times 480 \text{ V}/120 \text{ V} = 120 \text{ A}$$

In general, however, transformer secondary conductors are not considered to be protected by the overcurrent device on the primary side.

8-4 INSTALLATION RULES FOR CAPACITORS AND OTHER DISTRIBUTION EQUIPMENT

In addition to the primary distribution equipment supplying a premises, other units may be installed to perform ancillary functions. Capacitors are used in many cases for power factor correction for electrical systems or individual motors. Generators and storage batteries may be installed to supply power when there is a primary power failure. The entire system is sometimes protected from excessive current due to short circuits or lightning discharges by networks of resistors and reactors.

This section discusses the installation of capacitors and other equipment in an electrical system. The design of branch circuits for capacitors not covered in Part I of the Guide is also presented.

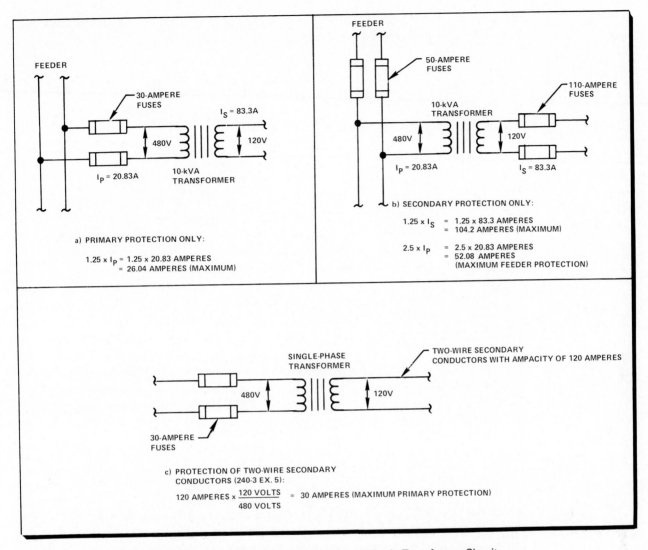

Figure 8-3. Location of Overcurrent Protection in Transformer Circuits

8-4.1 Capacitors

The Code provides specific rules for the installation and protection of capacitors **ARTICLE 460** other than surge capacitors or capacitors that are part of another apparatus. The chief use of the capacitors as discussed in the Code is to improve the power factor of an electrical installation or an individual motor.

The rules for capacitors operating under 600 volts are listed in Table 8-7. Since capacitors may store an electrical charge and hold a voltage that is present even when a capacitor is disconnected from a circuit, capacitors must be enclosed, guarded, or located so that persons cannot accidentally contact the terminals. In most installations, capacitors are installed out of reach or are placed in an enclosure accessible only to qualified persons. The stored charge of a capacitor must be drained by a **460-6** discharge circuit either permanently connected to the capacitor or automatically connected when the line voltage of the capacitor circuit is removed. The windings of a motor or a circuit consisting of resistors and reactors will serve to drain the capacitor charge.

The capacitor circuit conductors must have an ampacity of not less than 135% **460-8** of the rated current of the capacitor. This current is determined from the kilovolt-ampere or kilovar rating of the capacitor as for any load. A 100-kilovolt-ampere three-phase capacitor operating at 480 volts has a rated current of

$$\frac{100\ 000\ \text{kVA}}{1.73\ \times\ 480\ \text{V}} = 120.3\ \text{A}$$

The minimum conductor ampacity is then

$$1.35\ \times\ 120.3\ \text{A} = 162.4\ \text{A}$$

When a capacitor is switched into a circuit, a large inrush current results to charge the capacitor to the circuit voltage. Therefore, an overcurrent protective device for the capacitor must be rated or set high enough to allow the capacitor to charge. Although the exact setting is not specified in the Code, typical settings vary between 150% and 250% of the rated capacitor current.

In addition to overcurrent protection, a capacitor must have a disconnecting means rated at not less than 135% of the rated current of the capacitor unless the capacitor is connected to the load side of a motor-running overcurrent device. In this case, the motor disconnecting means would serve to disconnect the capacitor and the motor.

A capacitor connected to a motor circuit serves to increase the power factor and reduce the total kilovolt-amperes required by the motor-capacitor circuit. The power factor (pf) is defined as the true power in kilowatts divided by the total kilovolt-amperes or

$$\text{pf} = \frac{\text{kW}}{\text{kVA}}$$

where the power factor is a number between .0 and 1.0 (unity). A power factor less than one represents a lagging current for motors and inductive devices. The capacitor introduces a leading current that reduces the total kilovolt-amperes and raises the power factor to a value closer to unity. If the inductive load of the motor is completely balanced by the capacitor, a maximum power factor of unity results and all of the input energy serves to perform useful work (neglecting certain losses in the motor) since kilovolt-amperes = kilowatts in this case.

The capacitor circuit conductors for a power factor correction capacitor must have an ampacity of not less than 135% of the rated current of the capacitor. In addition, the ampacity must not be less than one-third the ampacity of the motor circuit conductors.

Figure 8-4(a) illustrates the proper conductor sizes for a 10-kilovolt-ampere capacitor connected to a 230-volt, three-phase motor supplied by No. 3/0 THW conductors. The rated current of the capacitor is 25.1 amperes, and conductors with a minimum ampacity of $1.35 \times 25.1 = 33.9$ amperes are required to satisfy the basic **TABLE 310-16** Code rule. But since the motor conductors can carry 200 amperes, the capacitor conductors must be rated at least $\frac{1}{3} \times 200$ amperes = 66.7 amperes. No. 4 THW conductors will meet this requirement.

The connection of a capacitor reduces current in the feeder up to the point of connection. If the capacitor is connected on the load side of the motor-running

Table 8-7. Installation Rules for Capacitors and Other Equipment

Application	Rule	Code Ref.
Enclosing and Guarding	Capacitors must be enclosed or guarded unless accessible only to qualified persons.	460-2(b)
Stored Charge	Capacitors must be provided with a means of draining the stored charge.	460-6
Over 600V	Special Code rules apply to capacitors operating at over 600 volts.	Article 460, Part B
Conductor Ampacity	a) The ampacity of capacitor circuit conductors must not be less than 135% of the rated current of the capacitor in any case. b) When connected to a motor, the ampacity of conductors connecting the capacitor must not be less than 1/3 the ampacity of the motor circuit conductors.	460-8(a)
Overcurrent Protection	Overcurrent protection is required in each ungrounded conductor unless the capacitor is connected on the load side of a motor-running overcurrent device. The setting must be as low as practicable. Note: The setting of the overcurrent device is normally less than 250% of the capacitor rating.	460-8(b)
Disconnecting Means	a) A disconnecting means is required for a capacitor unless it is connected to the load side of a motor-running overcurrent device. b) The rating must not be less than 135% of the rated current of the capacitor.	460-8(c)
Improved Power Factor	a) The total kilovar rating of capacitors connected to the load side of a motor controller must not exceed the value required to raise the no-load power factor to 1. b) If the power factor is improved, the motor-running overcurrent device must be selected based on the reduced current drawn, not the full-load current of the motor.	460-7 460-9
Grounding	Capacitor cases must be grounded except when the system is designed to operate at other than ground potential.	460-10
Generators	Generators and their associated equipment must conform to specific Code rules.	Article 445
Ampacity of Generator Conductors	The ampacity of phase conductors from the generator must not be less than 115% of the nameplate current rating of the generator. See exceptions in Code.	445-5
Resistors and Reactors	Specific Code rules cover the installation of resistors and reactors.	Article 470
Storage Batteries	Storage batteries must be installed according to applicable Code rules.	Article 480
Lightning Arresters	The location and method of installation of surge arresters are specified in the Code.	Article 280

overcurrent device, the current through this device is reduced and its rating must be based on the actual current, not on the full-load current of the motor. This case is illustrated in Figure 8-4(b).

8-4.2 Other Distribution Equipment
Code rules applicable to installation of other distribution equipment are listed in Table 8-7. Generators, resistors and reactors, storage batteries, and surge arresters are covered by separate Code articles.

ARTICLES 445, 470, 480, and 280

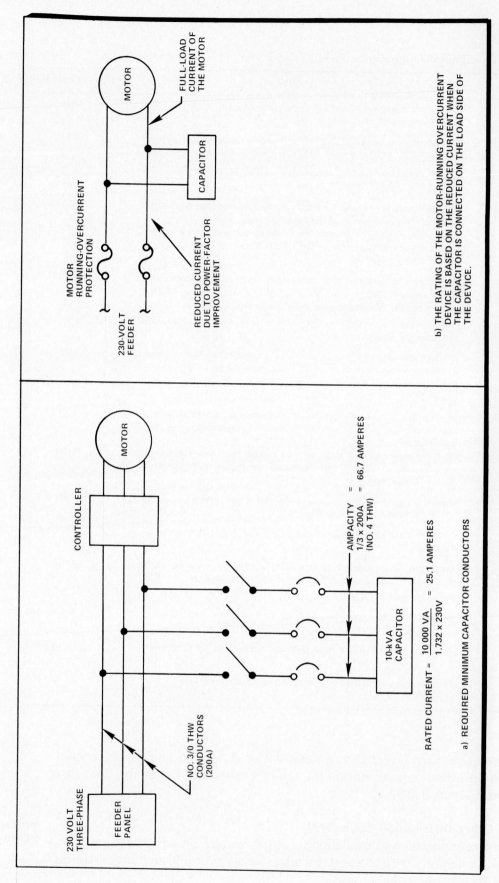

Figure 8-4. Capacitor Circuits for Power Factor Correction

FULL-LOAD CURRENT OF THE MOTOR

MOTOR RUNNING-OVERCURRENT PROTECTION

CAPACITOR

REDUCED CURRENT DUE TO POWER-FACTOR IMPROVEMENT

230-VOLT FEEDER

b) THE RATING OF THE MOTOR-RUNNING OVERCURRENT DEVICE IS BASED ON THE REDUCED CURRENT WHEN THE CAPACITOR IS CONNECTED ON THE LOAD SIDE OF THE DEVICE.

CONTROLLER

MOTOR

230 VOLT THREE-PHASE

FEEDER PANEL

NO. 3/0 THW CONDUCTORS (200A)

$$\text{AMPACITY} = 1/3 \times 200A = 66.7 \text{ AMPERES}$$
(NO. 4 THW)

10-kVA CAPACITOR

$$\text{RATED CURRENT} = \frac{10\,000 \text{ VA}}{1.732 \times 230\text{V}} = 25.1 \text{ AMPERES}$$

a) REQUIRED MINIMUM CAPACITOR CONDUCTORS

TEST CHAPTER 8

I. True or False *(Closed-Book)*

	T	F
1. The service-entrance equipment for a one-family dwelling if rated 200 amperes and installed indoors must be illuminated.	[ ]	[ ]
2. The working clearance in front of live parts operating at 120 volts must be at least 3 feet.	[ ]	[ ]
3. The minimum headroom around a motor control center must be 6½ feet.	[ ]	[ ]
4. On a switchboard with bus bars, phase B would have the highest voltage to ground in a high-leg delta arrangement.	[ ]	[ ]
5. A 200-ampere panelboard may supply a 200-ampere load.	[ ]	[ ]
6. A lighting and appliance branch-circuit panelboard has 10% or more of its overcurrent devices rated 30 amperes or less.	[ ]	[ ]
7. The terminal bar is not connected to the neutral bar in a panelboard not used as service equipment.	[ ]	[ ]
8. Transformers rated over 35 000 volts must be installed in a vault.	[ ]	[ ]
9. Potential transformers installed indoors require primary fuses.	[ ]	[ ]
10. Capacitor circuit conductors must have an ampacity at least 125% of the rated current of the capacitor.	[ ]	[ ]
11. A capacitor operating at 300 volts must be discharged to 50 volts or less within 1 minute after it is disconnected from the supply.	[ ]	[ ]
12. The nominal voltage of a lead-acid battery is 2.0 volts per cell.	[ ]	[ ]
13. If a generator operates at 115 volts or more, protection of live parts is required when the generator is accessible to unqualified persons.	[ ]	[ ]
14. A surge arrester for a 480-volt system requires a connecting conductor of No. 14 copper or larger.	[ ]	[ ]
15. A surge arrester must be connected on the load side of all connected apparatus.	[ ]	[ ]

II. Multiple Choice *(Closed-Book)*

1. The working clearance between live parts for a system operating at 480 volts is:
    (a) 3 feet
    (b) 3½ feet
    (c) 4 feet

    1. _____

2. Guarding by elevation for a 600-volt circuit requires an elevation of:
    (a) 8 feet
    (b) 6¼ feet
    (c) 8 feet 6 inches

    2. _____

3. Conduit or fittings must not rise into the bottom of a switchboard enclosure more than:
    (a) 8 inches
    (b) 3 inches
    (c) 10 inches

    3. _____

4. A three-phase panelboard with 40-ampere circuit breakers may contain a number of devices with a maximum of:
    (a) 42 poles
    (b) Any number of poles
    (c) 40 poles

    4. _____

5. A 200-ampere lighting and appliance branch-circuit panelboard may be protected by two main breakers with a combined rating of:
    (a) 200 amperes
    (b) 225 amperes
    (c) 160 amperes

    5. _____

6. To qualify as a lighting and appliance branch-circuit panelboard, the number of circuits rated 30 amperes or less with neutrals must be:
   (a) More than 10%
   (b) 42 or less
   (c) 10%

7. Panelboards equipped with 20-ampere snap switches must be protected not in excess of:
   (a) 150 amperes
   (b) 100 amperes
   (c) 200 amperes

8. A panelboard supplies six three-phase, 208-volt loads and one 20-ampere, four-wire, 208Y/120-volt circuit. Is main overcurrent protection required?
   (a) Yes
   (b) No
   (c) Cannot tell

9. A generator that according to the nameplate rating delivers 100 amperes requires conductors with an ampacity of at least:
   (a) 100 amperes
   (b) 125 amperes
   (c) 115 amperes

10. A 5-kilovolt-ampere, 240-volt, single-phase transformer requires primary overcurrent protection set at not more than:
    (a) 20 amperes
    (b) 30 amperes
    (c) 25 amperes

11. A 480/208Y/120-volt, 100-kilovolt-ampere transformer is tapped to a feeder protected at 200 amperes. Where is transformer protection required and what is its maximum rating?
    (a) Primary at 200 amperes
    (b) Secondary at 350 amperes
    (c) Primary and secondary

12. The rated current of a 20-kilovolt-ampere, three-phase, 230-volt capacitor is:
    (a) 87 amperes
    (b) 50 amperes
    (c) 63 amperes

13. The minimum ampacity of capacitor conductors connected to motor-circuit conductors of 300 amperes is:
    (a) 300 amperes
    (b) 135 amperes
    (c) 100 amperes

14. A power factor correction capacitor connected on the load side of the overload device of a motor reduces the line current by 20%. If the full-load current of the motor is 100 amperes, the setting of an overload device in the circuit is based on a current of:
    (a) 100 amperes
    (b) 125 amperes
    (c) 80 amperes

15. The conductor between a surge arrester and the line for installations operating at 1000 volts or more must be at least:
    (a) No. 14 copper
    (b) No. 6 copper
    (c) No. 8 copper

6. _____

7. _____

8. _____

9. _____

10. _____

11. _____

12. _____

13. _____

14. _____

15. _____

III. Fill in the Blanks (Closed-Book)

1. Live parts of equipment operating at _____ volts or more must be guarded.

2. The minimum distance between two 1000-volt buses is _____ feet if working space between them is provided.

3. The minimum elevation for a 1000-volt bus for guarding purposes is _____ .

4. The distance between the top of a switchboard and a nonfireproof ceiling must be at least _____ feet.

5. All sections of a switchboard must be _____ together.

6. A lighting and appliance branch-circuit panelboard with 42 devices must have more than _____ devices rated less than 30 amperes for which neutral connections are provided.

7. If primary protection is required for a 480-volt transformer, the device must be rated or set at not more than _____ percent of the rated primary current.

8. The minimum height of a doorsill in a transformer vault is _____ inches.

9. An installation with a kilovolt-ampere rating of 100 kilovoltamperes which delivers 75 kilowatts of power has a power factor of _____ .

10. A disconnecting means for a capacitor must be rated at least _____ percent of the rated current of the capacitor.

Installation
of Utilization Equipment

ARTICLE 100
DEFINITIONS*
Utilization equipment is designed to use electrical energy for heating, lighting, or other useful purposes. This chapter presents the installation rules for such equipment including appliances, heating equipment, and electric motors. The rules discussed here apply to these units when installed in typical occupancies. Any special considerations required in areas such as hazardous locations are not covered in this chapter.

Each appliance, heating unit, or motor requires a disconnecting means so that the equipment is disconnected from all ungrounded conductors when servicing is necessary. The type of disconnecting means required and its location are covered in detail in this chapter. Code requirements for grounding noncurrent-carrying metal parts of utilization equipment are also discussed.

9-1 APPLIANCES

ARTICLE 422
Appliances are used in dwellings and other occupancies to perform specific functions such as heating, cooling, and washing. The Code defines the rules for such appliances

---

*References in the margins are to the specific rules and tables in the *National Electrical Code.*

in terms of their characteristics as well as by the method of connection to the circuit supplying the appliance.

For the purpose of applying the Code rules discussed in this section, a distinction is made between the installation requirements of general-purpose appliances, heating appliances, and motor-driven appliances. The characteristics of each type of appliance, including the rating of a unit, determines the allowed method of connection to a branch circuit and the requirements for a disconnecting means. In addition, the grounding requirements for portable appliances differ somewhat from those for appliances which are normally fixed in place. Thus, the Code rules listed in Table 9-1 apply according to the type of appliance, its rating, and its mobility.

An appliance may be connected to its branch circuit by cord and plug or it may be permanently connected. The means of connection does not identify the appliance as either being fastened in place or portable, although most portable appliances are connected by cord and plug.

Certain appliances are permitted to be cord- and plug-connected provided all of the Code rules are met. Heating appliances require cords that are approved for the intended purpose. In a dwelling, appliances such as disposals, dishwashers, and trash compactors may be connected by approved *hard service* cords (Type S, SO,

422-8

Table 9-1. Installation of Appliances

Application	Rule	Code Ref.
Flexible Cords	a) Heating appliances—flexible cords used with electrically heated appliances must be of approved type as listed. b) Waste disposers—a disposal provided with a "hard service" cord with a grounding conductor is permitted to be installed as follows:	422-8(a), 422-8(b)
	1) The cord length must be between 18 inches and 36 inches 2) The receptacle must be accessible and located to avoid physical damage	422-8(d) (1), Table 400-4
	c) Dishwashers and trash compactors may be installed with "hard service" cords with a grounding conductor as follows:	422-8(d) (2),
	1) The length of the cord must be 3 to 4 feet 2) The receptacle must be accessible, located to avoid physical damage, and located in or near the space occupied by the appliance	Table 400-4
	Exception: Grounding is not required if the equipment is doubly insulated.	

	d) Other Appliances—flexible cords are permitted to:
	    1) Facilitate frequent interchange or prevent transmission of noise or vibration
	    2) Facilitate removal for maintenance
		422-8(c)

Grounding	Metal frames of appliances must be grounded under the conditions specified in the Code.	422-16, Article 250

	The Code provides rules for these specific appliances:
	    a) immersion heaters
	    b) heating appliances
	    c) flat irons
	    d) water heaters
	    e) infrared lamp industrial heating units
	    f) wall-mounted ovens, etc.	

Wall-Mounted Ovens	a) Wall-mounted ovens and counter-mounted cooking units are permitted to be connected by cord and plug or permanently connected.
	b) Separable connector or plug and receptacle in supply line to such units facilitates installation and servicing, but
	    1) cannot serve as required disconnecting means, and
	    2) must be approved for temperature of location.
		422-17

ST, STO, etc.) of short lengths when the receptacles are accessible and located so that damage to the cord is avoided. Wall-mounted ovens and other counter-mounted cooking units may also be connected by cord and plug or they may be permanently connected.

422-20   An appliance, no matter how it is connected, must have a means of disconnecting it from all ungrounded conductors of the circuit supplying the appliance. The types of disconnecting means permitted for appliances are summarized in Table 9-2.

422-21   For small permanently connected appliances rated at not over 300 voltamperes or ⅛ horsepower, any branch-circuit overcurrent device may serve as the disconnecting means. The branch-circuit overcurrent device may also serve as the disconnecting means for larger permanently connected appliances if it is accessible to the user.

      In addition to being accessible, a switch or circuit breaker disconnecting a motor-driven appliance rated more than ⅛ horsepower must normally be in sight from the motor controller and qualify as a motor disconnecting means. The disconnecting means is not required to be within sight of the appliance if a suitable unit switch is provided with the appliance.

Table 9-2. Disconnecting Means for Appliances

Application	Rule	Code Ref.
Disconnecting Means	A means must be provided to disconnect each appliance from all ungrounded conductors.	422-20
Permanently Connected Appliances	a) The branch-circuit overcurrent device may serve as disconnecting means for appliances rated not over 300 VA or 1/8 hp	422-21 (a)
	b) For higher rated appliances, the branch-circuit switch or circuit breaker may serve as disconnecting means if readily accessible	422-21 (b)
	c) A switch or circuit breaker may serve as the disconnecting means for a motor-driven appliance of more than 1/8 hp if:	422-26
	1) located within sight from motor controller, and 2) qualifies as motor-disconnecting means	Article 430, Part H
	Exceptions are provided for appliances with unit switches.	422-21, 422-26
Cord- and Plug-Connected Appliance	a) Separable connector or attachment plug and receptacle may serve as disconnecting means for cord-and-plug-connected appliances.	422-22 (a)
	b) Attachment plug and receptacle at rear of electric range permitted to be accessible through range.	422-22 (b)
	c) Rating and construction of such connectors, plugs, and receptacles must conform to applicable Code rules.	422-22 (d)
Unit Switches as Disconnecting Means	A unit switch may serve as the required disconnecting means if it has a marked "off" position and disconnects all ungrounded conductors provided that other means for disconnection are available in the following types of occupancies. a) multifamily dwellings—within dwelling unit or on same floor (may also supply lamps or other appliances) b) two-family dwellings—may be outside dwelling unit (may be individual switch) c) one-family dwellings—may be service disconnecting means d) other occupancies—may be branch-circuit switch or circuit breaker if accessible	422-24
Motor-Driven Appliances (More Than 1/8 hp.)	Motor driven appliances require a disconnecting means in sight from motor controller. Exception: Unless a unit switch is provided.	422-26, Exception

Appliances attached by cord and plug may be disconnected by a separable con-   **422-22**
nector or by an attachment plug and receptacle. If the receptacle is accessible, no
additional disconnecting means is required for the equipment.

A switch on the unit with a marked "off" position and that disconnects all the   **422-24**
ungrounded conductors may disconnect the appliance when other means such as the
branch-circuit switch or circuit breaker are provided to disconnect the circuit. This
other disconnecting means must be accessible to the user in the specified locations
of dwellings and other occupancies as summarized in Table 9-2.

9-2 FIXED ELECTRIC SPACE-HEATING EQUIPMENT

The Code rules for fixed electric space-heating equipment cover unit heaters, heating   **ARTICLE 424**
cables, duct heaters, resistance-type boilers, electrode-type boilers, and heating
panels. This section is primarily concerned with the general rules for installation of
any fixed electric space-heating equipment and the specific rules for electric space
heating cables.

9-2.1 General Rules for Installation
of Fixed Heating Equipment

The general rules for the installation of fixed heating equipment are presented
in Table 9-3. Installation of the units must be made in an approved manner unless
special permission is obtained to use methods not covered in the Code.

In addition to the requirements for installation, all fixed heating equipment,   **424-19**
including any associated motor controllers and supplementary overcurrent protective devices, must have a disconnecting means. This disconnecting means may be a
switch, circuit breaker, unit switch, or a thermostatically controlled switching device.
The selection and use of a disconnecting device are governed by the type of overcurrent protection and the rating of any motors that are part of the heating unit.

Table 9-3. General Rules Installation of Fixed Space Heating Equipment

Application	Rule	Code Ref.
Supply Conductors	Fixed electric space heating equipment with conductors having over 60°C insulation must be marked and marking must be visible after installation.	424-11
Exposed to Damage	Fixed electric heating equipment exposed to severe physical damage must be adequately protected.	424-12(a)
Damp or Wet Areas	Heaters installed in damp or wet conditions must be approved for the location and be installed so that water cannot enter or accumulate.	424-12(b)
Spacing From Combustible Material	Heating equipment must be installed with required spacing from combustible material or be acceptable for direct contact with such material.	424-13
Grounding	Exposed metal parts of fixed electric space heating equipment must be grounded as specified in the Code.	424-14, Article 250
Disconnecting Means	Means must be provided to disconnect heating equipment, including motor controllers and supplementary overcurrent protective devices, from all ungrounded conductors. Note: See Code for specific rules about the disconnecting means for heaters with supplementary overcurrent protection. Rules for other heating equipment are similar to those for any appliance.	424-19

424-22   In certain heating units, supplementary overcurrent protective devices other than the branch-circuit overcurrent protection are required. These supplementary overcurrent devices are usually used when heating elements rated more than 48 amperes are supplied as a subdivided load. In this case, the disconnecting means must be on the supply side of the supplementary overcurrent protective device and within sight of it. This disconnecting means may also serve to disconnect the heater and any motor controllers, provided the disconnecting means is within sight from the controller and heater, or it can be locked in the open position. If the motor is rated over ⅛ horsepower, a disconnecting means must comply with the rules for motor disconnecting means unless a unit switch is used to disconnect all ungrounded conductors.

A heater without supplementary overcurrent protection must have a disconnecting means that complies with rules similar to those for permanently connected
424-19(c)   appliances. A unit switch may be the disconnecting means in certain occupancies when other means for disconnection are provided as specified in the Code.

9-2.2 Electric Space-Heating Cables
ARTICLE 424,   Heating cables are used in dwellings and other occupancies to heat rooms or
PART E   entire areas. The cables are usually embedded in plaster or run between two layers of gypsum board in the ceiling. The Code rules for installation of heating cables are summarized in Table 9-4.
424-34   The nonheating leads that extend from the supply to the heating cables are
424-43   furnished in lengths of 7 feet or more. In any junction box at least 6 inches of free nonheating lead must be left within the box. The excess length of nonheating lead from heating cables must be embedded in the ceiling finish and not cut to length.
424-38   The cables must not extend beyond the room or area in which they originate.
424-39   Also, the heating elements must clear outlet boxes and junction boxes by at least 8 inches. A 2-inch minimum clearance is required from recessed fixtures or other openings.
424-44   Cables may also be installed in floors when the Code restrictions for such installations are followed. A maximum heating load of 16½ watts per linear foot of cable is allowed in floor installations.

Table 9-4. Installation Rules for Heating Cables

Application	Rule	Code Ref.
Lead Length	Heating cables must be furnished with nonheating leads at least 7 feet in length.	424-34
Area Restriction	a) Cables must not extend beyond room or area in which they originate. b) Cables must not be installed in closets unless used to control relative humidity and there are no obstructions such as shelves to the floor. See Code for exceptions and other restrictions.	424-38
Installation of Leads	a) Nonheating leads of cables must be installed using approved wiring methods such as conductors in raceway, Type UF, NMC, MI, or other approved conductors. b) 6 inches or more of free nonheating lead must be left within a junction box.	424-43

Table 9-4. Installation Rules for Heating Cables (continued)

Application	Rule	Code Ref.
	c) Excess leads of heating cable may not be cut. Excess must be embedded in plaster.	
Clearances	a) Wiring in Ceiling—wiring above heated ceiling must be spaced at least 2 inches above that ceiling. Wiring is to be considered as operating in 50° C ambient for purposes of ampacity computation.	424-36
	Exception: Wiring above 2 inches or more of thermal insulation requires no temperature correction.	
	b) Branch-Circuit Wiring in Exterior Wall—Wiring must not be subjected to excessive heating or damage.	424-37

	c) Clearances from Objects and Openings—Heating elements and cables must be separated from boxes and fixtures by at least:
	    1) 8 inches from edge of boxes used to mount surface fixtures
	    2) 2 inches from recessed fixtures or ventilating openings
		424-39

Installation in Concrete or Masonry	a) Heating cable not to exceed 16 1/2 watts/linear ft. b) Spacing between adjacent runs to be at least 1 in. on centers. c) Cables to be secured by nonmetallic means while concrete or finish is applied d) Spacing to be maintained between cable and metal in floor. (See exception.) e) Leads to be protected by approved means (conduit, etc.) f) Bushings to be used where leads emerge from slab.	424-44
Lead Marking	Heating cables must have permanently marked nonheating leads as follows:	424-35

Nominal Voltage	Color
120	yellow
208	Blue
240	Red
277	Brown

Application	Rule	Code Ref.
Splices	Embedded cables may be spliced only where necessary and in an approved manner, but lead length must not be altered.	424-40
Installation of Cables on Dry Board, in Plaster, or on Concrete Ceilings	a) Cables not permitted in walls b) Adjacent runs of cable not exceeding 2 3/4 watt/ft to be installed on at least 1 1/2-in. centers c) Cables to be applied to fire-resistant material d) All heating cable, splices, and 3-inch minimum nonheating lead at splice to be embedded in plaster e) Entire ceiling must have 1/2-inch thick finish of thermally noninsulating material f) Cables to be secured at least every 16 inches by approved means. (See exception.) g) In dry board installation, ceiling below cable to be covered with 1/2 in. or less of gypsum board and void between filled with approved material h) Cables to be free from conducting surfaces i) In dry board installation, cable to be run parallel to joist with clear space of 2 1/2 in. between adjacent runs centered under joist and with minimum crossings of joist.	424-41
Inspection and Tests	Heating cable installations must be inspected and approved before cables are covered or concealed.	424-45

424-35   *Heating Cables.* The nonheating leads of heating cable are color-coded with a
424-40 different color for each nominal voltage. Splices are permitted in embedded cables,
but the length of the cable must not be changed.

424-41   Heating cables on dry board, in plaster, or on concrete ceilings are subject to
restrictions listed in the Code. The rules are to prevent excessive heat buildup or
damage to the cables.

424-45   Cables must be inspected and tested before they are covered or concealed. The
tests are for continuity and insulation resistance.

Figure 9-1. Required Clearances and Wiring Locations for Heating Cables

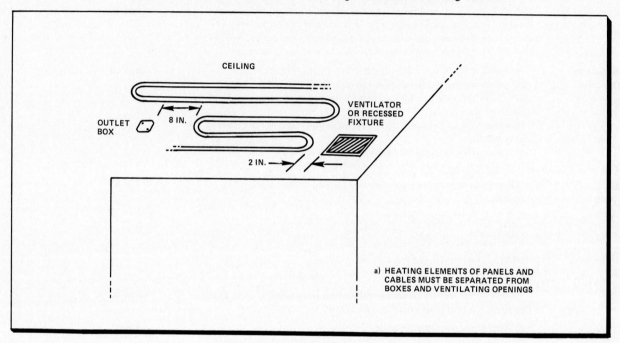

Clearance of Wiring from Heated Surfaces. Conductors run above heated ceilings are affected by the heat even though they are not part of the heating system. Since the wiring in such cases operates in a higher ambient than the nominal 30°C, the appropriate correction factor given in the Code tables of conductor ampacities must be applied. Figure 9-1 illustrates important Code rules for installing conductors near heated surfaces.

424-36, 424-37

9-2.3 Other Fixed Heating Equipment
The Code also provides specific installation rules for:

ARTICLE 424, PARTS F, G, H, and J

  a. Duct heaters
  b. Resistance-type boilers
  c. Electrode-type boilers
  d. Electric radiant heating panels

These units must be suitable for the purpose and installed in an approved manner. Normally, the manufacturer provides detailed instructions for the installation and operation of a specific unit.

9-3 MOTORS

Motors must be installed in a manner that allows them to be adequately cooled or ventilated and also that protects operators from shock hazard. The design of circuits and the selection of protective devices for motors are covered extensively in Part I of the Guide and are not repeated here.

ARTICLE 430

As indicated in Table 9-5 (page 274), the installation location must allow ventilation for the motor with enough space so that maintenance can be performed. In locations where combustible material, dust, or other flying material may be present, special precautions must be taken in selecting and installing motors.

430-14

9-3.1 Disconnecting Means
When maintenance is being performed, a suitable disconnecting means must be available to assure that the motor and its controller are disconnected from all sources of supply. A controller cannot serve as the disconnecting means since the controller must also be disconnected. This does not prohibit a single switch or circuit breaker from acting as both controller and disconnecting means as long as it may be disconnected from all ungrounded conductors itself and it meets other Code requirements.

ARTICLE 430, PART H
430-111

The disconnecting means must be within sight from the controller location (with exceptions) and should be within sight of the motor. Thus, the equipment must be within 50 feet and be visible from either location. If it is not practical to locate the disconnect within sight of the motor, a disconnecting means that can be locked open and is within sight of the controller is required. The switch must disconnect the motor from its source of supply, not just stop the motor. This type of installation is typical whenever a central motor control center is used to control a number of motors, many of which are not in sight from the control center.

ARTICLE 100
430-102

9-3.2 Guarding and Grounding
Exposed live parts of motors operating at 50 volts or more between terminals must be guarded as specified in Table 9-5. Suitable working space must also be provided around motors with exposed live parts even when guarded if maintenance has to be performed while the motor is operating.

430-132

Table 9-5. Installation Rules for Motors

Application	Rule	Code Ref.
Location	a) Motors must be located so that ventilation is provided and maintenance can be readily accomplished. b) Open motors must be located or protected so that sparks cannot reach combustible material. c) Suitable types of enclosed motors must be installed where dust or flying material may interfere with ventilation or cooling.	430-14, 430-16
Motor-Control Circuits	a) All conductors of a remote motor control circuit outside of the control device must be installed in a raceway or otherwise protected. b) The circuit must be wired so that an accidental ground in the control device will not start the motor. c) Motor control circuits require a disconnecting means to disconnect them from all sources of supply	430-73 430-74

Disconnecting Means	A motor disconnecting means must:
	    a) Be within sight from controller location. (See exceptions in Code.)
	    b) Disconnect both motor and controller
	    c) Indicate whether open (off) or closed (on) position
	    d) Be readily accessible
		Article 430,
		Part H

Use of Service Switch	The service switch may serve to disconnect a single motor if it complies with other rules for a disconnecting means.	430-106
Type of Disconnecting Means	The disconnecting means must be a motor-circuit switch rated in horsepower or a circuit breaker. (See exceptions for various motor applications and ratings.)	430-109
Over 600 V	Special installation rules apply to motors operating at over 600 volts.	Article 430, Part J

Protection of Live Parts	Exposed live parts of motors and controllers operating at 50 volts or more must be guarded by:
	    a) installation in a room, enclosure, or location so as to allow access by only qualified persons
	    b) elevation 8 feet or more above the floor
	Exception: Stationary motors as defined in the Code not operating at more than 150 volts.
		Article 430,
		Part K

Grounding Stationary Motors	As specified in the Code, stationary and portable motors operating at over 150 volts to ground, must have the frame grounded. Frames of stationary motors either supplied by metal-enclosed wiring or where in a wet or hazardous location must be grounded or suitably insulated from ground.	Article 430, Part L
Grounding Controllers	Controller enclosures must be grounded. Exceptions: Enclosures attached to ungrounded portable equipment or the lined covers of snap switches.	430-144

Grounding Method	Where required, grounding must be done in manner specified by Code, and:
	    a) Where Type AC cable or specified raceways are used, junction boxes to house motor terminals are required to be connected to armor or raceway in manner specified
	    b) A junction box may be placed up to 6 feet from motor if:
	        1) the leads to the motor are Type AC cable, armored cord, or are stranded leads in metal conduit or tubing not smaller than 3/8 inch
	        2) the armor or raceway is connected to both motor and box
	        3) if stranded leads are used, the maximum size is No. 10
		430-145

The frames of motors, the controller enclosure for other than ungrounded portable equipment, and junction boxes housing motor terminals supplied by metal raceways must be grounded as specified in the Code.

ARTICLE 430, PART L

9-4 MISCELLANEOUS UTILIZATION EQUIPMENT

The Code provides specific installation rules for other utilization equipment as follows:

a. Fixed outdoor electric de-icing and snow-melting equipment   **ARTICLE 426**
b. Fixed electric heating equipment for pipelines and vessels   **ARTICLE 427**
c. Air-conditioning and refrigerating equipment   **ARTICLE 440**

The rules for air-conditioning equipment are in addition to those for motors when electric motors are used as part of the equipment.

TEST CHAPTER 9

I. True or False *(Closed-Book)*

	T	F
1. Toasters may have live parts exposed to contact.	[ ]	[ ]
2. Flexible cords are permitted only to supply portable appliances.	[ ]	[ ]
3. Dishwashers may be connected with Type SP flexible cords.	[ ]	[ ]
4. A cord and plug may be the disconnecting means for a wall-mounted electric oven.	[ ]	[ ]
5. Unit switches may disconnect a heater if other means of disconnection are provided.	[ ]	[ ]
6. The nonheating leads of 120-volt heating cables are yellow.	[ ]	[ ]
7. Conductors located above a heated ceiling are considered to operate in a 40°C ambient.	[ ]	[ ]
8. In most cases, heating cables installed on ceiling boards must be secured at intervals not exceeding 16 inches.	[ ]	[ ]
9. Excess leads of heating cables must not be cut.	[ ]	[ ]
10. Heating panels for de-icing must not exceed a rating of 120 watts/square foot.	[ ]	[ ]
11. In general, enclosures for motor controllers must not be used as junction boxes.	[ ]	[ ]
12. A motor controller must open all conductors to the motor.	[ ]	[ ]
13. The branch-circuit overcurrent device may serve as the disconnecting means for motors rated less than ½ horsepower.	[ ]	[ ]
14. A disconnecting means for a 2300-volt motor must be capable of being locked in the open position.	[ ]	[ ]
15. The frames of stationary motors supplied by metal-enclosed wiring must be grounded.	[ ]	[ ]
16. Type AC cable must not be used to connect motors to their supply.	[ ]	[ ]
17. An attachment plug and receptacle may disconnect cord-connected household refrigerating equipment.	[ ]	[ ]
18. A three-phase air conditioner must be directly connected to a recognized wiring method.	[ ]	[ ]

II. Multiple Choice *(Closed-Book)*

1. The length of a Type S cord connecting a trash compactor must not exceed:
    (a) 18 inches
    (b) 4 feet
    (c) 36 inches

    1. _____

2. Screw-shell lampholders must not be used with infrared lamps rated over:
    (a) 300 watts
    (b) 600 watts
    (c) 150 watts

    2. _____

3. The branch-circuit overcurrent device may serve as the disconnecting means for motor-driven appliances rated not over:
    (a) ½ horsepower
    (b) 1 horsepower
    (c) ⅛ horsepower

    3. _____

4. A 20-ampere appliance must not be protected at more than:
    (a) 20 amperes
    (b) 25 amperes
    (c) 30 amperes

    4. _____

5. The minimum length of nonheating leads furnished for heating cables is:
    (a) 7 feet
    (b) 8 inches
    (c) 3 feet

    5. _____

6. The nonheating leads of 240-volt heating cable are:
    (a) Yellow
    (b) Blue
    (c) Red

    6. _____

7. Wiring above heated ceilings must be spaced above the ceiling at least:
    (a) 4 inches
    (b) 2 inches
    (c) 3 inches

7. _____

8. The separation between heating cables and an outlet box must be at least:
    (a) 8 inches
    (b) 2 inches
    (c) 6 inches

8. _____

9. Adjacent runs of heating cables not exceeding $2^3/_4$ watts/foot in ceilings must be spaced on centers of:
    (a) 1½ inches
    (b) 16 inches
    (c) 1 inch

9. _____

10. The branch-circuit conductors supplying a resistance heater rated at 16 amperes must have an ampacity of:
    (a) 16 amperes
    (b) 20 amperes
    (c) 32 amperes

10. _____

11. If the conductors for a 16-ampere heater are run above a heated ceiling, the smallest type RH conductor permitted is:
    (a) No. 12
    (b) No. 10
    (c) No. 8

11. _____

12. A disconnecting means that must be visible from a motor location must not be farther than:
    (a) 25 feet
    (b) 75 feet
    (c) 50 feet

12. _____

13. A 20-ampere ac snap switch may disconnect a 2-horsepower motor with a maximum full-load current of:
    (a) 16 amperes
    (b) 10 amperes
    (c) 20 amperes

13. _____

14. Live parts of motors must be guarded if they operate at over:
    (a) 150 volts
    (b) 50 volts
    (c) 100 volts

14. _____

15. The smallest size conduit allowed to enclose motor conductors from a junction box is:
    (a) ½ inch
    (b) ¾ inch
    (c) ⅜ inch

15. _____

III. Fill in the Blanks *(Closed-Book)*

1. The length of a cord supplying a waste disposer must be not less than _____ inches and not over _____ inches.

2. Metal frames of electrically heated appliances operating at over _____ volts must be grounded.

3. The maximum rating of the overcurrent device for industrial heating appliances is _____ amperes.

4. Heating equipment requiring supply conductors with over _____ °C insulation must be marked.

5. The maximum protection allowed for subdivided heating elements is _____ amperes.

6. The nonheating leads of 277-volt heating cable are _____ in color.

7. Conductors run above a heated ceiling need not be derated if above _____ inches or more of thermal insulation.
8. The clearance between ventilation openings and heating cables must be at least _____ inches.
9. A minimum of _____ inches of nonheating lead of a heating cable must be embedded in plaster in the same manner as the heating cable.
10. Among the approved conductors used in nonheating leads of cables are Types _____, _____, and _____.
11. In concrete floors, heating cables must be installed at least _____ inch on centers.
12. A general-use switch used to disconnect a motor must have an ampere rating at least _____ percent of that of the motor.
13. To guard live parts of motors by elevation, an elevation of at least _____ feet is required.
14. Insulating mats must be provided around motors requiring service of live parts at voltages of over _____ volts to ground.
15. A junction box required to house motor terminals may be separated from the motor by not more than _____ feet.

Special Equipment

The Code rules for special equipment supplement or modify the general rules for installation of conductors or equipment. These rules are not discussed in great detail in this Guide because they are of interest primarily to specialists and only a general acquaintance with them is needed by other electrical designers and electricians. Three types of special equipment do merit some discussion, however, because they are more accessible to unqualified persons than other equipment of this type covered by the Code. The special equipment discussed includes electric signs, data-processing systems, and swimming pool installations.

10-1 ELECTRIC SIGNS

The Code presents rules for the installation of conductors and equipment for electric signs and outline lighting. As defined by the Code, an electric sign is illuminated utilization equipment designed to convey information or attract attention; outline lighting is designed to outline or call attention to certain features of a building or other structure. In either case, the lighting is provided by incandescent lamps or

ARTICLE 100 DEFINITIONS*

*References in the margins are to the specific applicable rules and tables in the *National Electrical Code.*

electric-discharge tubing, such as neon lights. The important rules for the installation of signs and outline lighting are summarized in Table 10-1.

600-4   The Code requires that every sign of any type must be listed and installed in conformance with that listing. Presumably, however, a sign could be constructed at the location where it is to be installed and if the installation were in compliance with all applicable Code rules, the sign would be approved by the local inspection authority.

600-8   Outdoor signs must be weatherproof so that exposure to the weather will not affect their operation. The sign body material must be metal or other suitable non-combustible material having ample strength and rigidity. For the safety of persons involved in installation and maintenance, and to reduce any fire hazard, internal conductors and terminals must be enclosed in metal or other noncombustible material. Any cutouts, flashers, and other control devices must be enclosed in metal boxes with accessible doors.

Table 10-1. Installation Rules for Signs and Outline Lighting

Application	Rule	Code Ref.
Listing and Marking	Every electric sign must be listed and installed in conformance with that listing.	600-4, 600-7
	Signs and transformers must be marked with maker's name and other information such as rating data as specified in Code.	
Enclosures	a) Signs and outline lighting must be constructed of metal or other noncombustible material. Decorative wood allowed if more than 2 inches from lampholder or current-carrying parts.	600-8(d), 600-8(e)
	b) All steel enclosure parts must be galvanized or protected from corrosion.	600-8(f)
	c) Outdoor enclosures must be weatherproof and have ample drain holes of 1/4 to 1/2 inch diameter.	600-8(g)
	d) Conductors and terminals must be enclosed in metal or other noncombustible material. Control devices must be enclosed in metal boxes with accessible doors.	600-8(a), 600-8(b)
Clearances	a) Signs and outline lighting systems must have vertical and horizontal clearance from open conductors as for outside branch circuits and feeders.	600-10 Article 225
	b) The bottom of enclosures must be at least 16 feet above areas accessible to vehicles unless protected from damage.	
Branch Circuits	a) Circuits supplying lamps, ballasts and transformers must be rated 20 amperes or less.	600-6
	b) Circuits supplying only electric-discharge lighting transformers may be rated up to 30 amperes.	
Disconnecting Means	Each outline lighting system and each permanent sign requires a means to disconnect all ungrounded conductors. The disconnecting means must be within sight of the sign.	600-2
	Exception: An automatically controlled sign must have disconnecting means within sight from controller location. It must disconnect both sign and controller and be capable of being locked in open position.	
Control Devices	Switches, flashers, and other devices controlling transformers must be rated for purpose or have an ampere rating of twice that of the transformer.	600-2(b)
	Exception: AC general-use snap switches may control inductive loads as high as switch rating.	
Pull-Boxes	The supply wiring must terminate in the sign or transformer enclosure.	600-3
	Exception: In certain cases, signs, and transformer enclosures may be used as pull or junction boxes.	

Table 10-1. Installation Rules for Signs and Outline Lighting (continued)

Application	Rule	Code Ref.
Grounding	Signs and boxes must be grounded unless completely insulated from ground and other conducting surfaces and are inaccessible to unauthorized persons.	600-5

Portable Letters	Portable letters or similar displays used with fixed outdoor signs must have supply cords as follows:
	    a) A grounding type, weatherproof receptacle and attachment plug for each individual unit
	    b) All hard service (type S, SJ, SJO, etc.) 3-conductor cords with one conductor grounded
	    c) Cords at least 10 feet above ground level
		600-9

Installation (600V or Less)	a) Installation of Conductors:	
	1) Wiring methods—Conduit, electric metallic tubing, metal-clad cable, metal trough, and mineral-insulated metal-sheath cable	600-21(a)
	2) Insulation and Size—No. 14 general use conductors (with exceptions)	600-21(b)
	3) Exposed to weather—Exposed conductors, cables, or enclosures must be lead-covered or approved for the conditions	600-21(c)
	4) Number of Conductors in Raceway—Limited by Code.	Table 1, Chap. 9
	b) Lampholders—Lampholders must be unswitched and not miniature size. The screw-shell contact must be grounded in grounded circuits	600-22
	c) Protection of Wires—Wires within sign must be mechanically secure and protected by bushings when fed through enclosures	600-23, 600-24
Installation (Over 600V)	a) Installation of Conductors	600-31
	1) Wiring Method—Concealed conductors on insulators, metal type conduits, rigid nonmetallic conduits, and electrical metallic tubing, or type MC cable	
	2) Insulation and Size—No. 14 conductors identified for purpose and voltage	
	3) Bends in Conductors—Sharp bends to be avoided	
	4) Separation of Conductors in Insulators—Specified by Code according to voltage levels	
	5) Length of Sheathing and Insulation—Sheaths must extend beyond raceway. Insulation extension is specified according to voltage levels	
	b) Transformers	600-32
	1) Maximum Secondary Voltage—15,000 volts	
	2) Type—Identified for purpose with maximum rating of 4500 VA	
	3) Must be weatherproof or protected by enclosure	
	4) High-voltage windings must not be connected in series or parallel. See exceptions.	
	5) Must be accessible and, if not in sign, must have specified working spaces.	

Signs and outline lighting installations must be clear of open conductors and be **600-10** elevated at least 16 feet above areas accessible to vehicles, except when the installation is protected from damage at a lower height.

A branch circuit supplying lamps, ballasts, and transformers, or combinations, **600-6** is limited in its rating to 20 amperes. If the circuit supplies only electric-discharge lighting transformers, however, it may have a rating of up to 30 amperes. While the rating of a branch circuit is limited, no restrictions are placed on the number of branch circuits that may be used to supply a sign. Each fixed sign and each outline **600-2** lighting installation must have a disconnecting means that is externally operable, that opens all ungrounded conductors regardless of the number of branch circuits used, and that is within sight of the sign or outline lighting.

Switches, flashers, and other devices controlling the primary circuits of transformers may be subject to damage from arcing at the contacts, especially if gas discharge tubes are supplied. Such control devices, therefore, must be a type approved for the purpose or have at least twice the ampere rating of the transformer; except in the case of ac general-use snap switches which may control ac circuits with an inductive load as great as the rating of the switch.

ARTICLE 600, PART C

The wiring methods, types of lampholders, and type of transformers used in the construction of a sign are defined in detail by the Code according to the range of operating voltages of the lamps used. Generally, signs operating at 600 volts or less use either incandescent lamps or electric-discharge lamps. In most cases, signs operating at over 600 volts employ electric-discharge lamps only and the Code rules are more extensive because of the increased fire and shock hazard associated with high voltages (up to 15 000 volts).

10-2   DATA-PROCESSING SYSTEMS

ARTICLE 645
With the increasing use of computer equipment, data-processing rooms or areas are becoming more common. The Code is primarily concerned with the proper installation of power supply wiring, grounding of equipment, and other such provisions that will insure a safe installation as the Code rules listed in Table 10-2 indicate.

645-2
In many installations, data-processing equipment operates continuously. Thus, it would be expected that the branch circuits for such equipment would be required to have an ampacity of at least 125% of the connected load. The supply conductors are normally flexible cables or cords because it is not uncommon to move computer equipment frequently or to add new equipment to an existing installation.

Table 10-2. Installation Rules for Data Processing Systems

Application	Rule	Code Ref.
Conductors and Cables	Branch-circuit conductors must have an ampacity of not less than 125 percent of the total connected load.	645-2(a)
	Units may be connected by cables or flexible cords approved as part of the data processing system.	645-2(b)
Under Raised Floors	Power and communications supply cables are permitted under suitable raised floors provided: a) Branch-circuit conductors to receptacles employ rigid conduit, intermediate metal conduit, EMT, metal wireway, metal surface raceway, flexible metal conduit, liquid-tight flexible metal conduit, type MI cable, type MC cable, or type AC cable b) Ventilation in underfloor area is used for data processing equipment and the data processing area only	645-2(c)
Grounding	All exposed noncurrent-carrying metal parts must be grounded.	645-4
Disconnecting Means	a) In data processing room, the disconnecting means must 1) Disconnect all electronic equipment 2) Be controlled from locations readily accessible to operator at exits b) A disconnecting means must also be provided to 1) Disconnect the air conditioning system 2) Be controlled from location readily accessible to operator	645-3

Since there are so many interconnections for power, control, and communications, computer equipment is sometimes installed on raised floors with the cables running beneath the floor. The branch-circuit conductors in that case must be in conduit, a metal surface raceway, or a suitable cable. These conductors supply outlet boxes with several receptacle outlets to allow separate cord- and plug-connection to individual units.

The underfloor area can also be used to circulate air for computer equipment cooled from the bottom. Panels in the floor are removed under the equipment to allow air to enter the cabinets that house the equipment. In such an arrangement, fire spread is a real hazard and, therefore, the Code requires that the ventilation in the underfloor area be used for the data-processing equipment and data-processing area only. **300-22**
645-2(c)

A disconnecting means must be provided that will allow the operator to disconnect all data-processing equipment in the data-processing room. A disconnecting means must also be provided to disconnect the ventilation system serving the data-processing area. **645-3**

10-3 SWIMMING POOLS

The Code recognizes the potential danger of electric shock to persons in swimming pools, wading pools, and therapeutic pools or near decorative pools or fountains. This shock could occur from electric potential in the water itself or as a result of a person in the water or a wet area touching an enclosure that is not at ground potential. Accordingly, the Code provides rules for the safe installation of electrical equipment and wiring in or adjacent to swimming pools and similar locations. **ARTICLE 680**

This section discusses not only the general requirements for the installation of wiring and equipment in or adjacent to any permanently installed or storable pool, but is also discusses the specific requirements applicable only to permanently installed pools. These specific rules include installation of underwater fixtures and associated junction boxes and the requirements for grounding and bonding.

10-3.1 General Requirements for Pools

The general requirements for the installation of outlets, overhead conductors, and other equipment are summarized in Table 10-3. Many of these general rules are also illustrated in Figure 10-1 (see page 285), including the requirements for receptacles, lighting fixtures, and cord- and plug-connected equipment.

Receptacles with ground-fault circuit interrupter (GFCI) protection are permitted in the area between 10 feet and 20 feet from the inside wall of any type pool. In fact, at least one such receptacle is required to be installed in that area when a pool is permanently installed at a dwelling. Lighting outlets have similar requirements, but they are permitted to be within 5 feet of the inside wall of the pool if they are GFCI protected or rigidly installed at least 5 feet above the water level. Lighting fixtures extending over the water or near poolside must generally be at least 12 feet above the water level. Exceptions for existing fixtures or fixtures at indoor pools allow lower heights if certain conditions are met. **680-6**

In most cases, overhead conductors are not permitted to pass over diving platforms, observation stands, or the area extending 10 feet horizontally from the inside wall of the pool. An exception in the Code specifies clearances allowed for utility owned and maintained lines. **680-8**

Table 10-3. General Installation Rules for Equipment Installed in or near Pools

Application	Rule	Code Ref.
Receptacles	a) Receptacles must be at least 10 feet from inside wall of pool. b) At least one receptacle must be located in area 10 feet to 20 feet from inside wall of pool permanently installed at a dwelling. c) All receptacles in area 10 feet to 20 feet from inside wall of pool must have GFCI protection. Exception: A receptacle for a permanently installed water pump.	680-6(a)

Lighting Fixtures	a) Lighting fixtures over pool or the area 5 feet from pool must be 12 feet above water level.
	
	See Exceptions for existing fixtures and fixtures at indoor pools.
	b) Lighting fixtures in area 5 feet to 10 feet from inside wall must
	    1) Have GFCI protection, or
	    2) Be rigidly attached at least 5 feet above water level
	
	c) A cord-connected lighting fixture within 16 feet of the water surface must meet the requirements of other cord- and plug-connected equipment.
		680-6(b)

Switching Devices	Switching devices must be at least 5 feet from inside wall of pool unless separated from pool by a permanent barrier.	600-6(c)

Cord- and Plug-Connected Equipment	Fixed or stationary equipment rated 20 amperes or less (including lighting fixture within 16 feet of water) may be cord and plug connected. For other than storable pools, the cord must
	    1) Not exceed 3 feet
	    2) Have equipment grounding conductor not smaller than No. 12
	    3) Have grounding type plug
		680-7

Clearance of Overhead Conductors	Overhead conductors must not be placed over
	    a) Pool or area 10 feet horizontally from inside wall
	    b) Diving structures
	    c) Observation stands, towers, or platforms
	
	See exception for clearances permitted for utility owned and maintained lines.
		680-8

Transformers and GFCI	a) Transformers and enclosures must be approved for purpose. Transformers must be two-winding type with grounded metal barrier between windings b) A ground-fault circuit interrupter must be an approved type	680-5(a)
Pool Heaters	Electric water heaters for pools must not exceed 48 amperes in rating or be protected at more than 60 amperes.	680-9
Underground Wiring	Underground wiring is not permitted under the pool or under the area within 5 feet from pool (with exceptions).	680-10

10-3.2 Requirements for Permanently Installed Pools

680-4   A pool that is constructed in such a manner that it cannot be easily disassembled is considered a permanently installed pool. Since such pools usually have lighting fixtures and other equipment installed below the water level, there are extensive requirements for grounding and bonding. These installation rules for equipment installed in or adjacent to permanently installed pools are summarized in Table 10-4.

680-20   *Underwater Lighting Fixtures.* Lighting fixtures installed below the normal water level are either dry-niche fixtures or wet-niche fixtures. The dry-niche fixture is completely sealed. The wet-niche fixture and the metal-forming shell enclosing it are completely surrounded by pool water. As indicated in Table 10-4, the Code provides rules applicable to either type fixture and specific rules applicable to each type separately. Several of the specific rules for wet-niche fixtures are illustrated in Figure 10-2 (see page 288).

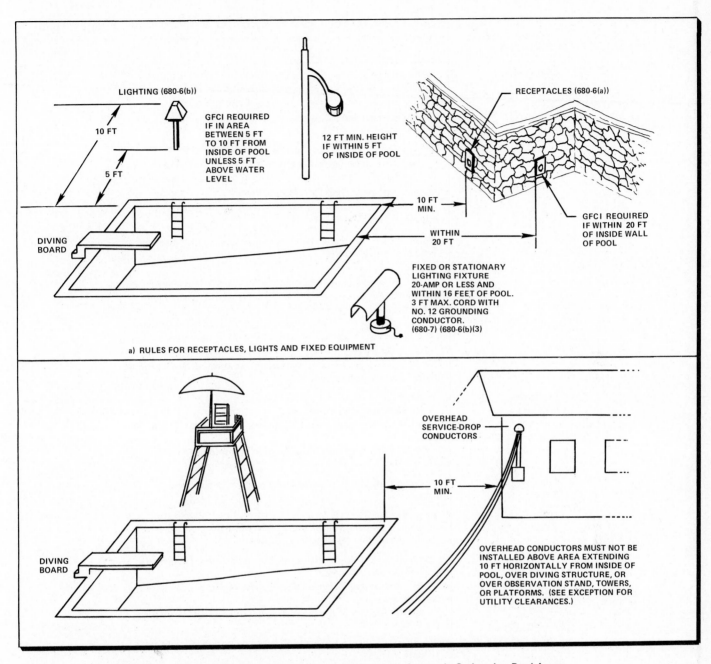

LIGHTING (680-6(b))

GFCI REQUIRED IF IN AREA BETWEEN 5 FT TO 10 FT FROM INSIDE OF POOL UNLESS 5 FT ABOVE WATER LEVEL

10 FT

5 FT

12 FT MIN. HEIGHT IF WITHIN 5 FT OF INSIDE OF POOL

RECEPTACLES (680-6(a))

10 FT MIN.

WITHIN 20 FT

GFCI REQUIRED IF WITHIN 20 FT OF INSIDE WALL OF POOL

DIVING BOARD

FIXED OR STATIONARY LIGHTING FIXTURE 20-AMP OR LESS AND WITHIN 16 FEET OF POOL. 3 FT MAX. CORD WITH NO. 12 GROUNDING CONDUCTOR. (680-7) (680-6(b)(3)

a) RULES FOR RECEPTACLES, LIGHTS AND FIXED EQUIPMENT

OVERHEAD SERVICE-DROP CONDUCTORS

10 FT MIN.

DIVING BOARD

OVERHEAD CONDUCTORS MUST NOT BE INSTALLED ABOVE AREA EXTENDING 10 FT HORIZONTALLY FROM INSIDE OF POOL, OVER DIVING STRUCTURE, OR OVER OBSERVATION STAND, TOWERS, OR PLATFORMS. (SEE EXCEPTION FOR UTILITY CLEARANCES.)

Figure 10-1. General Rules for Installation of Electrical Equipment in Swimming Pool Areas

Table 10-4. Installation Rules for Equipment Installed in or near Permanently Installed Pools

Application	Rule	Code Ref.
Underwater Fixtures	a) Under normal use, the fixture should present no shock hazard. Conductors on the load side of GFCI or a transformer must not occupy enclosures containing other conductors (with exceptions). b) Any underwater fixture operating at more than 15 volts must have GFCI protection. c) No fixture may operate at over 150 volts between conductors. d) Fixture lenses must be installed at least 18 inches below the normal water level. Exception: Approved types for lesser depths. e) Fixtures which depend on submersion must be protected from over-heating.	680-20(a), 680-5(c)

Wet-Niche Fixtures	a) Forming shell—The required and approved forming shell must be equipped with provisions for threaded conduit entries.
	
	 b) Wiring method
	     1) The supply conductors must be enclosed in rigid metal conduit or intermediate metal conduit of brass or other approved corrosion-resistant metal, or
	     2) Enclosed in rigid nonmetallic conduit with a No. 8 insulated, copper conductor connected to the shell for bonding purposes (the termination must be potted).
	
	 c) Terminations—Flexible cord terminations and grounding connections within a fixture must be potted.
		680-20(b)

Dry-Niche Fixtures	A dry-niche fixture must
	     a) Have provision for drainage
	     b) Accommodate an equipment grounding conductor for each conduit entry
	     c) Use approved rigid metal conduit, intermediate metal conduit, or rigid nonmetallic conduit (with exception).
		680-20(c)

Junction Boxes	a) Boxes and enclosures mounted above grade must be afforded protection.	680-21(c)
	b) Boxes and enclosures connected to a conduit that extends to a forming shell must be provided with at least one more grounding terminal than the number of conduit entries.	680-21(d)

Installation of Junction Boxes	A junction box or enclosure connected to a conduit that extends to a forming shell must be located
	     a) Not less than 8 inches above highest level of ground, pool deck, or water level
	     b) Not less than 4 feet from pool unless separated from pool by barrier.
	 (Note: An exception is made for flush deck boxes for lighting systems of 15 volts or less.)
	     c) Each box or enclosure must have provisions for threaded conduit entries.
		680-21

Bonding Required	The parts of a swimming pool and equipment that must be bonded together are
	     a) All metallic parts of the structure, including reinforcing metal
	     b) All forming shells
	     c) All metal fittings and all metal parts of pumps, etc.
	     d) All fixed metal parts, metal conduit, and metal pipes within 5 feet of pool and not separated by a permanent barrier. (See Code for exceptions.)
		680-22

Table 10-4. Installation Rules for Equipment Installed in or near Permanently Installed Pools (continued)

Application	Rule	Code Ref.
Bonding Method	All parts must be bonded with a solid copper conductor not smaller than No. 8 to a common bonding grid which may be reinforcing bars, metal walls of the pool, or a conductor.	680-22
Grounding	The equipment that must be grounded include a) Wet-niche fixtures and dry-niche fixtures b) Electrical equipment within 5 feet of pool c) Electrical equipment for recirculating system d) Junction boxes and transformer enclosures e) Panelboards and GFCI systems	680-24
Grounding Wet-Niche Fixtures Supplied by Flexible Cords	Wet-niche fixtures supplied by flexible cables must be grounded by a No. 16 or larger, insulated, copper equipment grounding conductor connected to a grounding terminal in the supply enclosure. An unspliced grounding conductor must ground the enclosure.	680-25(a) (1)
Grounding Other Equipment	All equipment (other than the underwater lighting fixture) must be grounded by methods recognized in the Code.	680-25(d)
Panelboards	The panelboards must have an equipment grounding conductor connected to the grounding terminal in the service.	680-25(c)
Size and Type of Equipment Grounding Conductor	Equipment grounding conductors must be No. 12 or larger, insulated copper conductors installed in rigid metal conduit, intermediate metal conduit, or rigid nonmetallic conduit. (See exceptions in Code.)	680-25

Junction Boxes and Enclosures. Lighting fixtures and other equipment at the **680-21** pool are supplied by conductors that originate at a junction box, transformer enclosure, or GFCI unit enclosure. In general, such junction boxes and enclosures are not permitted to be located within 4 feet of the pool and not less than 8 inches above the ground or pool deck level if the conductors supply wet-niche fixtures.

In addition to enclosing conductors and splices, the boxes and enclosures must also provide the means to establish the proper grounding and bonding connections in a circuit.

Grounding and Bonding. The primary requirement for equipment grounding and bonding is that all metal parts and equipment within 5 feet of the pool must be bonded together and grounded. Other equipment or devices associated with the electrical system of the pool must also be grounded.

In most cases, the smallest bonding conductor permitted is No. 8 solid copper and the smallest equipment grounding conductor is No. 12 copper. Wet-niche fixtures **680-25** supplied by flexible cords, however, must have an equipment grounding conductor in the cord that is not smaller than the supply conductors and not smaller than No. 16.

The requirements for grounding wet-niche fixtures and other equipment are shown in Figure 10-2. As shown, all of the grounding conductors must eventually terminate at the panelboard grounding terminal.

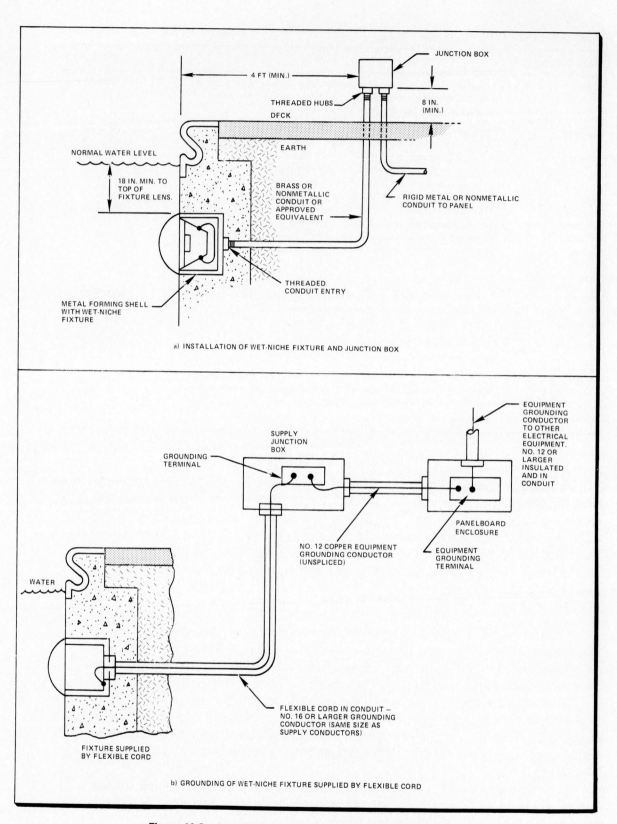

Figure 10-2. Installation and Grounding of Wet-Niche Fixtures

10-4 MISCELLANEOUS SPECIAL EQUIPMENT

The Code also provides specific rules for the installation of the special equipment listed below. The reader is referred to the Code text for further details.

Equipment	Code Article
Manufactured wiring systems	604
Office furnishings	605
Cranes and hoists	610
Elevators, dumbwaiters, escalators and moving walks	620
Electric welders	630
Sound-recording and similar equipment	640
Organs	650
X-ray equipment	660
Induction and dielectric heating equipment	665
Electrolytic cells	668
Electroplating	669
Metalworking machine tools and plastics machinery	670
Electrically driven or controlled irrigation machines	675
Integrated electric systems	685
Solar photovoltaic systems	690

TEST CHAPTER 10

I. True or False *(Closed-Book)*

	T	F

1. Electric signs using incandescent lamps must be marked with the number of lampholders.   [ ]   [ ]
2. The bottom of a sign must be at least 12 feet above an area accessible to vehicles.   [ ]   [ ]
3. Miniature lampholders may be used in outdoor signs.   [ ]   [ ]
4. Type NMC cable is permitted to connect data-processing equipment by underfloor wiring.   [ ]   [ ]
5. A disconnecting means must be provided for all data-processing equipment.   [ ]   [ ]
6. In a data-processing room the disconnecting means must be near the exit doors.   [ ]   [ ]
7. A storable swimming pool must not have any dimension over 12 feet.   [ ]   [ ]
8. Receptacles located within 20 feet of a swimming pool require GFCI protection.   [ ]   [ ]
9. GFCI protection is required for any branch circuit that supplies underwater fixtures operating at over 50 volts.   [ ]   [ ]
10. The minimum depth for any underwater lighting fixture is 4 inches below water.   [ ]   [ ]
11. Steel junction boxes may be used to supply underwater fixtures when they are connected by conduit to the forming shell.
12. The walls of a bolted or welded metal pool may serve as the common bonding grid.   [ ]   [ ]
13. In most cases, the equipment grounding conductors for poolside equipment must be run in conduit.   [ ]   [ ]
14. Fountains that have water common to a swimming pool do not need to comply with all of the swimming pool requirements.   [ ]   [ ]
15. All electrical equipment within 5 feet of the inside wall of a fountain must be grounded.   [ ]   [ ]

II. Multiple Choice *(Closed-Book)*

1. Flashers controlling a transformer in a sign must have a rating based on the ampere rating of the transformer of:
   (a) 100%
   (b) 200%
   (c) 125%

2. A branch circuit supplying lamps in a sign must not be rated more than:
   (a) 20 amperes
   (b) 15 amperes
   (c) 30 amperes

3. Wood used for decoration on a sign must not be closer to a lampholder than:
   (a) 6 inches
   (b) 2 inches
   (c) 3 inches

4. The minimum height of a cord supplying portable letters is:
   (a) 16 feet
   (b) 12 feet
   (c) 10 feet

5. In general, conductors for wiring a 120-volt sign must not be smaller than:
   (a) No. 14
   (b) No. 12
   (c) No. 16

6. Lampholders in a 480-volt sign must be:
   (a) Miniature
   (b) Unswitched
   (c) Screw-shell

1. _____

2. _____

3. _____

4. _____

5. _____

6. _____

7. Cords supplying portable gas tube signs must have a maximum length of:
   (a) 6 feet
   (b) 10 feet
   (c) 12 feet

7. _____

8. Which of the following wiring methods is not suitable for underfloor wiring in a computer room?
   (a) Type MI cable
   (b) Rigid conduit
   (c) Type NM cable

8. _____

9. Lighting fixtures above a pool must be installed at a minimum height of:
   (a) 5 feet
   (b) 12 feet
   (c) 10 feet

9. _____

10. In general, overhead conductors must not pass directly over a pool or horizontally within:
    (a) 5 feet
    (b) 18 feet
    (c) 10 feet

10. _____

11. A utility owned low-voltage service may pass directly over a pool at a minimum height of:
    (a) 18 feet
    (b) 14 feet
    (c) 16 feet

11. _____

12. A typical wet-niche fixture must be installed below the water line at least:
    (a) 12 inches
    (b) 15 inches
    (c) 18 inches

12. _____

13. The smallest bonding conductor permitted for pool equipment is:
    (a) No. 14
    (b) No. 12
    (c) No. 8

13. _____

14. The smallest permitted equipment grounding conductor from a junction box to a panelboard supplying poolside equipment is:
    (a) No. 12
    (b) No. 14
    (c) No. 8

14. _____

15. The maximum voltage between conductors for lighting fixtures in pools and fountains is:
    (a) 120 volts
    (b) 150 volts
    (c) 300 volts

15. _____

III. Fill in the Blanks (Closed-Book)

1. The maximum rating of a branch circuit supplying a transformer for a neon sign is _____ amperes.

2. Drain holes in outdoor signs must not be larger than _____ inches.

3. The height of a sign above a vehicle traffic lane must be at least _____ feet.

4. The maximum secondary voltage for a sign transformer is _____ volts.

5. The maximum length of a cord supplying a portable gas tube sign is _____ feet.

6. The disconnecting means in a data-processing room must disconnect all electrical equipment and the _____ system.

7. In dwellings, one receptacle is required near a pool not more than _____ feet from the inside walls.
8. To prevent shocks during relamping, branch circuits supplying underwater fixtures operating at more than _____ volts require GFCI.
9. Junction boxes and enclosures connected by conduit to underwater forming shell must be located _____ inches above ground and at least _____ feet from the pool.
10. Equipment within _____ feet of a pool must be bonded together and grounded.

Special Occupancies

Certain occupancies or locations because of their construction, use, or possible risk to safety of personnel, are termed *special occupancies* by the Code. Compliance with special rules is required for these locations to assure the safe installation of wiring and equipment.

An important group of special locations are called *hazardous* or *classified locations*. In these locations the possibility of fire or explosion exists because of the nature of the materials present. Normally, only certain areas of an industrial or commercial establishment are so classified where flammable vapors, combustible dust, or similar substances may be present in the air. This chapter discusses the various classifications and treats commercial garages and gasoline stations in detail as two representative examples of the hazardous locations covered by the Code.

Other occupancies, such as theaters, are considered as special occupancies by the Code since specific construction techniques may be required to provide the necessary safety. These locations are also mentioned in this chapter.

11-1 HAZARDOUS (CLASSIFIED) LOCATIONS

The Code *classifies* locations in which a fire or explosion hazard may exist. Equipment and installation methods used in these locations must insure safe performance and

ARTICLE 500*

*References in the margins are to the specific applicable rules in the *National Electrical Code*.

extra care should be exercised with regard to installation and maintenance. The Code rules are basically intended to confine arcing or sparking, restrict physical damage to equipment, and prevent the passage of hazardous vapors, dusts, or fibers to non-hazardous locations. Each room, section, or area of an occupancy that may contain hazardous materials must be considered individually to determine its classification.

As indicated in Table 11-1, the Code recognizes three major classifications. Each classification depends on the type of material present. Class I locations contain flammable gases or vapors. Class II locations contain combustible dust. Class III locations contain easily ignitable fibers or flyings.

In Class I and Class II locations the hazardous materials are further divided into groups. Groups A, B, C, and D apply to Class I locations and the gases or vapors in a particular group have various properties in common. Groups E, and G in Class II define the properties of combustible dusts.

Table 11-1. Definitions of Hazardous (Classified) Locations

Defintions	Class I	Class II	Class III
Class	Locations containing flammable gases or vapors in ignitible concentrations.	Locations containing combustible dust.	Locations containing easily ignitible fibers or flyings but not in ignitible concentrations.
Groups and Qualifying Properties of Equipment or Element	Chemical A—acetylene B—hydrogen, etc. C—ethylene, etc. D—gasoline, etc. Qualifying Properties • maximum explosion pressure • clearance of clamped joints • minimum ignition temperatures	Dust E—magnesium, etc. G—grain dust, etc. Qualifying Properties • tightness of joints • overheating effects • electrical conductivity of dust • ignition temperature	Fibers No groups established Examples: rayon cotton jute
Division 1	A location in which hazardous concentrations of gases or vapors exist: 1) under normal conditions 2) because of repair or leakage 3) due to faulty operation of equipment	A location in which combustible dust may be in the air in dangerous concentrations: 1) under normal conditions 2) due to faulty operation of equipment 3) due to the electrically conductive nature of the dust	A location in which ignitible fibers or flyings are: 1) handled 2) manufactured 3) used
Division 2	A location in which: 1) volatile liquids or gases are handled, processed or used but are normally confined 2) ventilation to prevent hazardous concentrations may fail 3) communication of hazardous concentrations from adjacent Class I, Division 1 location is possible	A location in which: 1) deposits of combustible dust would not interfere with operation of equipment 2) deposits of combustible dust might be ignited by failure of electrical equipment	A location in which easily ignitible fibers are either 1) stored 2) handled (except in process of manufacture)

Once the class of an area is determined, the conditions under which the hazardous material may be present determines the division. In Class I and Class II Division 1 locations, the hazardous gas or dust may be present in the air under normal operating conditions in dangerous concentrations. For instance, at a gasoline dispensing pump the interior of the dispenser and the immediate area are classified as Class I, Division 1. Other Division 1 conditions are outlined in Table 11-1.

In Division 2 locations the hazardous material is not normally in the air, but it might be released if there is an accident or if there is faulty operation of equipment. Storage areas for hazardous materials are frequently Division 2 locations.

Unless equipment is intrinsically safe, it must be approved for the class, group (if any), and division in which it is used as outlined in Table 11-2. In many cases, the equipment must be approved for the specific gas, vapor, or dust that is encountered.

The Code provides rules for specific occupancies with Class I locations including **ARTICLE 510** commercial garages, aircraft hangars, gasoline stations, and bulk storage plants. Areas in other occupancies for finishing processes and anesthetizing locations in health care facilities are also Class I locations. The general rules for installation in Class I locations are modified in some cases because of the characteristics of a specific occupancy. Examples of Class II locations include grain handling and storage plants, grinding and pulverizing rooms, coal or charcoal processing plants, and areas containing magnesium or aluminum dust. Class III locations include cotton mills, clothing manufacturing plants, woodworking plants, and textile mills. In the case of Class II and Class III locations, the Code does not provide rules for specific occupancies.

Table 11-2. General Installation Rules for Equipment in Hazardous Locations

Application	Rule	Code Ref.
Conduit	Conduit must be threaded with a standard 3/4 inch taper per foot. The joints must be made up wrench-tight or a bonding jumper is required.	500-2
Equipment	a) Approval—Equipment must be approved for both the class of location and the explosive properties of the gas, vapor, or dust. Division 1 equipment may be used in Division 2 locations of the same class and group. b) Markings—Equipment must be marked to show the class, group, and operating temperature, or temperature range. (See Code for exceptions.) c) Temperature—The temperature marking must not exceed the ignition temperature of the specific gas or vapor.	500-3

11-1.1 Class I Locations

Class I locations are characterized by the presence or possible presence of flam- **500-5** mable gases or vapors in quantities sufficient to produce explosive or ignitable **ARTICLE 501** mixtures. In these areas, special attention is paid to enclosing any device that might **ARTICLE 100** produce sparking or arcing. The enclosures for switches, relays, and similar equip- **DEFINITIONS** ment must be *explosionproof* to prevent ignition of gases or vapors surrounding the enclosures. Such enclosures must be strong enough to withstand an internal explosion and also prevent hot gases generated by the explosion from escaping until they are sufficiently cooled. One method used to prevent an external explosion is to allow the hot gases caused by an internal explosion to cool sufficiently before they escape from the enclosure by forcing them to travel a specified distance. This is accomplished by threaded joints or machined flanges (clamped joints) having a precise spacing. An alternative method is to use purged and pressurized enclosures, especially for large equipment such as switchboards.

Table 11-3 lists the equipment and devices commonly used in Class I locations, either Division 1 or Division 2. The required characteristics and the Code reference for each item are listed separately for each division.

501-4   The wiring methods in Class I locations are restricted as shown in the Table. In Division 1 locations, threaded joints of conduit must have at least five threads fully engaged. These conduits when connected to enclosures must, in most cases, contain seals to prevent passage of gases, vapors, or flames.

501-5(a)   Figure 11-1 shows a typical hazardous location with the seals required in conduit runs. Seals are necessary in each conduit run entering or leaving an enclosure for switches, circuit breakers, fuses, etc., or other apparatus that may produce arcs, sparks, or high temperature. Any unions or couplings between the seal and the enclosure must be explosionproof. An enclosure containing only terminals, splices, or

Table 11-3. Installation Rules for Equipment in Class I Locations

Equipment	Division 1	Code Reference	Division 2	Code Reference
Boxes, fittings	Explosion proof	501-4(a)	Not explosion proof unless current interrupting contacts are exposed	501-4(b)
Circuit Breakers	Class I enclosure	501-6(a)	Class I enclosures unless contacts are sealed or oil-immersed	501-6(b)
Flexible cords	Extra-hard usage with grounding conductor	501-11	Same as Division 1 (with exceptions)	501-11
Fuses	Class I	501-6(a)	Standard fuses in enclosures approved for the purpose	501-6(b) (3)
Generators	Class I, totally enclosed or submerged	501-8(a)	Class I if generator has sliding contacts, switching contacts or integral resistance devices.	501-8(b)
Lighting Fixtures	Explosion proof	501-9(a)	Protected from physical damage. Switches must be approved for Class I	501-9(b) (2)
Lighting Protection	Required in areas where lightning disturbances are prevalent	501-16(c)	Same as Division 1	501-16(c)
Motors	Class I, totally enclosed, or submerged	501-8(a)	Class I if motor has sliding contacts, switching contacts or integral resistance devices; otherwise may be general-purpose	501-8(b)
Distribution Panel	Explosion proof	501-6(a)	General purpose if not more than 10 devices protect circuits to fixed lamps	501-6(b)
Portable lamps	Explosion proof	501-9(a) (1)	Same as Division 1	501-9(b) (1)
Receptacles	Explosion proof (with exception)	501-12	Same as Division 1	501-12
Seals for conduit, etc.	Approved for purpose	501-5(c)	Same as Division 1	501-5(c)

taps requires its conduit entries to be sealed if the trade size is 2 inches or larger. When required, one seal per conduit entry is necessary unless the conduit run is 36 inches or less. As illustrated in Figure 11-1(d), one seal not farther than 18 inches from either enclosure is sufficient in that case.

Seals in conduit runs are also required when conduit leaves a hazardous area. The seal may be on either side of the boundary. A seal is not required in an unbroken run of conduit passing through the hazardous area if no fittings are placed within 12 inches of the hazardous boundary.

As shown in Figure 11-2, pendant fixtures must be protected from physical **501-9** damage. They must be supplied by threaded conduit, and stems longer than 12 inches must be braced or allow flexibility.

No live parts may be exposed in Class I locations and all exposed noncurrent- **501-15,** carrying parts must be grounded. Bonding must not rely on locknuts or bushings but **501-16** be accomplished by bonding jumpers or other approved means.

Table 11-3. Installation Rules for Equipment in Class I Locations (continued)

Equipment	Division 1	Code Reference	Division 2	Code Reference
Signaling Units (Alarms, etc.)	Class I, Division 1	501-14(a)	Class I, Division 1 unless contacts are immersed in oil, sealed, or not dangerous	501-14(b)
Switches	Class I enclosure	501-6(a)	Class I enclosure unless contacts are sealed or oil-immersed	501-6(b)
Transformers, Capacitors	Installed in approved vault if filled with burnable liquid Otherwise, in vaults or approved for Class I	501-2(a)	1) General purpose 2) Control—Class I enclosure if switched or have resistors	501-2(b), 501-7(b)
Control Transformers, and Resistors	Class 1, Division 1 enclosures	501-7(a)	1) Switching mechanism—in Class I Division 1 or general—purpose if other protection provided 2) Coils—general-purpose enclosures 3) Resistors—in Class I enclosure unless not dangerous.	501-7(b)
Utilization Equipment	Class I, Division 1	501-10(a)	1) Heaters approved for Class I, Division 1 2) Others approved for Class I, Division 2	501-10(b)
Wiring methods	1) rigid metal conduit, steel intermediate metal conduit, or Type MI cable 2) flexible connections – Class I (explosion proof)	501-4(a)	1) rigid metal conduit, steel intermediate conduit, Types MI, MC, MV, TC, SNM, PLTC cable or enclosed gasketed busways or wireways. 2) Flexible connections— flexible metal conduit, liquid tight flexible metal conduit or extra-hard usage cord	501-4(b)

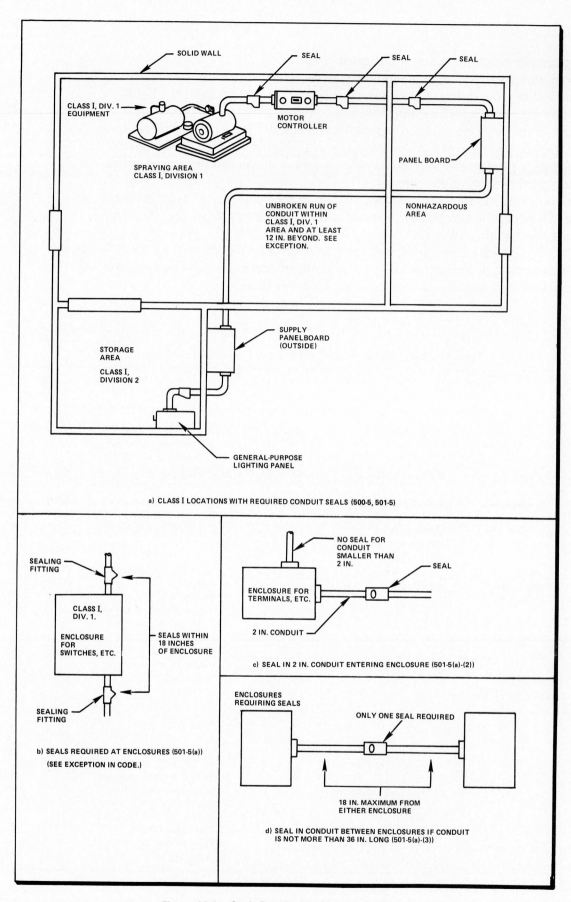

Figure 11-1. Seals Required in Class I Locations

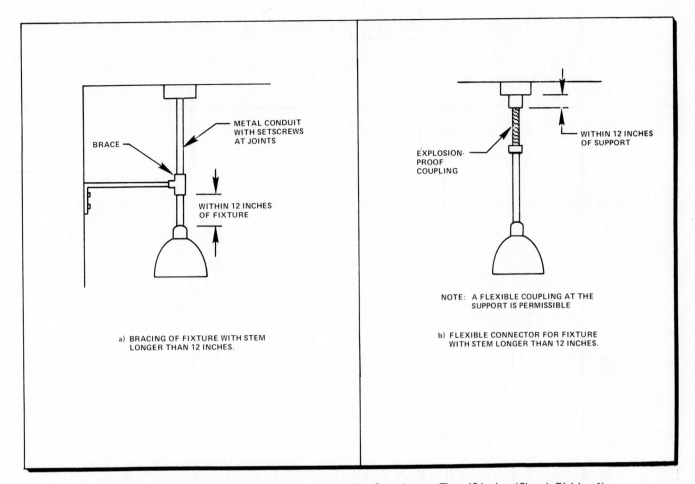

a) BRACING OF FIXTURE WITH STEM LONGER THAN 12 INCHES.

BRACE

METAL CONDUIT WITH SETSCREWS AT JOINTS

WITHIN 12 INCHES OF FIXTURE

EXPLOSION-PROOF COUPLING

WITHIN 12 INCHES OF SUPPORT

NOTE: A FLEXIBLE COUPLING AT THE SUPPORT IS PERMISSIBLE

b) FLEXIBLE CONNECTOR FOR FIXTURE WITH STEM LONGER THAN 12 INCHES.

Figure 11-2. Methods of Installing Pendant Fixture With Stem Longer Than 12 Inches (Class I, Division 1)

11-1.2 Class II Locations

Class II locations are hazardous because of the possible presence of combustible dust, electrically conductive dust, or accumulations of dust that might affect the operation of electrical equipment. The equipment used in such locations must exclude ignitable amounts of dust and not permit internal sparks or explosions to ignite external dust accumulations. Furthermore, the operating temperature of such equipment must be low enough so that it will not ignite layers of dust that may have been deposited on the equipment. **500-6** **ARTICLE 502**

Enclosures for switches, circuit breakers, and so forth must be *dust-ignition proof* and especially designed for Class II locations. Other enclosures and apparatus must meet the Code requirements summarized in Table 11-4. **502-1**

Seals in conduit are required to prevent the entrance of dust into dust-ignition proof enclosures. Figure 11-3 illustrates such methods for conduits between dust-ignition proof enclosures and general-purpose enclosures. A seal is not required if the Class II enclosure is more than 10 feet away horizontally or 5 feet above the general-purpose enclosure, or if the general-purpose enclosure is in an unclassified location. **502-5**

Table 11-4. Installation Rules for Equipment in Class II Locations

Equipment	Division 1	Code Reference	Division 2	Code Reference
Boxes, fittings	1) Class II boxes required if taps, joints or connections used or with conductive dust 2) Otherwise, dust tight with no openings	502-4(a)	Minimize the entrance of dust with tight covers and openings	502-4(b)
Circuit Breakers	Dust-ignition proof enclosures	502-6(a) (1)	Dust-tight enclosure	502-6(b)
Flexible cords	Extra-hard usage	502-4(a) (2), 502-12	Same as Division 1	502-4(b) (2), 502-12
Fuses	Dust-ignition proof enclosures	502-6(a) (1)	Dust-tight enclosure	502-6(b)
Generators	Class II, Division 1 or totally enclosed pipe-ventilated	508(a)	Dust-ignition proof or totally enclosed pipe-ventilated	502-8(b)
Lighting fixtures	Class II (165° C maximum exposed surface temperature)	502-11(a) (1)	Class II or minimize deposit of dust or lamps	502-11(b) (2)
Lightning Protection	Surge protection required	502-3	Same as Division 1	502-3
Motors	Class II, Division 1 or totally enclosed pipe-ventilated	502-8(a)	Dust-ignition proof or totally enclosed	502-8(b)
Panels (Distribution)	Dust-ignition proof	502-6(a)	Dust-tight enclosure	502-6(b)
Portable lamps	Class II	502-11(a)	Class II	502-11(b) (2)
Receptacles	Class II	502-13(a)	No live parts exposed when used	502-13(b)
Seals for conduit, etc.	Permanent and effective	502-5	Same as Division 1	502-5
Signaling units (alarms, etc.)	Class II enclosure unless contacts are immersed in oil or sealed	502-14(a)	Tight enclosures	502-14(b)
Switches	Dust ignition proof enclosures	502-6(a)	Dust-tight enclosure	506-6(b)
Transformers, Capacitors	1) Installed in vault if filled with burnable liquid 2) Otherwise, in vault or approved for Class II. 3) Not permitted when hazardous metal dust present	502-2(a)	1) Installed in vault if filled with burnable liquid. 2) Askarel over 25 kVA—provided with pressure-relief vents 3) Dry type—in vaults or in tight metal housings (600 volts or less)	502-2(b)
Utilization Equipment	Class II	502-10(a)	1) Heaters—Class II 2) Motors—dust ignition proof	502-10(b)
Ventilating Piping	Dust-tight throughout their length	502-9(a)	Tight to minimize the entrance of dust	502-9(b)
Wiring methods	Rigid or steel intermediate metal conduit, or Type MI cable	502-4(a)	Rigid or steel intermediate metal conduit; electrical metallic tubing; Types MI, MC, SNM cable or dust-tight wireways (See Code for others)	502-4(b)

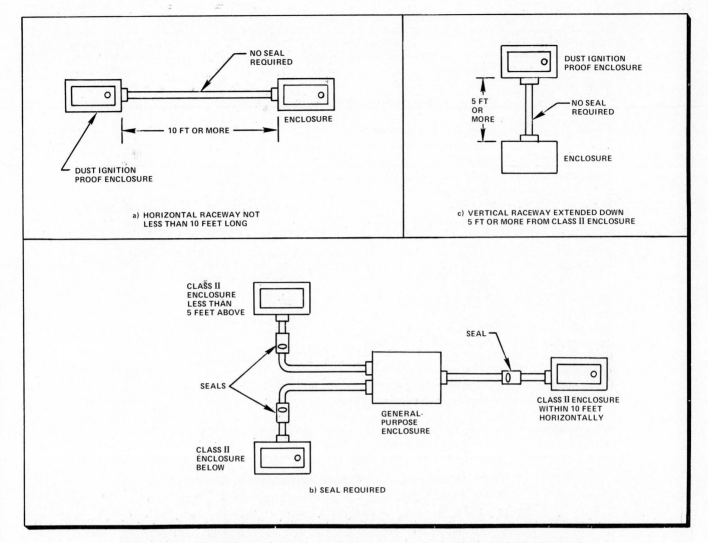

Figure 11-3. Required Sealing Between Class II Enclosures and General-Purpose Enclosures when Both are in a Class II Location

11-1.3 Class III Locations

Class III locations are potentially hazardous because of the presence of easily ignitible fibers or flyings. Class III, Division 1 locations usually are found where such material is manufactured or used in a manufacturing process. Division 2 areas are used to store easily ignitable fibers.

As with dusts, ignitable fibers are subject to combustion because of the excessive surface temperature of the equipment. Therefore, a maximum surface temperature is specified for equipment operating in Class III locations.

Table 11-5 lists the Code requirements for equipment and devices operating in Class III locations. In most cases, the Code does not distinguish between Division 1 and Division 2 of Class III locations. Since the fibers or flyings are normally large compared to dust particles, the enclosures for many devices are only required to be tight so that the entrance of the material is prevented. Enclosures for transformers, capacitors, and signaling units, however, must meet Class II standards.

500-7

503-1

ARTICLE 503

301

Table 11-5. Installation Rules for Equipment for Class III Locations

Equipment	Divisions 1 and 2	Code Reference
Boxes, fittings	Dusttight	503-3(a) (1)
Circuit breakers	Dusttight enclosures	503-4
Flexible cords	Extra-hard usage (same as Class II)	503-3(2), 503-10
Cranes	Meet Class III requirements	503-13
Fuses	Tight metal enclosure with no openings	503-4
Generators	Totally enclosed (with exception)	503-6
Lighting fixtures	Tight enclosures with no openings (165° C maximum temperature)	503-9(a)
Motors	Totally enclosed	503-6
Panels (distribution)	Dusttight	503-4
Portable lamps	Unswitched, guarded with tight enclosure for lamp	503-9(d)
Receptacles	No live parts exposed when used (same as Class II, Division 2)	503-11
Signaling units (alarms, etc.)	Suitable for Class III	503-12
Switches	Dusttight enclosures	503-4
Transformers, Capacitors	Same as Class II, Division 2	503-2
Utilization Equipment	1) Heaters—Class III 2) Motors—totally enclosed	503-8
Ventilating Piping	Tight to prevent entrance of fibers	503-7
Wiring methods	Rigid or intermediate metal conduit, rigid nonmetallic conduit, EMT, dust-tight wireways, or type MI, MC, or SNM cable	503-3

11-1.4 Comparison of Classified Locations (Division 1)

Table 11-6 provides a summary comparison of requirements and equipment specifications for the Division 1 locations of each class. Manufacturers of such equipment have their products approved for classified locations. In general, equipment used for one classified location may not be used in another unless specifically permitted by the Code.

11-2 SPECIFIC CLASS I LOCATIONS

ARTICLE 510 The Code presents installation rules for a number of specific occupancies that contain classified locations. These locations are Class I since the hazardous material present may be a flammable liquid, gas, or vapor. The rules for these locations modify the general rules in the Code and specify the extent of hazardous areas as well as any special rules for the specific occupancy.

Commercial garages and gasoline dispensing stations are selected for discussion in this section. The hazardous material is obviously gasoline or a similar fuel and the attendant vapors. Gasoline vapor is explosive in concentrations of approximately 1% to 8% by volume in air. The density of gasoline vapor is three to four times that of air; thus, areas close to or below the floor or ground level areas in areas in which gasoline is dispensed or otherwise used are normally Class I, Division I locations. Areas above the work area are usually either Division 2 locations or unclassified locations.

In any location where gasoline is dispensed, for example, from a pump, the rules for gasoline dispensers must be observed. Frequently, a garage serves as a gas station and a repair shop. Therefore, the rules for both garages and gasoline dispensing stations apply.

11-2.1 Commercial Garages

Commercial garages in which service and repair is made on vehicles propelled by a volatile flammable liquid such as gasoline have classified areas as specified in Figure 11-4. These rules normally do not apply to parking garages if they are adequately ventilated. **ARTICLE 511**

The main Class I, Division 2 hazardous area in garages includes the area up to 18 inches above the floor. Any pit or depression below the floor level is considered a Class I, Division 1 location. In either case, the enforcing agency may change the classification if it is determined that adequate ventilation is provided to remove any hazardous vapors. In each hazardous area, the installation rules for general Class I locations must be applied. The additional installation rules for electrical equipment in commercial garages are summarized in Table 11-7. **511-3**

Table 11-6. Comparison of Installation Requirements for Division 1 Hazardous Locations

Requirements	Class I	Class II	Class III
Wiring Method	• Rigid metal conduit • Steel intermediate conduit • Type MI cable	• Rigid metal conduit • Steel intermediate conduit • Type MI cable	• Rigid metal conduit • Rigid nonmetallic conduit • Intermediate metal conduit • EMT • Dust-tight wireways • Type MI, MC or SNM cable
Seals in Conduits Required	• Within 18 inches of enclosure Leaving Class I, Division 1 • Other cases	• Between dust-ignition proof enclosures and general-purpose enclosures (other means allowed)	Not required
Boxes, Fittings and joints	Explosionproof	Dust-ignition proof or dust-tight	Dusttight
Enclosures for Meters, Switches, etc.	Explosionproof, or Purged and Pressurized Enclosures	Dust-ignition proof	Dusttight
Motors	• Explosionproof • Totally-enclosed • Submerged	• Dust-ignition proof • Totally-enclosed	Totally enclosed (with exceptions)
Lighting Fixtures	• Explosionproof	Approved for Class II, Division 1	Enclosed
Portable Lamps	Explosion proof	Approved for Class II	Enclosed
Flexible Cords	Extra-hard usage	Extra-hard usage	Extra-hard usage
Receptacles and Plugs	Explosionproof	Approved for Class II	No live parts exposed during use
Bonding	Jumpers required	Jumpers required	Jumpers required

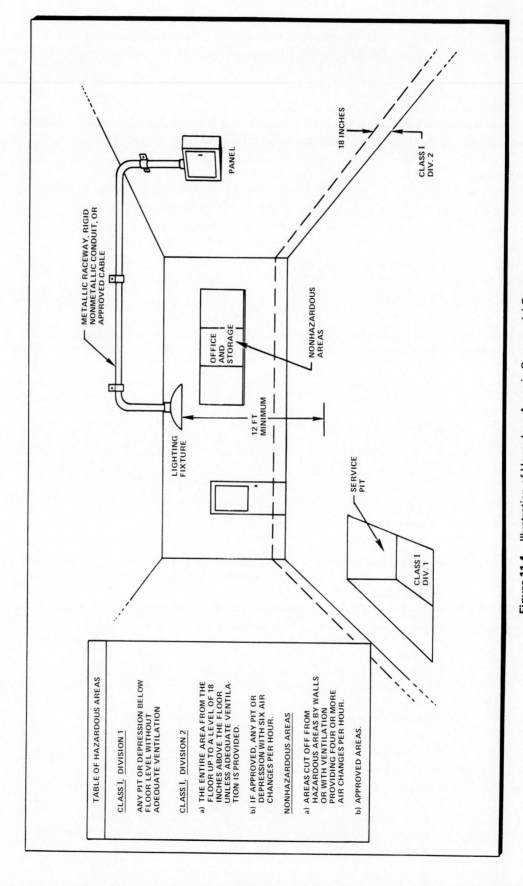

Figure 11-4. Illustration of Hazardous Areas in Commercial Garages

TABLE OF HAZARDOUS AREAS

CLASS I, DIVISION 1

ANY PIT OR DEPRESSION BELOW FLOOR LEVEL WITHOUT ADEQUATE VENTILATION

CLASS I, DIVISION 2

a) THE ENTIRE AREA FROM THE FLOOR UP TO A LEVEL OF 18 INCHES ABOVE THE FLOOR UNLESS ADEQUATE VENTILATION IS PROVIDED.

b) IF APPROVED, ANY PIT OR DEPRESSION WITH SIX AIR CHANGES PER HOUR.

NONHAZARDOUS AREAS

a) AREAS CUT OFF FROM HAZARDOUS AREAS BY WALLS OR WITH VENTILATION PROVIDING FOUR OR MORE AIR CHANGES PER HOUR.

b) APPROVED AREAS.

METALLIC RACEWAY, RIGID NONMETALLIC CONDUIT, OR APPROVED CABLE

PANEL

LIGHTING FIXTURE

OFFICE AND STORAGE

NONHAZARDOUS AREAS

12 FT MINIMUM

18 INCHES

CLASS I DIV. 2

SERVICE PIT

CLASS I DIV. 1

304

Some locations in a garage are considered nonhazardous but the installation **511-6,** Code rules in Table 11-7 must nevertheless be followed. In particular, the wiring methods that may be used above hazardous locations are restricted and equipment which may produce sparks, arcs, or hot metal, including lighting fixtures, must be located at least 12 feet above the floor unless constructed in a manner to eliminate any hazard.

Table 11-7. Installation Rules for Electrical Equipment in Commercial Garages

Application	Rule	Code Ref.
Equipment in Hazardous Areas	Wiring and equipment in hazardous areas must conform to the rules for Class I locations.	511-4, 511-5
Equipment not in Hazardous Areas		
Arcing Equipment	Equipment that may produce arcs or sparks and is located less than 12 feet above the floor level must be totally enclosed or constructed to prevent escape of sparks or hot metal.	511-7(a)
Battery Charging	Battery chargers must not be located in hazardous areas.	511-8
Lighting Fixtures	Lighting over vehicle lanes must be at least 12 feet above the floor unless totally enclosed or constructed to prevent escape of sparks or hot metal.	511-7(b)
Portable Lamps	Portable lamps must have a handle, hook, and guard and be unswitched type.	511-3(f)
Wiring Methods	Wiring methods used above hazardous areas must be metallic raceways, rigid nonmetallic conduit, or type MI, TC, SNM, or MC cable. Note: Cellular metal floor raceways are permitted in certain locations.	511-6(a)
Miscellaneous Equipment	Electric vehicle charging equipment must be equipped with suitable extra-hard service flexible cords with connectors. No connections permitted within hazardous location. See Code for additional requirements.	511-9
GFCI Protection	All 125-volt single phase, 15- and 20-ampere receptacles for connection of hand tools, portable lights, and similar equipment must have GFCI protection.	511-10

11-2.2 Gasoline Dispensing and Service Stations

Some of the hazardous locations in a gasoline service station are summarized in **ARTICLE 514** Table 11-8. Not only the dispenser, but also fill pipes or vent pipes and pits below grade constitute a possible hazard. The specific hazardous areas around outdoor dispensing pumps are shown in Figure 11-5.

As listed in Table 11-9, equipment and wiring in a service station must comply with specific Code rules, some of which apply even outside hazardous areas. The **514-4** wiring above hazardous areas, for example, must comply with the requirements for such wiring in a commercial garage.

Any circuit to the dispenser including any grounded neutral, must be discon- **514-5** nected completely. All conductors are required to be disconnected to insure that no live conductors are present during repair of the dispenser as a result of accidental polarity reversal or the accidental application of a voltage to an improperly grounded neutral.

Seals are required at the dispenser and in all conduits at the boundary of the **514-6** hazardous area. This applies to both horizontal and vertical boundaries.

Table 11-8. Hazardous Locations (Gasoline Dispensing and Service Stations)

Location	Outdoor Dispenser	Overhead Dispenser	Tank Vent Pipes and Fill-Pipes	Below Grade In Lube Room With Dispenser
Class I, Division 1	a) The space within dispenser up to 4 feet from base b) The space within a nozzle boot c) Any space below grade level	The volume within enclosure and 18 inches in all directions not cut off by ceiling or wall	The spherical volume within 3-foot radius from vent pipe discharge point Any space below grade level within the classified area	The space within pit or space below grade in lube room without ventilation
Class I Division 2	Any area in outside location within 20 feet horizontally and 18 inches above ground around dispenser Space above Division I area within dispenser (Indoor dispensers have other requirements)	a) Two feet horizontally in all directions beyond Division 1 area and extending to grade below b) The area above grade to 18 inches for a horizontal distance of 20 feet measured from point below edge of dispenser	a) The area in outside location within 10 feet horizontally of fill pipe and upward 18 inches from ground (for loose fill) b) The volume between 3-foot and 5-foot radius from vent pipe discharge point c) The cylindrical volume below Division 1 and 2 locations extending to ground for vent pipe that does not discharge upward	a) The space within entire lube room up to 18 inches above floor or grade extending 3 feet horizontally b) The space within 3 feet of any dispensing point c) Any pit with ventilation

306

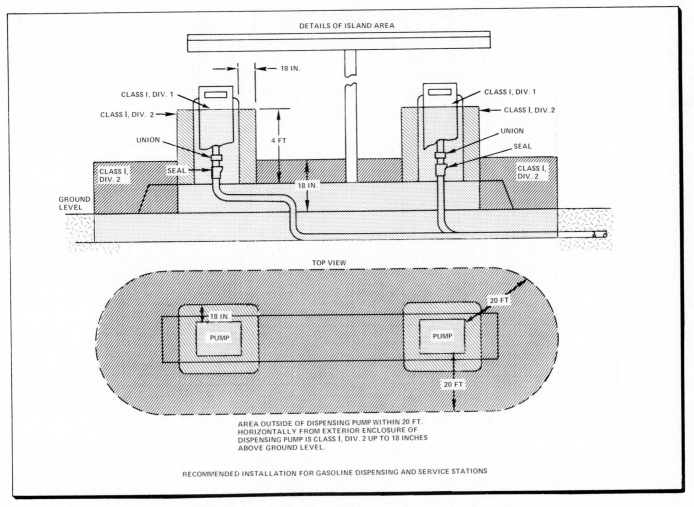

DETAILS OF ISLAND AREA

18 IN.

CLASS I, DIV. 1

CLASS I, DIV. 2

UNION

4 FT

SEAL

CLASS I, DIV. 2

GROUND LEVEL

18 IN.

CLASS I, DIV. 1

CLASS I, DIV. 2

UNION

SEAL

CLASS I, DIV. 2

TOP VIEW

18 IN.

PUMP

PUMP

20 FT

20 FT

AREA OUTSIDE OF DISPENSING PUMP WITHIN 20 FT. HORIZONTALLY FROM EXTERIOR ENCLOSURE OF DISPENSING PUMP IS CLASS I, DIV. 2 UP TO 18 INCHES ABOVE GROUND LEVEL.

RECOMMENDED INSTALLATION FOR GASOLINE DISPENSING AND SERVICE STATIONS

Figure 11-5 Hazardous Areas Near Gasoline Dispensing Pumps

Table 11-9. Installation Rules for Electrical Equipment in Service Stations

Application	Rule	Code Ref.
Equipment in Hazardous Locations	Wiring and equipment in hazardous locations must conform to the rules for Class I locations. Exception: wiring below grade.	514-3 514-6(a)
Equipment above Hazardous Locations	Wiring and equipment above hazardous locations must conform to the rules for such equipment in commercial garages.	514-4
Gasoline Dispenser	a) A disconnecting means must be provided for each circuit leading to or through a dispensing pump to disconnect all conductors including the grounded neutral, if any.	514-5
	b) An approval seal is required in each conduit entering or leaving a dispenser.	514-6(a)
Grounding	Metal portions of all noncurrent-carrying parts of dispensers, raceways, etc., must be grounded.	514-7

Underground Wiring	Underground wiring must be installed as follows:
	    a) Within 2 feet of ground level—in rigid metal or steel intermediate conduit.
	    b) Buried 2 feet or more—in rigid metal conduit, steel intermediate conduit, or rigid nonmetallic conduit.
	Exception: Type MI cable is permitted in some cases.
		514-8

11-3 OTHER SPECIAL OCCUPANCIES

In addition to classified locations, other special occupancies listed below are covered in the Code. The installation rules for each occupancy or location are presented in the appropriate Code article.

Location	Article
Health care facilities	517
Places of assembly	518
Theaters	520
Motion picture and television studios	530
Motion picture projectors	540
Manufactured buildings	545
Agricultural buildings	547
Mobile homes and mobile home parks	550
Recreational vehicles and recreational vehicle parks	551
Floating buildings	553
Marinas and boatyards	555

TEST CHAPTER 11

I. True or False *(Closed-Book)*

		T	F
1.	Equipment approved for a Division 1 location may be used in a similar Division 2 location.	[ ]	[ ]
2.	Gasoline is a Class I, Group D substance.	[ ]	[ ]
3.	Explosionproof equipment may be used in Class II locations.	[ ]	[ ]
4.	The sealing compound in a completed conduit seal must be at least ¾ inch thick in Class I locations.	[ ]	[ ]
5.	Open squirrel-cage motors without brushes are permitted in Class I, Division 1 locations.	[ ]	[ ]
6.	Double locknuts as a bonding means for conduit is not permitted in Class I locations.	[ ]	[ ]
7.	No transformer may be installed where dust from magnesium is present.	[ ]	[ ]
8.	A method to prevent dust from entering through a conduit into a dust-ignition proof enclosure is always required.	[ ]	[ ]
9.	Portable lamps must not be used in a Class II location.	[ ]	[ ]
10.	Stockrooms in commercial garages are normally not classified areas.	[ ]	[ ]
11.	In some cases, lighting fixtures may be placed over vehicle lanes at a height of less than 12 feet.	[ ]	[ ]
12.	Any area within 20 feet around a gasoline pump is a Class I, Division 1 location.	[ ]	[ ]
13.	The spherical volume within 3 feet around the discharge of a gasoline tank vent pipe is a Class I, Division 1 location.	[ ]	[ ]
14.	Interiors of spray booths are Class I, Division 2 locations.	[ ]	[ ]
15.	In an anesthetizing location the area up to 5 feet above the floor is considered a Class I, Division 2 location.	[ ]	[ ]
16.	Type NM cable may be used for wiring in a place of assembly that has a fire-rated construction.	[ ]	[ ]
17.	In a marina the space below a gasoline dispenser and 4 feet horizontally is a Class I, Division 1 location.	[ ]	[ ]

II. Multiple Choice *(Closed-Book)*

1. Conduit in hazardous locations must be made up wrenchtight or:
   - (a) Sealed
   - (b) Bonded
   - (c) Tapered

   1. _____

2. Locations where fibers are stored are:
   - (a) Class II, Division 2
   - (b) Class II, Division 1
   - (c) Class III, Division 2

   2. _____

3. Liquidtight flexible metal conduit is allowed for motor hookups in:
   - (a) Class I, Division 1
   - (b) Class I, Division 2
   - (c) Underground in gasoline stations

   3. _____

4. Seals in conduit supplying a panelboard must be placed near the enclosure within:
   - (a) 18 inches
   - (b) 36 inches
   - (c) 24 inches

   4. _____

5. A conduit to a splice box must be sealed if the trade size is:
   - (a) ½ inch or larger
   - (b) 1 inch or larger
   - (c) 2 inches or larger

   5. _____

6. In a Class I, Division 2 location a general-purpose enclosure may house circuit breakers protecting circuits of fixed lamps if the number of devices is not more than:
     (a) 12
     (b) 20
     (c) 10

6. _____

7. In a Class I location a pendant fixture with a rigid stem longer than 12 inches must be braced:
     (a) 12 inches from the top
     (b) Not more than 12 inches from the fixture
     (c) On the stem

7. _____

8. A dust-ignition proof enclosure supplied by a horizontal raceway does not require a raceway seal if the conduit length is at least:
     (a) 10 feet
     (b) 5 feet
     (c) 12 feet

8. _____

9. Motors in Class II locations must be approved or be:
     (a) Totally enclosed
     (b) Totally enclosed pipe-ventilated
     (c) Explosionproof

9. _____

10. When a flexible cord supplies a pendant fixture in a Class II location, it must be approved for:
     (a) Extra-hard usage
     (b) Class II
     (c) Hard usage

10. _____

11. In Class III locations, pendant fixtures must be braced or have a flexible connector if the stem is longer than:
     (a) 6 inches
     (b) 24 inches
     (c) 12 inches

11. _____

12. In a garage, unenclosed equipment that may produce arcs or sparks must be placed at least what height above the floor?
     (a) 18 inches
     (b) 12 feet
     (c) 10 feet

12. _____

13. The Class I, Division 1 location in a gasoline dispenser extends from the base upward for:
     (a) 18 inches
     (b) 3 feet
     (c) 4 feet

13. _____

14. The hazardous area near a gasoline fill-pipe with loose fill extends horizontally:
     (a) 10 feet
     (b) 20 feet
     (c) 5 feet

14. _____

15. The disconnecting means for a 120/240-volt circuit for a gasoline dispenser must disconnect:
     (a) The ungrounded conductors
     (b) All conductors
     (c) The neutral

15. _____

III. Fill in the Blanks (Closed-Book)

1. Explosionproof enclosures are used in _____ locations.

2. Class I substances are classified by _____ _____, _____ _____, and _____ .

3. Threaded joints in Class I, Division 1 locations must have at least _____ threads fully engaged.

4. Whenever seals are required, they must be no closer than _____ inches from the enclosure.

5. An unbroken run of conduit through a Class I area does not require sealing if the first fitting is more than _____ inches beyond the boundary.

6. Totally enclosed motors in Class I, Division 1 locations must have no external surface with an operating temperature in excess of _____ percent of the ignition temperature of the hazardous substance involved.

7. Telephones in Class I locations require _____ enclosures.

8. Any heating appliance used in a Class II, Division 2 location must be_____ .

9. Type MI and _____ cables are approved wiring methods in Class III locations.

10. Receptacles in Class III locations must meet the requirements for Class _____, Division _____ locations.

11. Flexible cords for pendant lighting in a commercial garage must be approved for _____ _____ usage if the fixture is above any hazardous area.

12. Any pit without ventilation in a commercial garage is a _____ , _____ _____ location.

13. The Class I, Division 1 area for an overhead gasoline dispensing unit includes the enclosure and extends _____ inches in all directions from the enclosure.

14. A sealing fitting at a gasoline dispenser must be the _____ fitting after the conduit emerges from the ground or concrete.

15. When PVC conduit is used underground at a service station, it must be buried at least _____ feet.

General
Electrical Theory

*Part III provides practical electrical theory discussions.
The material presented is consistent
with the types of general theory questions
included in the Master Electrician's examination
and with the depth of understanding needed
to perform the required calculations
for electrical system design.*

12

Review of Electrical Theory

This chapter presents a brief review of electrical theory consistent with the level presented in the Guide and in the various examinations. The areas of direct-current theory, conductors, alternating-current theory, and equipment in alternating-current circuits are discussed. Important formulas for problem solving in these areas are listed in Appendix A for convenient reference. The test in this chapter requires the application of these formulas to solve problems typical of many electrician's examinations.

12-1 DIRECT-CURRENT THEORY

The calculation of current and voltage in a direct-current circuit containing only resistive elements is determined by Ohm's law. This basic law states that the voltage across a resistive element is equal to the resistance of the element times the current through the element or

$$E = I \times R \quad \text{(volts)}$$

where E is the voltage, I is the current in amperes, and R is the resistance in ohms.

The power dissipated by the resistance R is given by

$$P = I^2 \times R = \frac{E^2}{R} \quad \text{(watts)}$$

When resistive elements are connected in a circuit, it is possible to combine the resistances to obtain a total resistance for the circuit. Then Ohm's law applies to the entire circuit if R in the equation above is taken to be the total resistance. Various formulas for combining resistive elements are given in Table A-1 of Appendix A. Review problems for direct-current circuits are contained in Part I of the test for this chapter.

12-2 CONDUCTORS

A conductor used in electrical circuits has a resistance that increases with its length and decreases with its area. The relationship is expressed as

$$R = \rho \, \frac{L}{A} \quad \text{(ohms)}$$

where L is the length in feet, A is the area in circular mils, and ρ is the resistivity which depends on the material and the temperature.

The circular mil area is the square of the diameter of the wire in mils. Thus, a wire of diameter one-thousandth of an inch (.001 inch) has an area of 1 circular mil (cmil). The true area in square inches is given by

$$A \, (\text{sq in.}) = A \, (\text{cmil}) \times \frac{\pi}{4} \times \frac{1 \, \text{in.}^2}{1 \, 000 \, 000 \, (\text{mils})^2}$$

$$= A \, (\text{cmil}) \times .7854 \times 10^{-6} \, \text{in.}^2$$

This formula is useful in converting between the area of a rectangular bus bar and that of a circular conductor since the current-carrying capacity of either depends on the area. Thus, they can be compared.

For large conductors, the circular mil area is measured in thousands of circular mils or MCM. Thus, a 1000 MCM conductor has an area of 1 000 000 cmils, a diameter of 1000 cmils or 1 inch, and an area of .7854 square inches.

The resistivity of the conductor depends on the material, either aluminum or copper, in most applications. The units of resistivity are ohms per circular mil-foot, which measures the resistance of a conductor 1 circular mil in area and 1 foot long. Typical values at 25°C (77°F) are as follows:

    a. Aluminum   17.7 ohms − cmil/foot
    b. Copper   10.5 ohms − cmil/foot

These values apply at the temperature stated and would increase with increasing temperature. The values at 75°C (167°F) are approximately 12.6 for copper and 20.6 for aluminum.

TABLE 8, CHAPTER 9 Many tables list the resistance of conductors in ohms per 1000 feet. Thus, if the length of a given size conductor is known, its resistance may be determined from the appropriate table. The code values are given at 75°C (167°F).

Selected formulas are listed in Table A-2 of Appendix A to aid in the solution of problems presented in Part II of the test for this chapter.

12-3 ALTERNATING-CURRENT THEORY

The alternating-current voltage or current waveform is characterized by its amplitude, frequency, and phase angle. The amplitude is the value of the wave at its maximum or peak and the frequency is the number of times the wave repeats its full cycle during 1 second. If this occurs 60 times per second, the frequency is 60 hertz (Hz). The period of the wave is the time duration of each cycle or the reciprocal of the frequency given by

$$T = \frac{1}{f} \quad \text{(seconds)}$$

where T is the period and f is the frequency in hertz.

The phase angle determines the shift in time of the wave from a reference point, with 360° being a shift of one entire cycle. In circuits that contain inductance or capacitance the voltage and current waveforms are not in phase but are shifted by a phase angle that depends on the properties of the circuit.

An important result of alternating-current theory is that a sinusoidal waveform such as the common 60-Hz power distribution wave can be represented by its root mean square (rms) value. This value is obtained by dividing the peak value by the square root of 2. A voltage wave with a peak of 162.6 volts has a root mean square value of $162.6 \div 1.414 = 115$ volts. The alternating-current waveform is then characterized by its root mean square value, frequency, and phase angle.

A brief review is given in this section of inductors and capacitors, impedance, power, and power factor in ac circuits. Table A-3 in Appendix A contains summary formulas for the problems in Part III of the chapter test.

12-3.1 Inductors and Capacitors

In addition to resistors, ac circuits often contain inductors and capacitors. These elements are characterized by their inductive reactance and capacitive reactance, respectively. When an ac voltage is applied to either circuit element, the voltage and current waveforms are not in phase. However, the root mean square magnitude of the current I is related to the root mean square value of the voltage V across an inductor or capacitor by the expression

$$I\,\text{(amperes)} = \frac{V\,\text{(volts)}}{X\,\text{(ohms)}}$$

where X is the reactance of the element as given by formulas in Table A-3. This reactance is measured in units of ohms.

12-3.2 Impedance

Impedance, abbreviated $\vec{Z}$, is the ratio of the voltage to the current in an ac circuit. Since the voltage and current may be out of phase, a phase angle is associated with the impedance. However, the magnitude of $\vec{Z}$, written Z, depends only on the resistance and reactance of the circuit elements as shown in Table A-3 and is the ratio of the root mean square voltage to the root mean square current.

12-3.3 Power and Power Factor

Power is the rate of doing work. The average power is associated only with resistive elements in a circuit. Capacitors and inductors store and release energy during each cycle of the ac wave, but they do not absorb or generate power on the average. However, their effect in a circuit is determined by their rating in voltamperes reactive or VARS. This value is calculated as follows:

$$\text{VARS} = I^2 \times X$$

where I is the root mean square current through the element and X is the reactance. The rating in VARS for a capacitor or inductor is mathematically similar to the power in a resistor in watts.

To combine the power in watts and the voltamperes reactive to determine the "apparent" power or voltampere requirement of a circuit, the power triangle is drawn. The adjacent side represents power in watts, and the opposite side at 90° is the reactive component. The hypotenuse becomes the voltampere component. Figure 12-1 shows the impedance, voltage, and power relationships in an ac circuit and defines the phase angle between the power and the voltampere component for a series circuit consisting of a resistor and an inductor. The reactive component in the power triangle is drawn down to indicate that the current has a lagging phase angle with respect to the voltage. In a capacitive circuit the reactive component would be drawn upward by convention.

The power factor is the cosine of the phase angle which is calculated as

$$\text{pf} = \cos \theta = \frac{\text{power (watts)}}{\text{voltamperes}}$$

In the circuit shown in Figure 12-1 the power factor is given by

$$\text{pf} = \frac{R}{Z}$$

since the same current flows in each element.

When the power factor of a circuit is known, the power can be calculated from the voltampere product for the circuit. If the root mean square voltage and current are known, then

$$P\,(\text{watts}) = V \times I \times \text{pf}$$

for a single-phase circuit. In a balanced three-phase circuit

$$P\,(\text{watts}) = \sqrt{3} \times V \times I \times \text{pf}$$

12-3.4 Power Factor Correction

To improve the power factor of a load, inductance or capacitance must be added to the circuit to reduce the reactive voltampere rating of the circuit. The ideal power factor is unity since the input voltampere rating is then converted entirely to

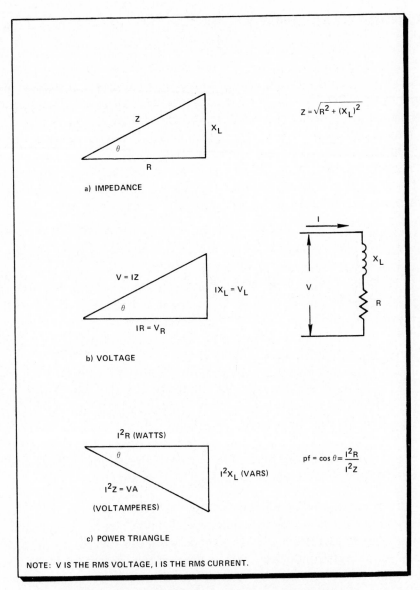

$$Z = \sqrt{R^2 + (X_L)^2}$$

a) IMPEDANCE

b) VOLTAGE

$$pf = \cos\theta = \frac{I^2 R}{I^2 Z}$$

c) POWER TRIANGLE

NOTE: V IS THE RMS VOLTAGE, I IS THE RMS CURRENT.

Figure 12-1. Relationships in Alternating-Current Circuits

power in watts. This is not always possible, but an increase in the power factor of the circuit is desirable. The formulas of Table A-3 give correction calculations in terms of the power and the new desired power factor of the circuit.

12-3.5 Energy (watt-hours)

The electrical energy supplied to a circuit is measured as the power consumed over a specified period of time. This energy is measured in watt-hours or kilowatt-hours as follows:

$$\text{watt-hours} = \text{power in watts} \times \text{time in hours}$$

The kilowatt-hour is the energy resulting from the power of 1 kilowatt operating for 1 hour.

In certain circuits the units of watts, voltamperes, or voltamperes reactive may be too small for convenient measurement. These units in thousands become kilowatts (kW), kilovoltamperes (kVA), and kilovoltamperes reactive (kVAR), respectively.

12-4 EQUIPMENT IN AC CIRCUITS

When equipment is supplied by an ac circuit, the calculation of the voltage, current, or voltampere requirement is based on the characteristics of the circuit and the equipment. A given single-phase or three-phase load draws a current at its rated voltage that is easily determined by basic ac theory. However, the voltage at the load may be less than the source voltage because of the voltage drop in the supply conductors. When the voltage drop is expected to be significant i.e., more than several percent of the supply voltage, it should be calculated as shown in this section.

The specific equipment considered in this section includes transformers and motors. The important formulas are summarized in Table A-4 and problems related to equipment in ac circuits are given in Part IV of this chapter's test.

12-4.1 Loads

Each item of ac equipment represents a load for the supply circuits whose capacity must be great enough to supply every unit. The rating of the load is normally specified in watts for purely resistive loads or in voltamperes for loads with reactive characteristics. The total load for a circuit or system determines the required rating of the generating equipment and any transformers used to stepup or stepdown the voltage of the system. The size or rating of conductors and overcurrent protective devices is based on the current drawn by the loads.

In single-phase and balanced three-phase circuits the current in amperes is calculated by using the formulas presented in Table A-4. The line current for each circuit depends only on the voltage and the voltampere rating of the load.

If the load of a three-phase circuit is unbalanced, the individual phases of the circuit draw currents that depend on the load of each phase. The calculation of the load currents involves ac theory beyond the scope of this Guide, but the calculation is treated in several of the references mentioned in Appendix C.

12-4.2 Voltage Drop in Conductors

Since a conductor has a resistance that depends on its size and length, the current flowing in a conductor causes a voltage drop between the source and the load. This voltage drop for a single conductor is the product of the current times the resistance of the conductor as given in Table A-4.[1] This drop is not, however, the voltage drop for the entire circuit since at least two conductors are required for any circuit, with three or four required in three-phase circuits.

The voltage drop for a circuit is defined as the difference between the voltage of the source and voltage at the load expressed as

$$V_{drop} = V_{source} - V_{load}$$

The percentage voltage drop in terms of the source voltage is

$$V_{drop} (\%) = \frac{V_{drop}}{V_{source}} \times 100$$

[1]The voltage drop calculations given in this section must be modified for any of the following conditions: (1) the power factor of the load is not 1.0, (2) the inductance of the circuit cannot be neglected, or (3) the ac resistance of the conductors must be considered. The *American Electricians Handbook* listed in the Bibliography explains procedures in these cases.

Thus, a two-wire, 230-volt source supplying 220 volts at the load has a 10-volt or 4.3% voltage drop.

In two-wire, single-phase circuits the voltage drop is calculated as

$$V_{drop} = 2 \times I \times R$$

where R is the resistance of one of the supply conductors. The value of R is determined from tables of conductor resistance based on the length of the conductor. The formula also may be used to determine the voltage drop for three-wire, single-phase circuits with a balanced load. In these circuits there is no current in the neutral conductor and hence no voltage drop from line-to-neutral.

Three-wire, three-phase circuits with a balanced load have a voltage drop determined as

$$V_{drop} = .866 \times 2 \times I \times R$$

where the factor .866 ($\sqrt{3}/2$) is due to the three-phase circuit and the other values represent the two-wire voltage drop in a single-phase circuit. Rewritten, the formula becomes

$$V_{drop} = \sqrt{3} \times I \times R$$

where I is the line current and R is the resistance of *one* of the conductors supplying the three-phase load. Thus, a 480-volt, three-phase circuit supplied by conductors with 1-ohm(Ω)resistance which carry 1 ampere has a voltage drop of

$$V_{drop} = 1.732 \times 1\,A \times 1\,\Omega = 1.732\,V$$

and a percentage voltage drop of .36%.

In a balanced, three-phase, four-wire circuit the line-to-neutral voltage drop is

$$V_{drop} = .5 \times 2 \times I \times R = I \times R$$

or one-half of the two-wire drop in a single-phase circuit.

12-4.3 Transformers

Transformers are used in ac systems to change the voltage level of a distribution system or service when required. Depending on the turns ratio of the primary and secondary windings, the voltage may be increased or decreased. Since the voltampere rating of the primary is equal to that of the secondary in an ideal transformer, the ratio of primary to secondary current must change as indicated by the formulas in Table A-4.

12-4.4 Motors

There is an extensive literature covering the characteristics and use of motors in ac circuits. From this wealth of information only a few basic principles concerning speed, efficiency, and power factor are presented here.

The speed of a synchronous motor is determined by the frequency of the ac voltage waveform and the number of poles the motor contains. The speed is proportional to frequency and decreases with the number of poles as shown in Table A-4.

In order to calculate the power output of a motor, the input power and the efficiency must be known. These quantities are related by the equation

$$P_{\text{out}} = P_{\text{in}} \times \eta$$

where η is the efficiency. Normally, the power input is measured in watts and the output power is measured in horsepower where 1 horsepower equals 746 watts. When a motor has a power factor, the input power is the voltampere input times the power factor.

The power factor of a motor may be improved by adding a capacitor to the circuit to cancel the lagging reactive component of the motor load. Once the reactive component is determined, the value of the capacitor to correct or increase the power factor is found as discussed in Section 12-3.

12-4.5 Electrical Diagrams

The reader should be familiar with the wiring diagrams and elementary or one-line diagrams for transformers, motors, and motor controllers. The wiring diagrams show the complete electrical system and the connection of each element. Since the complete diagram is very detailed, interpretation is sometimes difficult and the simpler elementary diagram is used. The elementary diagram is intended to yield a clear picture of the operation of a circuit, but it is not detailed enough to be followed when a circuit is being wired.

TEST CHAPTER 12

I. Basic dc Theory

1. What is the current in a series circuit consisting of resistors of 20 ohms, 10 ohms, and 30 ohms supplied by a 100-volt source?
2. A 20-ohm and a 15-ohm resistor are in series with a 120-volt source. What is the voltage across the 15-ohm resistor? What power is dissipated by the 15-ohm resistor?
3. Two 600-watt heaters are connected in series to a 120-volt source. What is the current in the circuit and what is the resistance of each heater?
4. What is the equivalent resistance of a 20-ohm and a 40-ohm resistor in parallel?

5. In the circuit shown below find the current from the 120-volt source. The values given are in ohms.

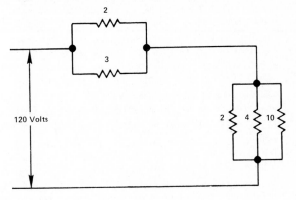

II. Conductors

1. What is the circular mil area of a No. 10 conductor whose area is .008155 square inches?
2. Calculate the dc resistance of 100 feet of copper wire .102 inches in diameter at 75°C.
3. It is required to replace a .75-inch × 1 inch bus bar with a conductor that carries the same current. Determine the MCM size of the conductor.
4. Calculate the dc resistance of 400 feet of No. 12 solid copper conductor at 75°C.

5. What is the dc resistance of four No. 1/0 aluminum conductors in parallel for a length of 1000 feet?
6. Calculate the ratio of the square mil area to the circular mil area for a conductor ¼ inch in diameter.
7. What is the diameter of a conductor 15 625 circular mils in area?

III. ac Theory

1. What is the period of a 60-Hz ac waveform?
2. In a 60-Hz ac waveform a phase shift of 60 degrees represents what shift in time?
3. A coil without resistance connected to a 150-volt, 60-Hz supply draws 3 amperes. Calculate the reactance and the inductance.
4. A coil with 5 ohms of resistance and 12 ohms of reacance is connected to a 120-volt supply. What is the current in the coil?
5. Compute the power factor of the coil in problem 4.
6. A resistor of 20 ohms and a capacitance of 31.8 ohms reactance are connected in series to a 200-volt ac source. Calculate (a) the current, (b) the voltage drop across the resistor, and (c) the power delivered by the source.

7. A capacitor connected to 240-volt source draws 120 amperes. What is the rating of the capacitor in VARS?
8. A 50-kilowatt load has a lagging power factor of .8. If the line voltage is 240-volts, find (a) the kilovolt-amperage for the load and (b) the current.
9. A balanced, three-phase load connected to a 208Y/120-volt supply has a power factor of .75 and draws 50 amperes. Calculate the power of the load.
10. A 10-kilowatt load operates continuously. What is the energy consumption in one day?
11. A 25-kilowatt load has a lagging power factor of .6. Determine the kilovolt-ampere reactive rating of the load and the value of the capacitive reactance to increase the power factor to .80.

IV. Equipment in ac Circuits

1. Two 50-kilowatt, 240-volt, single-phase loads are connected to a circuit. First, determine the currents drawn by each load if the power factors are 1.0 and .8, respectively. Next, combine the two loads and calculate the overall power factor, required kilovolt-amperes, and current.

2. A balanced 30-kilowatt load is connected to a 480-volt, three-phase supply. If the power factor is .75, find the supply current.

3. Calculate the voltage drop and percentage voltage drop if No. 10 copper conductors supply loads 1000 feet away. The individual loads have the following characteristics: (a) single-phase, 200-volt load with resistance of 18 ohms; (b) three-phase, 230-volt load of 2 kilowatts; (c) 2-kilowatt, 277-volt load supplied by a 480Y/277-volt circuit.

4. What size copper conductor is required to limit the voltage drop of a 50-ampere, 120-volt, single-phase load to less than 3% if the load is 300 feet away.

5. Determine the primary and secondary rated current for a 480/208/120-volt, 300-kilovolt-ampere transformer.

6. Determine the speed of a four-pole ac motor connected to a 60-hertz supply. What would the speed be if the supply frequency were 120 hertz?

7. A 460-volt, three-phase motor has a power factor of .8 and an efficiency of .85. If the motor draws 75 amperes, determine the horsepower delivered by the motor.

8. Using the appropriate Code tables, determine the efficiency of the following motors: (a) 1-horsepower, 115-volt, single-phase; (b) 5-horsepower, 115-volt, single-phase; (c) 1-horsepower, 460-volt, three-phase; (d) 100-horsepower, 460-volt, three-phase.

9. A 460-volt, three phase, 50-horsepower motor has a power factor of .7. Determine the required rating of the capacitor to raise the power factor to 1.0.

IV

Examinations

Part IV presents general review examinations to test the reader's understanding of the material presented in the Guide. The examinations include instructions and time limits representative of typical Master Electrician's examinations given around the country for both city and state licenses. Complete solutions to all examination problems are contained in Appendix B.

FINAL EXAMINATION NO. 1

Total Time: 4 Hours

Only the Code Book May Be Used
for the Open-Book Portion

I. *(Closed-Book)* **Test Time: 1 Hour**
True or False

		T	F
1.	No. 10 conductors may be connected by screws and nuts with upturned lugs.	[ ]	[ ]
2.	Medium-base lamps may be used on branch circuits of 30 amperes.	[ ]	[ ]
3.	EMT shall be supported at least every 15 feet.	[ ]	[ ]
4.	The wiring for a discontinued outlet in a cellular metal floor raceway shall be removed from the raceway.	[ ]	[ ]
5.	The minimum length for a pull box with 2-inch conduits for a straight pull is 16 inches.	[ ]	[ ]
6.	Wall-mounted ovens and counter-mounted cooking units are permitted to be cord-and plug-connected.	[ ]	[ ]
7.	Wiring beneath the dispenser in a gas station is considered Class II, Division 1 wiring.	[ ]	[ ]
8.	A motor disconnecting means located within 75 feet of the motor and visible is "in sight."	[ ]	[ ]
9.	An emergency storage battery shall be capable of supplying $87\frac{1}{2}\%$ of system voltage for at least 90 minutes.	[ ]	[ ]
10.	A lighting and appliance panelboard is allowed 42 overcurrent devices including the mains.	[ ]	[ ]
11.	A lighting and appliance panelboard has 10% of its overcurrent devices rated 30 amperes or less with neutral connections.	[ ]	[ ]
12.	The disconnecting means in a data-processing room shall disconnect:		
	(a) Ventilation system	[ ]	[ ]
	(b) Receptacles for computer equipment	[ ]	[ ]
	(c) Lighting	[ ]	[ ]
	(d) Emergency system	[ ]	[ ]
13.	The recommended grounding method is to bond the neutral terminal to the service equipment at the service.	[ ]	[ ]
14.	The highest voltage-to-ground in an industrial establishment for electric discharge lighting circuits is 240 volts.	[ ]	[ ]
15.	The total equivalent bend in conduit between outlet and outlet, fitting and fitting, or outlet and fitting shall not exceed $360°$.	[ ]	[ ]
16.	Conduit bound to a box by double locknuts is considered bonded in an acceptable manner.	[ ]	[ ]

Multiple Choice

1. An auxillary gutter may extend:
   (a) 10 feet
   (b) 20 feet
   (c) 30 feet
   (d) 50 feet

   1. _____

2. The minimum swimming pool grounding wire shall be:
   (a) No. 12
   (b) No. 6
   (c) No. 10
   (d) No. 8

   2. _____

327

3. The unit of electrical resistance to current is:
   (a) Henry
   (b) Ohm
   (c) Mho
   (d) Farad
4. Vertical conductors of No. 8 copper must be supported at the top and every:
   (a) 100 feet
   (b) 200 feet
   (c) 70 feet
   (d) 50 feet
5. It is possible to find the grounded wire of a circuit with:
   (a) An ohmmeter
   (b) A voltmeter
   (c) An ammeter
6. The wire with the highest temperature rating is:
   (a) RH
   (b) MI
   (c) THHN
   (d) RHW
7. A 240-volt, 100-ampere motor operating at 208 volts draws:
   (a) 100 amperes
   (b) 105 amperes
   (c) 112 amperes
   (d) 115 amperes
8. The field of an alternator is excited by:
   (a) dc current
   (b) ac current

3. _____

4. _____

5. _____

6. _____

7. _____

8. _____

Fill in the Blanks

1. How many thermal overload units are needed for a three-phase motor? _____

2. A 2400-watt load at 120 volts has what resistance?

   _____

3. How shall the conductor having the higher voltage to ground be identified on a four-wire, delta-connected secondary? _____

4. How many amperes does a 100-kilowatt load with a .8 power factor draw in a 480-volt, three-phase circuit? _____

5. A megger is used to measure? _____

6. A 50-kilovolt-ampere load operates at .5 power factor. How many kilowatt-hours are used in 10 hours?

   _____

Diagrams

1.

Connect the three-phase transformer for a 480-volt delta primary to a 208Y/120-volt secondary.

2. Connect three start-stop units for a motor.

START

STOP

Local Ordinance Questions

This part depends on the local ordinance. Usually there will be five or more questions from the Electrical Code of the city or state giving the examination.

II. Problems (Open-Book) Test Time: 3 Hours

1. Design the wiring system for a 12-unit apartment complex consisting of the following types of units:
   (a) Six 1800-square foot units equipped with the following:
      (1) 5-kilowatt, 240-volt hot water heater
      (2) Six 2-kilowatt, 240-volt space heaters
      (3) Four 4.9-ampere, 240-volt air conditioners
      (4) 4.5-kilowatt, 240-volt dryer
      (5) 120-volt washer
      (6) 12-kilowatt range
   (b) Four 1200-square foot units equipped with the following:
      (1) 5-kilowatt, 240-volt hot water heater
      (2) Six 2-kilowatt, 240-volt space heaters
      (3) Four 4.9-ampere, 240-volt air conditioners
      (4) 4.5-kilowatt, 240-volt dryer
      (5) 120-volt washer
      (6) 10-kilowatt range
   (c) Two 900-square foot units equipped with the following:
      (1) 8¾-kilowatt range
      (2) No washer (not permitted)
      (3) Four 2-kilowatt, 240-volt space heaters
      (4) Two 4.9-ampere, 240-volt air conditioners
   Determine (1) the required number and ratings of branch circuits for each type of unit, (2) the feeder (THW aluminum) and conduit size for each type of unit, and (3) the main service by using both the standard and optional methods.

2. A new installation of 400 four-tube, 40-watt fluorescent fixtures is planned. Find the required number of branch circuits if the circuits are 277 volts and the power factor of the units is .80. How large must the feeder conductors supplying the lighting panelboard be?

3. The following loads are supplied by a 480-volt, three-phase feeder: Two 60-horsepower motors, two 30-horsepower motors, four 10-horsepower motors, four 3-horsepower motors, and a 10-kilowatt heating load. Determine the ampacity of the feeder conductors. What size standard disconnecting means would be used?

4. The 230-volt, single-phase motor loads on a feeder are as follows:

Two 10-horsepower	Three 1-horsepower
Two 5-horsepower	Three ¾-horsepower
Four 3-horsepower	Three ⅓-horsepower

   Find the feeder size and overcurrent protection required if the branch circuits are protected by nontime-delay fuses.

5. What size conduit is required to contain four No. 4/0, four No. 2, and four No. 6 THW conductors?

6. A small store is served by a 120/240-volt service. The store contains the following:
   (a) 40 foot long foot show window
   (b) Twenty-one 200-watt floor lamps
   (c) 21 receptacles (continuous)
   (d) Twenty-one 150-watt lamps
   Determine (1) the required number of branch circuits if 20-ampere circuits are used and (2) the service rating and conductor size.

7. What is the size of conduit required and the ampacity of the conductor if eight No. 8 type THHN copper conductors are to be enclosed in conduit in an ambient of 60°C.

FINAL EXAMINATION NO. 2

Total Time: 6 Hours

*Only the Code Book May Be Used
for the Open-Book Portion*

I. Problems *(Closed-Book)* Test Time: 2 Hours

1. Define the following:
   (a) Feeder
   (b) Appliance
   (c) Branch circuit
   (d) Overload
   (e) Vented power fuse
2. What is the minimum height of a 120/240-volt service drop over the following:
   (a) Sidewalk
   (b) Residential driveway
   (c) Public streets
3. What is the color of the finish for the high voltage conductor in a high-leg delta feeder?
4. What are the requirements for lighting outlets in a one-family dwelling?
5. Which receptacles must have GFCI protection in one-family dwellings?
6. Plug fuses may be used to protect 120/240-volt, single-phase circuits. True or false?
7. An aluminum grounding conductor must not be used outside within _____ inches of the earth.
8. Metal water pipes used in the grounding electrode system must have a length of _____ feet or more.
9. Select the size aluminum conductors allowed in a single-phase residential service rated at (a) 100 amperes (b) 125 amperes (c) 150 amperes (d) 200 amperes.
10. Conductors of a 120-volt circuit may occupy the same raceway as a 4160-volt circuit if the low-voltage conductors are insulated for 5000 volts. True or false?
11. If six 300 MCM THW conductors are enclosed in conduit, what is the ampacity as a percentage of the rated ampacity?
12. Rigid metal conduit of size 1 inch must be supported at least every _____ feet if a straight run with threaded couplings is used.
13. In a wireway, the cross-sectional area of conductors must not exceed _____ percent of the area of the wireway and the area of splices must not exceed _____ percent.
14. A device box with an area of 12 cubic inches may hold up to _____ No. 14 conductors.
15. Busways may extend _____ feet from their supply if the current rating is at least _____ of the rating of the overcurrent device next back on the line.
16. Auxiliary gutters must not extend farther than _____ feet and contain more than _____ current-carrying conductors that are not derated.
17. Conductors for wiring lighting fixtures must not be smaller than _____.
18. A room air conditioner must be rated less than _____ amperes and _____ volts, single-phase. A flexible cord supplying a 208-volt unit must not exceed _____ feet in length.
19. The ampacity of conductors supplying a capacitor must be at least _____ percent of the rated current of the capacitor. What is the setting of the overcurrent device?
20. A location in which electrically conductive dusts are present is a Class I, Division 1 area. True or false?
21. An area in which rayon is manufactured is classified as:
    (a) Class II, Division 2
    (b) Class III, Division 1
    (c) Class III, Division 2
    (d) Class I, Division 1
22. Type MI cable may be used in all hazardous locations. True or false?
23. At a gasoline pump the first fitting used as the conduit supplying the pump leaves the ground must be a _____ _____. What is the next fitting?
24. Near a gasoline dispenser the area up to _____ inches from the base and extending _____ feet is a Class I, Division 2 location.
25. In the following circuit, determine the quantities listed:
    (a) The reactance of the inductor
    (b) The current the ammeter reads
    (c) The voltage $V1$ across the resistor
    (d) The voltage $V2$ across the coil
    (e) The impedance of the circuit

26. What is the horsepower output of a 75-horsepower, three-phase motor with 85% efficiency and a .80 power factor if the voltage is 230 volts and the motor draws 192 amperes?

27. A motor with eight poles runs at 900 revolutions per minute is connected to a 60-hertz source. What will the speed be if the source frequency is 120 hertz?

28. A delta-wound, three-phase, 240/480-volt motor has nine terminals. Connect these terminals for (a) high-voltage operation and (b) low-voltage operation.

29. A motor controller has two remote start-stop stations and one jog station. Draw the one-line diagram.

II. Problems *(Open-Book)* Test Time: 4 Hours

1. A 100-unit apartment building is supplied by a 12470Y/7200-volt service. Each apartment contains the following:
   (a) 2000 square feet of living area
   (b) 12-ampere, 120-volt dishwasher
   (c) 6-ampere, 120-volt disposal
   (d) 7.2-kilowatt counter-mounted cooking unit
   (e) 7.2-kilowatt wall-mounted oven
   (f) 5-kilowatt, 240-volt clothes dryer
   (g) Washing machine
   (h) 9.6-kilowatt, 240-volt electric heating
   (i) 1½-horsepower, 240-volt air conditioner
   The complex also contains a central 30-ampere, 480-volt, three-phase air conditioner for the recreation areas. The apartments are supplied by a single-phase, 7200/240/120-volt transformer and the motor is supplied by a separate 7200/480-volt, three-phase transformer. Calculate the following:
   (a) The feeder capacity for each apartment unit by using the standard method.
   (b) The main-service conductor ampacity and the main fuse size by using the optional method for the apartment building load.
   (c) The conductor ampacity and the fuse size for the motor circuit

2. A 140 000-square foot department store is supplied by a 120/240 volt, high-leg delta circuit. The store contains the following:
   (a) 40 duplex receptacles
   (b) 80 feet of show window
   (c) 80-horsepower, 230-volt, three-phase motor
   (d) Outside sign of 16 amperes
   (e) Six hundred 100-watt fixtures for general illumination

   Calculate the service capacity and transformer kilovolt-ampere rating.

3. Design the branch circuits and feeder supplying the following 460-volt, three-phase motors:
   (a) 25-horsepower, squirrel-cage motor with service factor 1.15, code letter F
   (b) Two 30-horsepower, wound-rotor motors

4. A 12 470-volt switchboard has exposed buses. The elevation of the buses must be at least _____ feet.

5. What is the impedance at 60 hertz of a 500 MCM THW aluminum conductor in a steel raceway that is 800 feet long?

6. A 230-volt, three-phase squirrel-cage motor with a service factor of 1.2 has a power factor of .7 and draws 104 amperes. A capacitor is added to raise the power factor to 1.0. Find the following:
   (a) The capacitor rating in VARS
   (b) The rating of the required overload device for the motor if the capacitor is connected to the load side of the device
   (c) The branch-circuit conductor ampacities for the motor circuit and the capacitor

Appendices

Important Formulas

Table A-1 DC Circuits

Description	Formula	Comments
Ohm's Law	E = I × R volts E in volts I in amperes R in Ohms	
Resistors in series	$R_T = R_1 + R_2$ ohms	
Resistors in parallel	$R_T = \dfrac{R_1 R_2}{R_1 + R_2}$ ohms	
Power in resistor	$P = I^2 R = E^2/R$ watts	
Power in DC circuit with only resistors	P = E × I watts	

Table A-2 Conductors

C mil area	A cmil = d²	D IN MILLS
Square mil area	A sq. mil = ℓ^2	L IN MILLS
Conversion A cmil to A sq. mil	A sq. mil = A cmil $\times \dfrac{\pi}{4}$	$\dfrac{\pi}{4}$ = .7854
Sq. in. area	A sq. in. = $\dfrac{\pi}{4}$D² = .7854D²	D IN INCHES
Conversion A sq. in. to A cmil	A sq. in. = A cmil $\times$ .7854 $\times$ 10⁻⁶	
Resistivity (at 25°C)	ρ AL = 17.7 ohms-cmil/ft. ρ CU = 10.4 ohms-cmil/ft.	1 CMIL 1 FT
Resistance	R = $\rho \dfrac{L(ft)}{A\ cmil}$ ohms	A (CMIL) L

Table A-3 AC Circuits

Description	Formula	Comments
AC waveform of frequency f H_z	$T = 1/f$ seconds $V_{rms} = V_P/\sqrt{2}$ volts	
Inductor	$X_L = 2\pi f L$ ohms (ℓ in henrys) $V = I \times X_L$	
Capacitor	$X_C = \dfrac{1}{2\pi f C}$ ohms (C in farads) $V = I \times X_C$	
Impedance (magnitude)	$Z = \dfrac{V}{I}$	
Impedance (magnitude) Series R-L	$Z = \sqrt{R^2 + X_L{}^2}$	
Impedance (magnitude) Series R-C	$Z = \sqrt{R^2 + X_C{}^2}$	

Table A-3 AC Circuits (continued)

Description	Formula	Comments
Impedance (magnitude) Series R-L-C	$Z = \sqrt{R^2 + (X_L - X_C)^2}$ (Resonance when $X_L = X_C$)	
Impedance (magnitude) Parallel R-L	$Z = \dfrac{RX_L}{\sqrt{R^2 + X_L^2}}$ (or X_C for a capacitor)	
Impedance (magnitude) Parallel L-C	$Z = \dfrac{X_L X_C}{X_L - X_C}$	
Impedance (magnitude) Parallel R-L-C	$Z = \dfrac{RX_L X_C}{\sqrt{X_L^2 X_C^2 + R^2 (X_L - X_C)^2}}$	
Volt-Amperes (single-phase)	$VA = V \times I$	
Three-phase Balanced	$VA = \sqrt{3}\, V \times I$	
Power factor (leading power factor)	$Pf = \dfrac{Watts}{Volt\text{-}Amperes}$ $\cos \theta = pf$	

338

Table A-3 AC Circuits (continued)

Description	Formula	Comments
Power	$P = VI \times pf$ ($P = \sqrt{3}\, VI \times pf$; three-phase)	 pf = POWER FACTOR
Energy	watts $\times$ time	measured in watt hours or kilowatt hours
VARS	$VARS = I^2 X$ or $\dfrac{V^2}{X}$	
Power triangle	$VA = \sqrt{(watts)^2 + (VARS)^2}$ $Watts = \sqrt{(VA)^2 - (VARS)^2}$ $VARS = \sqrt{(VA)^2 - (Watts)^2}$	
Trigonometric relations	$\theta = \cos^{-1}(pf)$ $Watts = VA \times \cos\theta$ $VARS = VA \times \sin\theta$	
Power factor correction from $pf_1 = \cos\theta_1$ to $pf_2 = \cos\theta_2$	To correct add X VARS $X = Watts \times (\tan\theta_1 - \tan\theta_2)$ or $VAR_1 - VAR_2$	

Table A-4 Equipment in AC Circuits

Description	Formula	Comments
Single-phase load current (two wire or three wire)	$I = \dfrac{\text{Power (watts)}}{\text{V (volts)} \times \text{pf}}$	LOAD IN WATTS, pf = POWER FACTOR
Three-phase balance current (three wire or four wire)	$I = \dfrac{\text{Power (watts)}}{\sqrt{3} \times \text{V (volts)} \times \text{pf}}$ amperes	LOAD IN WATTS, pf = POWER FACTOR
Voltage drop	$V_{drop} = V_{source} - V_{load}$ $\% \, V_{drop} = \dfrac{V_{drop}}{V_{source}} \times 100$	
Voltage drop for single conductor	$V_{drop} = I \times R$ ($R = \rho \dfrac{L}{A}$ or use Code tables)	
Voltage drop in single-phase circuit	$V_d = 2 \times I \times R$	R = RESISTANCE OF ONE CONDUCTOR
Voltage drop in three-phase three wire circuit	$V_d = \sqrt{3} \times I \times R$	R = RESISTANCE OF ONE CONDUCTOR
Voltage drop in three-phase four-wire circuit	$V_d = I \times R$ (line-to-neutral drop)	$I_N = 0$

340

Description	Formula	
Single-Phase Transformer	$V_p \times I_p = V_S \times I_S$ volt amperes	
Three-Phase Transformer	$\sqrt{3}V_p \times I_p = \sqrt{3} V_S \times I_S$ Note: The type of winding does not affect calculation	 V AND I ARE LINE VALUES NOT PHASE VALUES
Motor Synchronous Speed	$\text{speed} = \dfrac{120 \times f}{\text{number of poles}}$ f = frequency of ac supply in Hz	
Efficiency Single-Phase	$\text{hp out} = \dfrac{V \times I \times pf \times \eta}{746 \text{ W/hp}}$	 pf = POWER FACTOR η = EFFICIENCY
Efficiency Three-Phase	$\text{hp out} = \dfrac{\sqrt{3}\,V \times I \times pf \times \eta}{746 \text{ W/hp}}$	η is efficiency, pf is power factor $\eta = \dfrac{\text{hp out}}{\text{hp in}}$

B

Answers

CHAPTER 1 ANSWERS

Chapter 1 Test

1, 2, 3. See local ordinances
4. (a) True (Article 90)

(b) True (Article 90)
5, 6, 7. Contact local agency

CHAPTER 2 ANSWERS

2-1 Quiz *(Closed-Book)*

1. Refer to Figure 2-1
2. Contact utility company
3. Service; branch-circuits

2-2 Quiz *(Closed-Book)*

1,2. Refer to Table 2-2
3, 4, 5. See text
6. Refer to Figure 2-3
7. Refer to Table 2-1
8. No. 6 (250-94)

9. Refer to Code section 230-42
10. Refer to Code Article 100

2-2 Quiz *(Open-Book)*

1. Load (amperes) $= \dfrac{46\ 000\ \text{VA}}{240\ \text{V}} = 191.7$ A
    No. 3/0 THW (Table 310-16)
    2-inch conduit (Table 3A, Ch. 9)
2. Load = 192 amperes as before
    No. 3/0 THHN (Table 310-16)
    1½-inch conduit (Table 3B, Ch. 9)

3. (a) Load (amperes) = $\dfrac{10\,000\ \text{VA}}{240\ \text{V}}$ = 41.7 A

  Requires 60-ampere service (230-41)
  No. 6 THW (Table 310-16)
  1-inch conduit (Table 3A, Ch. 9)

 (b) Load (amperes) = $\dfrac{20\,000\ \text{VA}}{240\ \text{V}}$ = 83.3 A

  No. 4 THW (Table 310-16)
  1-inch conduit (Table 3A, Ch. 9)

 (c) Load (amperes) = 83.3 amperes
  Requires 100-ampere service (230-42)
  No. 3 THW (Table 310-16)
  1¼-inch conduit (Table 3A, Ch. 9)
  (Note 3, Table 310-16 ignored)

 (d) Load (amperes) = $\dfrac{100\,000\ \text{VA}}{240\ \text{V}}$ = 416.7 A

  600 MCM THW (Table 310-16)
  3-inch conduit (Table 3A, Ch. 9)

4. The ampacity is 175 amperes (Table 310-16)
  Load = 175 A × 240 V = 42 000 VA

5. (a) 800 amperes (240-3, 240-6)
 (b) 150 amperes (240-3, 240-6)

6. (a) No. 8 jumper [250-79(c)]
 (b) No. 8 grounding conductor (Table 250-94)
 (c) No. 8 bonding jumper [250-80(a)]

7. Load (amperes) = $\dfrac{40\,000\ \text{VA}}{\sqrt{3}\ \times\ 480\ \text{V}}$ = 48.1 A

  No. 8 THW (Table 310-16)

2-3 Quiz *(Closed-Book)*

1, 2, 3. Refer to Table 2-3
4. Refer to Code Article 100
5. See text
6. Demand load = .5 × 100 kW = 50 kW

2-3 Quiz *(Open-Book)*

1. (a) Line load = $\dfrac{30\,000\ \text{VA}}{240\ \text{V}}$ = 125 A

 (b) No. 1 THW (Table 310-16)
 (c) Area = 3 × .2027 in.² = .6081 in.² (Table 5, Ch. 9)
 (d) 1¼ "-conduit (Table 3A, Ch. 9)

2. (a) Line load = $\dfrac{30\,000\ \text{VA}}{\sqrt{3}\ \times\ 480\ \text{V}}$ = 36 A

 (b) No. 8 THW (Table 310-16)
 (c) Area = 4 × .0598 in.² = .2392 in.² (Table 5, Ch. 9)
 (d) 1-inch conduit (Table 3A, Ch. 9)

3. (a) Line load = $\dfrac{100\,000\ \text{VA}}{240\ \text{V}}$ = 417 A

 (b) Neutral ampacity = 200 A + .7 (417 − 200A)
  = 351.9 A (220-22)

 (c) Ungrounded conductors: 600 MCM (Table 310-16)
  Neutral: 500 MCM (Table 310-16)

4. Neutral ampacity = 200 A + .7(1000 A − 200 A) =
  760 A (220-22)

2-4 Quiz *(Closed-Book)*

1. Refer to Code Article 100
2. No. 14 [210-19(c)]
3. 15, 20, 30, 40, and 50 amperes (210-3)
4. Yes, as an individual branch circuit (210-3)

5. Load (amperes) = $\dfrac{10\,000\ \text{W}}{240\ \text{V}}$ = 41.7 A

  A 50-ampere circuit would be used

2-4 Quiz *(Open-Book)*

1. Load (amperes) = $\dfrac{20\,000\ \text{VA}}{240\ \text{V}}$ = 83.3 A

  USE: (1) No. 4 Type THW copper conductors for individual branch circuit (Table 310-16)
   (2) Standard 90-ampere overcurrent device (240-6)
   (3) No. 8 copper equipment grounding conductor
    (Table 250-95)
   (4) 1-inch conduit to enclose three No. 4 and one
    No. 8 THW conductors (Table 3A, Ch. 9)

2. Load (amperes) = $\dfrac{20\,000\ \text{VA}}{\sqrt{3}\ \times\ 480\ \text{V}}$ = 24 A

  Use: (1) No. 10 Type THW conductors (Table 310-16)
   (2) 25- or 30-ampere overcurrent protective device
    (240-6)
   (3) No. 10 THW equipment grounding conductor
    (250-95)
   (4) ¾-inch conduit to enclose five No. 10 Type
    THW conductors (Table 3A, Ch. 9)

CHAPTER 2 TEST

I. (True or False)

1. F (Article 100)
2. F (230-41 exceptions)
3. T (230-42) [250-23(b)]
4. T[230-79(a)]
5. F (Article 100)
6. T (Table 1, Ch. 9)
7. F (230-79)
8. F (210-21a)

9. T (210-3)
10. F (230-42)

II. (Multiple Choice)

1. (b) (230-42)
2. (c) (220-22)
3. (b) $\sqrt{3}$ × 2400 V
4. (b) [250-79(c)]

5. (b) (220-22)
6. (d) (Article 100)
7. (b)
8. (b) (210-3)
9. (a) (210-22)

III. (Problems)

1. Load (amperes) $= \dfrac{92\,000 \text{ VA}}{240 \text{ V}} = 383$ A

   (a) 500 MCM type THHN (Table 310-16)
   (b) Neutral current $= 200 \text{ A} + .7 \times 183$ A
       $= 328$ A
       Use 350 MCM THHN for neutral
   (c) 400-amperes standard (240-3, 240-6)
   (d) No. 1/0 (Table 250-94)

2. Load (amperes) $= \dfrac{20\,000 \text{ VA}}{\sqrt{3} \times 208 \text{ V}} = 55.5$ A
   Use four No. 4 Type THW aluminum (Table 310-16)
   Use 1¼-inch conduit (Table 3A, Ch. 9)

3. (a) Load (amperes) $= \dfrac{100\,000 \text{ VA}}{240 \text{ V}} = 416.7$ A,
       600 MCM THW (Table 310-16)
   (b) Load (amperes) $= \dfrac{100\,000 \text{ VA}}{\sqrt{3} \times 208 \text{ V}} = 277.5$ A,
       300 MCM THW

   (c) Load (amperes) $= \dfrac{100\,000 \text{ VA}}{\sqrt{3} \times 408 \text{ V}} = 120.3$ A,
       No. 1 THW
   (d) Same as (c)—three conductors only

4. (a) Feeder demand $= 20 \text{ kW} \times .5 = 10$ kW
       Load (amperes) $= \dfrac{10\,000 \text{ VA}}{240 \text{ V}} = 41.7$ A
   (b) Service demand $= .35 (20 \text{ kVA} + 20 \text{ kVA}) = 14$ kVA
       Load (amperes) $= \dfrac{14\,000 \text{ VA}}{240 \text{ V}} = 58.3$ A
   (c) Use No. 8 THW conductors for feeders and No. 6 THW for service (Table 310-16)

5.

Conductors	*Area*	(Table 5, Ch. 9)
Six No. 10 TW	$6 \times .0224 =$	.1344 in.²
Three No. 14 THW	$3 \times .0206 =$	.0618 in.²
Two No. 12 THHN	$2 \times .0117 =$	.0234 in.²
		.2196 in.²

Since 40% area must be greater than .2196 in.², use 1-inch conduit (Table 4, Ch. 9)

CHAPTER 3 ANSWERS

3-1.1 Quiz *(Closed-Book)*

1. (a) (220-2)
2. True [210-19(a)]
3. True [210-23(a)]
4. 10 ft × 180 VA/5 ft = 360 VA [220-3(e)]
5. 200 W/ft × 20 ft = 4000 W [220-3(c)], Exception 1]
6. Lighting is continuous (Article 100)
7. Only fixed units [210-23(b)]

3-1.1 Quiz *(Open-Book)*

1. Load = 200 VA/ft × 30 ft = 6000 VA [220-3(c)]
   $\dfrac{6000 \text{ VA}}{120 \text{ V} \times 15 \text{ A}} = 3.3$, or 4 circuits
2. Load = 1.25 × 3.5 VA/ft² × 5000 ft² = 21 875 VA [220-2(b)]
3. $\dfrac{21\,875 \text{ VA}}{20 \text{ A} \times 120 \text{ V}} = 9.1$, or 10 circuits
4. Lighting:
   $\dfrac{1.25 \times 10,000 \text{ VA}}{20 \text{ A} \times 120 \text{ V}} = 5.2$, or 6 circuits
   Show windows:
   $\dfrac{200 \text{ VA/ft} \times 10 \text{ ft}}{20 \text{ A} \times 120 \text{ V}} = .83$, or 1 circuit
   Receptacles:
   $\dfrac{1.25 \times 180 \text{ VA/unit} \times 100 \text{ units}}{20 \text{ A} \times 120 \text{ V}} = 9.37$, or 10 circuits

   Fluorescent lights:
   $\dfrac{1.25 \times 8000 \text{ VA}}{30 \text{ A} \times 120 \text{ V}} = 2.78$, or 3 circuits

   Total seventeen 20-ampere circuits, three 30-ampere circuits

3-1.2 Quiz *(Closed-Book)*

1. See Article 100
2. See Figure 3-4
3. (c) (430-6)
4. (a) (430-32)
5. A motor-circuit switch or a circuit breaker (430-109)
6. F (430-83)
7. F [430-32(c)(4)]

3-1.2 Quiz *(Open-Book)*

1. FLA = 54 A (Table 430-150)
   Ampacity = 1.25 × 54 A = 67.5 A minimum (430-22)
   Relay set at 1.15 × 54 A = 62.1 A maximum [430-32(a)]
2. 1.25 × 54A = 67.5A [430-32 (a)]
3. FLA = 9.6 A (Table 430-150)
   Ampacity = 1.25 × 9.6 A = 12 A (430-22)
   No. 14 THW conductors (Table 310-16)
4. FLA = 34 A (Table 430-150)
   Ampacity = 1.25 × 34 A = 42.5 A (430-22)
   Overload = 1.25 × 34 A = 42.5 A (430-32)

Branch-circuit protection = $1.5 \times 34\,A = 51\,A$
(Table 430-152)
(The motor is a wound-rotor motor)
Controller and disconnecting means ratings: 25 horsepower
Use conduit as the equipment grounding conductor or
    No. 8 aluminum (Table 250-95)
Thus, three No. 6 THW aluminum conductors (Table
    310-16) and a 1-inch conduit (Table 3A, Ch. 9) are
    required.

5. FLA = 50 A (Table 430-148)
   Ampacity: $1.25 \times 50\,A = 62.5\,A$ (430-22)
   Overload: $1.15 \times 50\,A = 57.5\,A$ (430-32)
   Branch-circuit protection: $2.5 \times 50\,A = 125\,A$ (Table 430-152, Code letter F motor)
   Controller and disconnecting means ratings: 10 horsepower
   Equipment grounding conductor: No. 6 copper (Table 250-95)
   Thus, two No. 4 TW copper conductors (Table 310-16)
       and a 1-inch conduit with equipment ground (Table 3A, Ch. 9) are required.

3-1.3 Quiz *(Closed-Book)*

1. See Article 100, Definitions
2. A water heater of 120 gallons or less capacity [422-14(b)]
3. F (Table 220-19, Note 4)
4. (b) (Table 220-19, Note 1)
5. F [220-2 (c) (1)]
6. $1.25 \times 10\,A + 20\,A = 32.5\,A$ [210-22(a)]

3-1.3 Quiz *(Open-Book)*

1. (a) 8 kilowatts (Table 220-19, column A)
   (b) 8 kilowatts (Table 220-19, column A)
   (c) $8\,kW + (2 \times 400\,W) = 8800\,W$ (Table 220-19, Note 1)
   (d) $5\,kW \times .8 = 4\,kW$ (Table 220-19, column C)
2. FLA = 8 A (Table 430-148)
   Ampacity = $1.25 \times 8\,A = 10\,A$; use 15-ampere circuit
   Disconnecting means: ac snap switch (422-26, 430-109)
       exception 2)

3. $\text{Load (amperes)} = \dfrac{6000\,W}{240\,V} = 25\,A$

   No. 10 Type THW aluminum conductors (Table 310-16)
   Use a 30-ampere circuit
4. Load = $9\,kW + 8\,kW = 17\,kW$ (Table 220-19, Note 4)
   Demand = $8\,kW + 400\,W\,(17 - 12) = 10\,kW$

   $\text{Load (amperes)} = \dfrac{10\ 000\,W}{240\,V} = 41.7\,A;$

   use No. 8 type THW copper conductors (Table 310-16)

5. $\text{Load (amperes)} = \dfrac{10\ 000\,W}{\sqrt{3} \times 480\,V} = 12.0\,A$

   Overcurrent protection: 15 A
   No. 14 MI conductors [Table 310-16, 210-19 (c)]

3-1.4 Quiz *(Closed-Book)*

1. $20\,A \times 1.25 = 25\,A$ [424-3(b)]

2. 15, 20, or 30 amperes [424-3(a)]
3. (a) [424-3 (b)]

3-1.4 Quiz *(Open-Book)*

1. $\text{Load (amperes)} = 1.25 \times \dfrac{10\ 000\,W}{240\,V} = 52\text{ minimum}$

   [424-3(b)]
2. FLA = 17 A (Table 430-148)

   $\text{Heater current} = \dfrac{5000\,W}{240\,V} = 20.8\,A$

   $\text{Load (amperes)} = 1.25 \times (17\,A + 20.8\,A) = 47.3\,A$
   [424-3(b)]
   No. 8 THW copper conductors (Table 310-16)
   Overcurrent protection: 50 A standard (240-6)
   Equipment grounding conductor: No. 10 (Table 250-95)

3. $\text{Load (amperes)} = 1.25 \times \dfrac{4\text{ units} \times 1200\,W/\text{unit}}{240\,V} = 25\,A$

   (424-3)
   This requires two 15- or 20-ampere circuits or one 30-ampere circuit. [424-3(a)]

3-1.5 Quiz *(Closed-Book)*

1. (a) $1.25 \times 100\,A = 125\,A$ (440-32)
   (b) $1.75 \times 100\,A = 175\,A$ [440-22(a)]
   (c) $1.15 \times 100\,A = 115\,A$ [440-12(a)(1)]
   (d) $1.25 \times 100\,A = 125\,A$ [440-52(a)(3)]
2. (b) [440-41(a)]
3. $1.25 \times 12\,A = 15\,A$ [440-62(b)]
4. $15\,A/2 = 7.5\,A$ [440-62(c)]
5. Six times rated-load current [440-12(c)]
6. Ampere rating and horsepower rating [440-12(a)]

3-1.5 Quiz *(Open-Book)*

1. Ampacity = $1.25 \times 20\,A = 25\,A$ (440-32)
   Overload protection: $1.4 \times 20\,A = 28\,A$ maximum
       [440-52(a)]
   Branch-circuit protection: $1.75 \times 20\,A = 35\,A$ maximum
       (440-22)
   Disconnecting means: $1.15 \times 20\,A = 23\,A$ minimum
       (440-12)
   Thus, No. 10 THW copper conductors (Table 310-16);
       35 A circuit breaker for overcurrent protection and
       disconnecting means; overload relay set at 28 amperes.
2. Current rating = 14 A (Table 430-150)
   Equivalent locked rotor rating = 84 A (Table 430-151)
   The 10-horsepower controller may control a motor-compressor with the above ratings [440-41(a)]
3. (a) Disconnecting means
       Current rating: $1.15 \times 124\,A = 142.6\,A$ [440-12(a)(1)]
       Locked rotor equivalent: 100 horsepower (based on
           730 amperes) (Table 430-151)
       Full-load equivalent: 100 horsepower (based on 124
           amperes) (Table 430-150)

Thus, required horsepower rating is 100 horsepower [440-12(a)(2)]

(b) Controller

Current rating: 124 amperes (full-load current) [440-41(a)] and 730 amperes (locked-rotor current)

4. Based on 750-ampere locked-rotor current, the equivalent horsepower rating is 125 horsepower (Table 430-151) [440-12(a)(2)]

3-1.6 Quiz *(Closed-Book)*

1. 50% of branch-circuit rating [210-23(a)]
2. (a) Appliances and fixed lighting units [210-23(b)]
   (b) Appliances
3. Load is 8 amperes
   A 20-ampere circuit is required (16 amperes or greater) (440-62)
4. 125% of motor full-load current (430-25) plus the current of the other load
5. (a) [430-110(c)(2)]

3-1.6 Quiz *(Open-Book)*

1. Motor: FLA = 50 A (Table 430-148)

   Heater: $I = \dfrac{1000\ W}{240\ V} = 4.2\ A$

   Total Load = 1.25 × 50 A + 4.2 A = 66.7 A (430-25)
   Use No. 4 THW copper conductors (Table 310-16)

3-1.7 Quiz *(Closed-Book)*

1. (b) (430-22)
2. 16 amperes [220-3(a)]
3. Code table [Table 430-148]
4. (c) (Table 430-152)
5. 180 voltamperes [220-3(c)]
6. 225 voltamperes [220-3(a), 220-3(c)(4)]

3-1.7 Quiz *(Open-Book)*

1. (a) 9.8 amperes (Table 430-148)
   (b) 4.9 amperes (Table 430-148)
   (c) 9.6 amperes (Table 430-150)
   (d) 4.8 amperes (Table 430-150)
2. FLA = 14 A (Table 430-150)
   Ampacity = 1.25 × 14 A = 17.5 A
3. 700% of FLA (Table 430-152)
   7 × 14 A = 98 A
   Standard Size is 100 amperes (240-6)

4. $\text{Ampacity} = \dfrac{50\ 000\ W}{240\ V} = 208.3\ A$

The circuit rating must be 1.25 × 208.3 A = 260.4 A or greater if the overcurrent protective device is not rated for 100% continuous duty.

5. Use 4/0 THW copper conductors (Table 310-16)

6. $\text{Number} = \dfrac{50\ 000\ VA}{20\ A \times 240\ V} = 10.4\ \text{or}\ 11\ \text{circuits}$

7. $\text{Number} = \dfrac{1.25 \times 180\ VA/unit \times 100\ units}{15\ A \times 120\ V} = 12.5,$

   or 13 circuits [220-2(c)] (See Problem 8)

8. $\text{Load per unit} = \dfrac{1.25 \times 180\ VA}{120\ V} = 1.87\ A/unit$

   $\text{Units/circuit} = \dfrac{15\ A}{1.87\ A/unit} = 8.02,\ \text{or}\ 8\ \text{per circuit}$

   thus, 13 circuits must be provided to supply 100 outlets.

9. (a) $\text{Lighting load} = 1.25 \times \dfrac{2000\ VA}{120\ V} = 20.8\ A$

   Use two 20-ampere circuits

   (b) Motor full-load current is 34 amperes (Table 430-150)
   Conductor ampacity = 1.25 × 34 A = 42.5 A
   Requires three No. 8 THW copper conductors
   Fuse can be rated as high 3 × 34 A = 102 A (Table 430-152)
   Use standard 110-ampere fuse
   A 25-horsepower rated controller is required

   (c) The fluorescent lighting load per unit is
   1.25 × 1.7 A = 2.1 A [210-22(b)]
   Ten such units require a 21-ampere supply. A 30-ampere branch circuit could be used if the lampholders were heavy-duty type [210-21(a)]

   (d) (1) $\text{Unit load (continuous)} = 1.25 \times \dfrac{600\ VA}{120\ V} = 6.25\ A$

   [220-3(c)]
   (2) Units per 30-ampere circuit is

   $\dfrac{30\ A}{6.25\ A} = 4.8,\ \text{or}\ 4\ \text{per circuit}$

   Requires three 30-ampere circuits

3-2.1 Quiz *(Closed-Book)*

1. (b) (220-12)
2. 180 voltamperes [220-3(c)]
3. Circuits supplying household electric ranges (220-22)
4. (b) (220-22)
5. The current drawn by the load and whether the load is continuous or not (220-10)

3-2.1 Quiz *(Open-Book)*

1. Lighting: 10 000 ft² × 3.5 VA/ft² × 1.25
   Receptacles: 100 × 180 A × 1.25

	Line	Neutral
	43 750 VA	35 000 VA
	22 500 VA	18 000 VA	[220-3(c)]
Total load =	66 250 VA	53 000 VA

   Load in amperes = 66 250 VA ÷ 240 V = 276 A
   Neutral load: 200 A + .7(20 A) = 214 A (220-22)
   THW copper conductors: Use two 300 MCM; one 4/0
   (neutral) (Table 310-16)

2. (a) Incandescent load (amperes) = $\dfrac{104\ 000\ \text{VA}}{240\ \text{V}}$ = 433 A

   Neutral load = 200 + (.7 × 233 A) = 363 A
   (220-22)

   (b) Fluorescent load current: $\dfrac{52\ 000\ \text{VA}}{240\ \text{V}}$ = 217 A

   (c) Total neutral: 363 + 217 A = 580 A
   (d) Total load: 433 A + 217 A = 650. THW alumi-
       num conductors: The line conductors must be larger
       than 2000 MCM or equivalent (Table 310-16)

3. The overcurrent protective device must be set at 276 A or
   greater. A standard 300 A device could be used. The
   connected load is 35 000 VA (lighting) + 18 000 VA
   (receptacles) or 220.8 A, thus allowing 4/0 THW copper conductors. However, 300 MCM THW copper conductors would be used with the 300 A device. [220-10(b)]
   (240-3).

3-2.2 Quiz *(Closed-Book)*

1. The feeder overcurrent protective device is selected based
   on the rating of the largest branch-circuit protective device and the full-load currents of any other motors connected to the feeder. (430-62)
2. The feeder load is the sum of the full-load current of the
   motors plus the current drawn by any other loads, plus
   25% of the full-load current of the motor with the
   largest full-load current. (430-24)

3-2.2 Quiz *(Open-Book)*

1. FLA = 50 amperes for each 10-horsepower, 230-volt
   motor (Table 430-148)
   Branch-circuit protective device: 7 × 50 A = 350 A
   maximum (Table 430-152)
   Feeder ampacity: 50 A + 50 A + 50 A + (.25 × 50 A) =
   162.5 A (430-24)
   THW copper conductors: Two No. 2/0 (Table 310-16)
   Feeder protective device: 350 A + 50 A + 50 A = 450 A
   maximum (430-62)

2.
   Compressor
   5-horsepower, 480-volt, three-phase motor

FLA	Branch-circuit protection
25 A	1.75 × 25 A = 43.8 A (440-22)
7.6 A;	2.5 × 7.6 A = 19 A (Table 430-150, Table 430-152)

   Feeder ampacity: 25 A + 7.6 A + (.25 × 25 A) = 38.9 A
   (430-240, 440-33)
   THW aluminum conductors: Three No. 8 (Table 310-16)
   Feeder protective device: 43.8 A + 7.6 A = 51.4 A maximum (430-62, 240-6) Use 50 ampere standard breaker

3-2.3 Quiz *(Closed-Book)*

1. T (220-15)
2. T (220-14)
3. (c) (Table 220-20)
4. Five (Table 220-18)
5. 70% (220-22)

3-2.3 Quiz *(Open-Book)*

		Line	
1. 7½-horsepower, 230-volt motor:	=	40 A	(Table 430-148)
Heater: $\dfrac{10\ 000\ \text{W}}{240\ \text{V}}$	=	41.7 A	
Neglect 12-ampere air conditioner			(220-21)
25% of largest motor: .25 × 40 A	=	10 A	(430-25)
Feeder current		91.7 A	

THW copper conductors: Two No. 3   (Table 310-16)
Since the branch-circuit protective device is not
specified, the rating of the feeder protective device
cannot be selected.

2. $.3 \times 30 \times 5\,kW = 45\,kW$ (Table 220-18)
3. The load for twenty-five 10-kilowatt electric ranges is 40 kilowatts (Table 220-19)
   The neutral load is $.7 \times 40\,kW = 28\,kW$ (220-22)
4. Ranges:

	Line	Neutral	
	40 kW	28 kW	(Table 220-19, 220-20)
Dryers (240 volts):	45 kW	31.5 kW	
Total load:	85 kW	59.5 kW	(Table 220-18)

Current at 240 V = 354 A 247.9 A
THW copper conductors: Two 500 MCM, one 250 MCM (neutral) (Table 310-16). Feeder protective device: 400 A standard (240-6)

3-2.4 Quiz *(Closed-Book)*

1. (a) $115\,V \times 16\,A = 1840\,VA$
   (b) $230\,V \times 28\,A = 6440\,VA$
   (c) $\sqrt{3} \times 460\,V \times 10\,A = 7967\,VA$
2. If there is no feeder demand factor, the feeder load is equal to the sum of the branch-circuit loads for lighting loads, receptacle loads, and certain nonmotor driven appliance loads
3. The motors must meet the conditions of Code rule (430-112)

3-2.4 Quiz *(Open-Book)*

		Line	Neutral	
1. Lighting:	$1.25 \times \dfrac{14\,000\,VA}{\sqrt{3} \times 480\,V}$	21.0 A	16.8 A	(220-10)
30-horsepower, three-phase motor:		40 A	—0—	(Table 430-
25% of largest motor		10 A	—0—	150)
	Total load	71 A	16.8 A	

		Line	Neutral
2. Range: $8\,kW + (2 \times 400\,W) =$		8800 W	6160 W
4.5-kilowatt water heater:		4500 W	—0—
5-kilowatt dryer:		5000 W	3500 W
A/C: $6 \times 7\,A \times 240\,V$		10 080 W	—0—
25% of largest motor:			
$.25 \times 7\,A \times 240\,V$		420 W	—0—
Lights:		5000 W	5000 W
	Total	33 800 W	14 660 W
Load (amperes) at 240 volts		140.8 A	61 A
Conductors THW aluminum		No. 3/0	No. 4 neutral

NOTE: This is not intended to be a feeder for a dwelling as discussed in Chapter 4.

3-2.5 Quiz *(Closed-Book)*

1. Doubled
2. Four (Tables 310-16 to 310-18, Note 8)
3. $kVA_{in} = kVA_{out}$

4. Power factor $= \dfrac{kW}{kVA}$
5. Line-to-line = 240 V
   Line-to neutral = 120 V, 120 V, 208 V

3-2.5 Quiz *(Open-Book)*

1. $\dfrac{Vp}{Vs} = \dfrac{Is}{Ip}$ $\dfrac{Ip}{10\,A} = \dfrac{100\,V}{1000\,V}$

   $Ip = 1$ ampere

   $kVA = 1000\,VA \times \dfrac{1\,kVA}{1000\,VA} = 1\,kVA$

2. $I = \dfrac{50\,000\,W}{240\,V \times .8} = 260.4\,A$

3. (a) $I_F = \dfrac{40\,000\,VA}{\sqrt{3} \times 480\,V \times (.85)} = 56.6\,A$

   (b) Power per phase $= \dfrac{40\,000\,VA}{3} = 13.33\,kVA$

4. $I_F = \dfrac{831\,380\,VA}{\sqrt{3} \times 480\,V} = 1000\,A$

The ampacity of the conductors must be reduced to 80% since there are four conductors in the conduit. (Note 8 to Tables 310-16 through 310-19). The ampacity of the circuit must be

$$\dfrac{1000\,A}{.8} = 1250\,A$$

Number of 500 MCM conductors $= \dfrac{1250\,A}{380\,A} = 3.29,$

or 4 per phase in 4 conduits

CHAPTER 3 TEST

I. (True or False)

1. F (210-3)
2. T [210-21(a)]
3. F [210-22(c)]
4. F [210-23(b)]
5. T (Article 100)
6. T (220-22)
7. T (Table 220-19)
8. T (422-20)
9. F [424-3(a)]
10. T [440-22(a)]

II. (Multiple Choice)

1. (b) [210-22(c)]
2. (b) (Table 310-16)
3. (a) (210-3)
4. (b) [220-3(c)]
5. (c) [220-3(c) Exception No. 3]
6. (b) (430-22)
7. (b) [430-32(a)]
8. (a) (430-24)
9. (b) [430-62(a)]
10. (c) (310-4)

III. (Problems)

1. Each receptacle is a load of 180 voltamperes [220-3(c)] As a noncontinuous load, the current per receptacle at 120 volts is

$$\frac{180 \text{ VA}}{120 \text{ V}} = 1.5 \text{ A}$$

The number per circuit is

$$\frac{20 \text{ A}}{1.5 \text{ A/receptacle}} = 13.3, \text{ or } 13 \text{ receptacles}$$

2. (a) FLA = 5.8 amperes (Table 430-148)
   Ampacity: $1.25 \times 5.8 \text{ A} = 7.25 \text{ A}$
   No. 14 THW copper conductors (Table 310-16)
   Branch-circuit protection: $3 \times 5.8 \text{ A} = 17.4 \text{ A}$
   maximum (Table 430-152)
   Use either a 15-ampere or 20-ampere circuit (No. 12 conductors)

   (b) Load $= 1.25 \times \dfrac{2000 \text{ W}}{240 \text{ V}} = 8.33 \text{ A}$ [424-3(b)]

   No. 14 THW copper conductors (Table 310-16)
   Use a 15-ampere circuit

3.

Lighting load:	$\dfrac{99\ 720 \text{ VA}}{\sqrt{3} \times 480 \text{ V}}$	*Line* 120 A	*Neutral* 120 A	
50-horsepower motor		65 A	—0—	(Table 430-150)
30-horsepower motor		40 A	—0—	
25% of largest motor:		16.25 A		
		241.25 A	120 A	

THW aluminum conductors: three 350 MCM; No. 1/0 neutral

4. Load (amperes) $= \dfrac{303\ 300 \text{ W}}{\sqrt{3} \times 480 \text{ V}} = 364.8 \text{ A}$

(Assuming the load is not continuous)
Demand load $= .65 \times 364.8 \text{ A} = 237.13 \text{ A}$ (Table 220-20)

5. (a) Load $= 8 \text{ kW} + 400 \text{ W} (17 - 12) = 10 \text{ kW}$ (Table 220-19 Note 1)
   (In this case, an excess of 5 kilowatts over 12 kilowatts was used since the calculated excess was 4.6 kilowatts and was rounded off to 5 kilowatts)
   (b) Demand load $= 20 \text{ kW}$ (Table 220-19)

6.

Load	*Line*	*Neutral*	
Lighting load:	40 000 W	40 000 W	
Range load (neutral at 70%):	25 000 W	17 500 W	(Table 220-19)
Total	65 000 W	57 500 W	

Load (amperes) = 270.8 A
Neutral $= 200 \text{ A} + .7(39.6 \text{ A}) = 227.7 \text{ A}$ (220-22)
THW conductors: 300 MCM, No. 4/0 neutral
Main overcurrent: 270.8 amperes, use 300-ampere standard (230-90, 240-6)
Disconnecting means: 270.8 amperes minimum (230-79)
Grounding electrode conductor: No. 2 copper (250-94)
Conduit: $2 \times 300 \text{ MCM THW} = 2 \times .5581 = 1.1162 \text{ in}^2$
No. 4/0 $\quad\quad\quad\quad\quad\quad\quad\quad\quad\quad\quad\quad\quad\quad \dfrac{.3904 \text{ in}^2}{1.5066 \text{ in}^2}$

(Table 5, Ch. 9)
Use 2½-inch conduit (Table 4, Ch. 9)

7. Ampacity of No. 3 conductor: 100 amperes (Table 310-16)
   Ampacity of four No. 3 conductors: $4 \times 100 \text{ A} = 400 \text{ A}$
   Reduced ampacity for 12 conductors in conduit:
   $.7 \times 400 \text{ A} = 280 \text{ A}$ (Table 310-16, Note 8)

8.

Load	Line A	Line B	Line C	Neutral	Code Refs
Lighting:					
$1.25 \times \dfrac{10\ 000\ \text{W}}{\sqrt{3} \times 208\ \text{V}}$	34.7 A	34.7 A	34.7 A	27.8 A	
15-horsepower motor (42 A + 4.2 A) 208 V, three-phase	46.2 A	46.2 A	46.2 A	—0—	(Table 430-150)
5-horsepower motor (28 A + 2.8 A) 208 V, single-phase	30.8 A	30.8 A	—0—	—0—	(Table 430-148)
25% of largest motor = .25 × 46.2 A	11.6 A	11.6 A	11.6 A		
Current:	123.3 A	123.3 A	92.5 A	27.8 A	

THW Aluminum conductors: No. 2/0, No. 2/0, No. 1, No. 10 (Table 310-16)

Note: The 208-volt motors draw 10% more current than the corresponding 230-volt motors.

9. (a) Branch circuits:
   (1) 25-horsepower squirrel-cage
       FLA = 34 A (Table 430-150)
       Ampacity: 1.25 × 34 A = 42.5 A (430-22)
       No. 8 THW copper conductors (Table 310-16)
       Branch-circuit protection: 3 × 34 A = 102 A, or 110-ampere standard fuse (Table 430-152, 240-6)
       Overload: 1.15 X 34 A = 39.1 A maximum (430-32)
       Controller: 25 horsepower (430-83)
       Disconnecting means: 25-horsepower motor-circuit switch (430-110)
       Conduit enclosing three No. 8 THW conductors: ¾ inch (Table 3A, Ch. 9)
   (2) 30-horsepower wound rotor:
       FLA = 40 A
       Ampacity: 1.25 × 40 A = 50 A (Table 430-150)
       No. 8 THW copper conductors (Table 310-16)
       Branch-circuit protection: 1.5 × 40 A = 60 A fuse (Table 430-152)
       Overload: 1.15 × 40 A = 46 A maximum (430-32)
       Controller: 30 horsepower (430-83)
       Disconnecting means: 30-horsepower motor-circuit switch (430-110)
       Conduit enclosing three No. 8 THW conductors: ¾ inch (Table 3A, Ch. 9)
   (b) Feeder:
       Ampacity: 1.25 × 40 A + 34 A = 84 A (430-24)
       No. 4 THW copper conductors (Table 310-16)
       Feeder overcurrent protection: 110 A + 40 A = 150 A maximum (430-62)

9. (cont.)
   Conduit enclosing three No. 4 THW conductors: 1 inch (Table 3A, Ch. 9)
   (The conduit is used as the equipment grounding conductor)

IV. (Special Problems)

1. $I_p = \dfrac{100\ 000\ \text{VA}}{\sqrt{3} \times 480\ \text{V}} = 120.28\ \text{A}$

2. (a) Ampacity:
      The circuit ampacity is 25 A × .8 = 20 A (Table 310-16, Note 8, Note 10)
   (b) Branch-circuit overcurrent device:
      The conductors must be protected at 20 amperes or less

3. $I = \dfrac{15\ 000\ \text{W}}{\sqrt{3} \times 208\ \text{V} \times (.75)} = 55.5\ \text{A}$

   $\text{kVA} = \sqrt{3} \times 208\ \text{V} \times 55.5\ \text{A} = 20\ \text{kVA}$

4. $I_s = \dfrac{207\ 846\ \text{W}}{\sqrt{3} \times 240\ \text{V}} = 500\ \text{A}$ (secondary)

   $I_p = \dfrac{240\ \text{V}}{480\ \text{V}} \times 500\ \text{A} = 250\ \text{A}$ (primary)

Check:

$P_{\text{in}} = \sqrt{3} \times 480\ \text{V} \times 250\ \text{A} = 207\ 846\ \text{W}$

5. FLA = 15.2 A with unity power factor (Table 430-150)

   $\text{Load} = \dfrac{15.2\ \text{A}}{.80} = 19\ \text{A}$

   Ampacity = 1.25 × 19 A = 23.75 A (430-22)
   No. 10 THW copper conductors (Table 310-16)

CHAPTER 4 ANSWERS

4-1.1 Quiz *(Closed-Book)*

1. Outside dimensions [220-3(b)]
2. (a) General lighting circuits (220-4)
   (b) Small appliance circuits
   (c) Laundry circuit
3. 8 kilowatts (Table 220-19, Note 4)
4. 3 watts [Table 220-3(b)]
5. 20 amperes (220-4)

4-1.1 Quiz *(Open-Book)*

1. Yes, if circuit conductors are rated 50 amperes [422-27(e)]
2. (a) Lighting (use 20-ampere circuits):

$$\frac{3 \text{ VA/ft}^2 \times 1600 \text{ ft}^2}{20 \text{ A} \times 120 \text{ V}} = 2.0$$

   Use two 20-ampere circuits with No. 10 conductors [220-3(a)]
   (b) Small appliance and laundry circuits: Use three 20-ampere circuits with No. 10 conductors [220-4(b)] [220-4(c)]
   (c) 12-kilowatt range: Load given as 8 kilowatts (220-19) Load (amperes) = 8000 W ÷ 240 V = 33.3 A; use 40-ampere, 120/240-volt circuit with No. 8 THW aluminum conductors (40 A); required ampacity = .7 × 40 A = 28 A; use No.10 for neutral conductor (210-19)

3. (a) Lighting: area = 40 ft by 40 ft = 1600 ft²
   16 ft by 16 ft = <u>256 ft²</u>
   1856 ft²

   No. of 20-ampere circuits =

$$\frac{3 \text{ VA/ft}^2 \times 1856 \text{ ft}^2}{20 \text{ A} \times 120 \text{ V}} = 2.3, \text{ or 3 circuits}$$

   Use three 20-ampere circuits with No. 12 conductors [220-3(b)]
   Small appliance and laundry: use three 20-ampere circuits with No. 12 conductors [220-4(b)] [220-4(c)]
   (b) 12-ampere, 230-volt compressor: 1.25 × 12 A = 15 A (440-32)
   Use 15- or 20-ampere, 240-volt circuit
   (c) 12-kilowatt range: Load given as 8 kilowatts (220-19) Load (amperes) = 8000 W ÷ 240 V = 33.3 A
   Use 40-ampere, 120/240-volt circuit with No. 8 conductors (50 A)
   Neutral ampacity = .7 × 40 A = 28 A; use No. 10 conductor
   (d) 5-kilowatt, 240-volt space heating:

$$1.25 \times \frac{5000 \text{ W}}{240 \text{ V}} = 26.0 \text{ A}$$

   Use 30-ampere, 240-volt circuit with No. 10 conductors

4-1.2 Quiz *(Closed-Book)*

1. 4500 watts (220-16)
2. 5000 watts (220-18)
3. (b) (Table 220-11)
4. 100 amperes (230-41)

4-1.2 Quiz *(Open-Book)*
See Detailed Calculations in Appendix B

4-1.3 Quiz *(Closed-Book)*

1. Nameplate rating (220-30)
2. True
3. (b)
4. See Code section 220-30
5. 40% of Nameplate rating (220-30)

4-1.3 Quiz *(Open-Book)*
See Detailed Calculations in Appendix B

4-1.4 Quiz *(Closed-Book)*

1. 60 A × 208 V = 12 480 VA
2. 8000 VA + .4*(X* − 8000) VA = 12 480 VA

4-1.4 Quiz *(Open-Book)*

1. (a) Existing loads are
   Lighting: 3 VA/ft² × 2000 ft² = 6 000
   Small appliance: 2 × 1500 VA = 3 000
   Range (10 kW) <u>10 000</u>
   Total 19 000 VA
   (b) Water heater is subject to 40% demand and may have connected load of 24 000 VA − 19 000 VA = 5000 VA maximum
2. Maximum added air-conditioning load (*L*) is
   L + 8000 W + .4(19 000 − 8000) W = 14 400 W
   L = 2000 W

4-1.5 Quiz *(Closed-Book)*

1. The service must be at least 100 amperes. The optional calculation (220-30) usually results in a smaller service than the standard calculation when there is a large heating and appliance load because of the application of demand factors to central heating and other loads
2. (a) Required branch circuits are

   (1) Lighting: $\dfrac{3 \text{ VA/ft}^2 \times 1000 \text{ ft}^2}{20 \text{ A} \times 120 \text{ V}} = 1.25$, or 2 circuits

   (2) Two 20-ampere small appliance circuits
   (3) One 20-ampere laundry circuit
   A total of five 20-ampere circuits are required
   (b) Service load is
   Lighting: 3 VA/ft² × 1000 ft² = 3 000
   Small appliances: 2 × 1500 VA = 3 000
   Laundry = <u>1 500</u>
   7 500 VA

   Application of demand factors:

   3000 VA @ 100% = 3000
   4500 VA @ 35% = <u>1575</u>
   Demand Load = 4575 VA

$$\text{Load (amperes)} = \frac{4575 \text{ VA}}{240 \text{ V}} = 19.1 \text{ A}$$

Minimum service size is 60 amperes (230-41) (For a 3-wire service)

4-1.5 Quiz *(Open-Book)*

See Detailed Calculations in Appendix B

4-2.1 Quiz *(Closed-Book)*

1. (b) (220-11)
2. 5000 watts (220-18)
3. Five (220-18)

4-2.1 Quiz *(Open-Book)*

See Detailed Calculations in Appendix B

4-2.2 Quiz *(Closed-Book)*

1. Three (Article 100-Definitions) (220-32)
2. False [220-32(b)]

4-2.2 Quiz *(Open-Book)*

See Detailed Calculations in Appendix B

4-2.4 Quiz *(Closed-Book)*

1. Connected load is 400 kilovolt-amperes
   Demand load is:

	Standard		*Optional*
3 000 VA @ 100% =	3 000		.28 × 400 kVA = 112 000 VA
117 000 VA @ 35% =	40 950		
280 000 VA @ 25% =	70 000		
	113 950 VA		

   There is no significant difference in demand loads

4-2.4 Quiz *(Open-Book)*

See Detailed Calculations in Appendix B

4-3 Quiz *(Open-Book)*

See Detailed Calculations in Appendix B

CHAPTER 4 TEST

I. (True or False)

1. F[220-3(b)]
2. F[220-3(a), Exceptions]
3. F [210-52(e), Exception 1, 2]
4. T [423-3(b)]
5. T (220-10)
6. T (220-18)
7. F (220-30)
8. T [220-32(a)-(3)]

II. (Multiple Choice)

1. (c) [422-27(e)]

2. (b) (Table 220-11)
3. (b) (Table 220-30)
4. (c) (220-31)
5. (a) (220-30)
6. (a) (220-30)
7. (c) [250-23(b)], [230-41(c)]
8. (a) [210-19(b)]

III. (Problems)

See Detailed Calculations in Appendix B

CHAPTER 5 ANSWERS

5-1 Quiz *(Closed-Book)*

1. (b) (240-21, Exception 3)
2. Yes; the neutral may not be smaller than the grounding electrode conductor [250-23(b)] [230-41(c)]
3. Rated primary current is 10 000 kVA ÷ 480 V = 20.8 A The maximum primary feeder overcurrent device setting is 2.5 × 20.8 A = 52 A; maximum size of secondary side overcurrent device is 1.25 × (10 000 kVA ÷ 240 V) = 52 A [450-3(b)]. Maximum standard size is 60 A.

5-1 Quiz *(Open-Book)*

See Detailed Calculations in Appendix B

5-2 Quiz *(Closed-Book)*

1. (b) (Table 220-20)
2. (c) (Table 220-40)
3. (a), (c) (Table 220-11)

5-2 Quiz *(Open-Book)*

See Detailed Calculations in Appendix B

5-3 Quiz *(Closed-Book)*

1. 3[550-13, 551-10(d)]
2. 16 000 [550-22(a)]
3. 35[550-13, 551-10(d)]
4. (a) (555-5)
5. (b) (645-2)
6. (c) [550-3(l)]

5-3 Quiz *(Open-Book)*

1. *See Detailed Calculations in Appendix B*
2. 20 × 16 000 VA × .25 = 8 000 VA (550-22)
3. 30 amperes (551-12)
4. 30 × 3 600 VA × .25 = 27 000
   20 × 2 400 VA × .26 = 12 480
   Total load   39 480 VA (551-44)
5. 5 × 30 A × .9 = 135 A (555-5)

CHAPTER 5 TEST

I. (True or False)

1. T[Table 220-3(b)]
2. T (384-14)
3. T (210-4)
4. T [210-6(a), Exception 1]
5. F (See text)
6. F [215-2]
7. T (555-3)
8. F (220-34)
9. F (550-22)
10. T (Table 220-20)

II. (Multiple Choice)

1. (b) [422-27(e)]

2. (c)$1.25 \times 30 \times 600$ VA $= 22\ 500$ VA [220-3(c)] (220-10)
3. (a)[Table 220-3(b)]
4. (a) (240-3, Exception 1)
5. (b) [460-8(a)]
6. (c) [600-6(a)]
7. (a) [450-3(b)]
8. (a)[Table 220-3(b)]
9. (c) (Table 220-41)
10. (c) (Table 220-11)

III. (Problems)

See Detailed Calculations in Appendix B

CHAPTER 6 TEST

I. (True or False)

1. T (210-4, Exception 2)
2. F (210-5 Exceptions)
3. T [210-8 (a)(1)]
4. F [210-70(a) and Exception 1]
5. T [230-24(a)]
6. F [230-24(b)]
7. F (230-82)
8. T [230-71(a)]
9. T (250-32, 230-63)
10. T (230-75)
11. F [250-61(a)]
12. F [250-81(a)]
13. F (250-84)

II. (Multiple Choice)

1. (b) [210-52(a)]
2. (c) [210-52(a)]
3. (a) (210-52)
4. (d) [210-8(a)]

5. (d) [210-70(a)]
6. (a) (210-62)
7. (c) [230-24(b)]
8. (a) [230-42(c)]
9. (c) [250-5(b)]
10. (a) [250-79(c)]

III. (Fill in the Blanks)

1. 120 (210-6)
2. 5½ (210-52)
3. 12 [210-52(b)]
4. 12 [230-24(b)]
5. 10 [230-50(b)]
6. Raintight (230-54)
7. 24 [230-54(c)]
8. 8 [250-83(c)]
9. 8 [250-83(c)(3)]
10. 2 [250-83(d)]

CHAPTER 7 TEST

I. (True or False)

1. T (240-3)
2. F (240-21, Exception 3)
3. T [310-12(a)]
4. F (310-15, Note 8)
5. T (Chapter 9, Tables 8, 9)
6. T (400-8)
7. T (402-6)
8. F [300-4(a)(1)]
9. F (300-14)
10. T (110-14)
11. T (310-4)
12. F [300-19(a)]

13. T [300-5(a)]
14. F [300-5(d)]
15. T (318-3)
16. F (320-14)
17. T (326-1)
18. T (336-25)
19. T (348-1)
20. F [349-4(e)]
21. T (374-8(b)]
22. T [250-76]
23. F [240-51(b)]
24. F [370-6(a)(1)]

25. T (370-19)
26. F [380-14(a)(2)]
27. T [410-8(a)]

II. (Multiple Choice)

1. (b) [200-6(b)]
2. (d) (240-6)
3. (a) [240-50(a)]
4. (b) (240-61)
5. (c) (422-16) (250-42, Exception 2)
6. (a) [250-60(b)]
7. (b) (250-95)
8. (c) [300-5(a)]
9. (a) [300-5(a), Exception 4)
10. (b) (310-15)
11. (a) [342-3(a)]
12. (b) (346-10)
13. (c) (346-12, Exception 1)
14. (b) [348-5(b)]
15. (a) (374-5)
16. (b) [370-6(b)]
17. (c) (370-6)
18. (b) [380-14(b)(2)]
19. (c) (402-5)
20. (a) (410-15)

III. (Fill in the Blanks)

1. Surface (Article 100)
2. AWG (110-6)
3. Hexagonal [240-50(c)]
4. Load [240-50(e)]
5. Interchangeable [240-53(b)]
6. Six [250-79(e)]
7. No. 6 [250-92(b)(2)]
8. 12 [300-5 (a), Exception 1]
9. No. 8 (310-3)
10. 100 (310-15, Note 3)
11. 4½ (320-6)
12. 6 [333-12(a)]
13. ¾ [346-7(b)]
14. 4, 360° (346-11)
15. No. 1/0 (356-4)
16. ¹⁵⁄₁₆ (370-14)
17. 8 [370-18(a)(1)]
18. 1000 (374-6)
19. Blades [380-6(e)]
20. Grounded (410-47)

CHAPTER 8 TEST

I. (True or False)

1. F [110-16(e), Exception]
2. T [Table 110-16(a)]
3. F [110-16(f)]
4. T [384-3(f)]
5. T (384-13)
6. F (384-14)
7. T (384-27)
8. T (450-21)
9. T [450-3(c)]
10. F [460-8(a)]
11. T [460-6(a)]
12. T [480-2(c)]
13. F (445-6)
14. T (280-21)
15. T (280-21)

II. (Multiple Choice)

1. (c) [Table 110-16(a)]
2. (a) [110-17(a)(4)]
3. (b) (384-10)
4. (b)
5. (a) [384-16(a)]

6. (a) (384-14)
7. (c) [384-16(b)]
8. (a) [384-16(a)]
9. (c) (445-5)
10. (b) [450-3(b), (1) Exception 1]
11. (b) [450-3(b)(2)]
12. (b)
13. (c) [460-8(a)]
14. (c) (460-9)
15. (b) (280-23)

III. (Fill in the Blanks)

1. 50 [110-17(a)]
2. 5 [Table 110-34(a)]
3. 8 ft, 6 in. [Table 110-34(e)]
4. 3 (384-8)
5. Bonded [384-3(c)]
6. 4 (384-14)
7. 125 [450-3(b)(1)]
8. 4 [450-43(b)]
9. .75
10. 135 [460-8(c)]

CHAPTER 9 TEST

I. (True or False)

1. T (422-2)
2. F [422-8(c)]
3. F [422-8(d)(2)]
4. F (422-17)
5. T [424-19(c)]
6. T (424-35)
7. F (424-36)
8. T [424-41(f)]
9. T [424-43(c)]
10. T [426-24(b)]
11. T (430-10)
12. F (430-84)
13. F (430-109, Exception 1)
14. T (430-127)
15. T (430-142)
16. F (430-145)
17. T (440-13)
18. T (440-60)

II. (Multiple Choice)

1. (b) [422-8(d)(2)]
2. (a) [422-15(b)]
3. (c) [422-21(b)]
4. (c) [422-27(e)]
5. (a) (424-34)
6. (c) (424-35)
7. (b) (424-36)
8. (a) (424-39)
9. (a) [424-41(b)]
10. (b) [424-3(b)]
11. (b) (424-36)
12. (c) ARTICLE 100
13. (a) (430-109, Exception 2)
14. (b) (430-132)
15. (c) [430-145(b)]

III. (Fill in the Blanks)

1. 18, 36 [422-8(d) (1)]
2. 150 (422-16)
3. 50 [422-27(c)]
4. 60 (424-11)
5. 60 [424-22(b)]
6. Brown (424-35)
7. 2 (424-36)
8. 2 (424-39)
9. 3 [424-41(d)]
10. UF, NMC, MI [424-43(a)]
11. 1 [424-44(b)]
12. 200 [430-109, Exception 2]
13. 8 [430-132(c)]
14. 150 (430-133)
15. 6 [430-145(b)]

CHAPTER 10 TEST

I. (True or False)

1. T [600-7(a)]
2. F [600-10(b)]
3. F (600-22)
4. F [645-2(c)(2)]
5. T (645-3)
6. T [645-3(a)]
7. F (680-4)
8. T [680-6(a)]
9. F [680-20(a)(1)]
10. T [680-20(a) (3) [Exception]
11. F [680-21(a)(2)]
12. T [680-22(b)(2)]
13. T [680-25(d)]
14. F (680-50)
15. T (680-54)

II. (Multiple Choice)

1. (b) [600-2(b)]
2. (a) [600-6(a)]
3. (b) [600-8(d)]
4. (c) [600-9(c)]
5. (a) [600-21(b)]
6. (b) (600-22)
7. (b) [600-37(c)]
8. (c) [645-2(c)(2)]
9. (b) [680-6(b)(1)]
10. (c) [680-8(a)]
11. (a) (680-8, Exception)
12. (c) [680-20(a)(3)]
13. (c) [680-22(b)]
14. (a) [680-25(d)]
15. (b) [680-41(b)]

III. (Fill in the Blanks)

1. 30 [600-6(a)]
2. ½ [600-8(g)]
3. 16 [600-10(b)]
4. 15 000 [600-32(a)]
5. 10 [600-37(c)]
6. Ventilation (645-3)
7. 15 [680-6(a)]
8. 15 [680-20(a)]
9. 8, 4 [680-21(a)(4)]
10. 5 (680-22, 680-24)

CHAPTER 11 TEST

I. (True or False)

1. T[500-3(a)]
2. T(Table 500-3)
3. F (502-1)
4. F [501-5(c)(3)]
5. F (501-8)
6. T [501-16(b)]
7. T [502-2(a)(3)]
8. T (502-5)
9. F (502-11)
10. T[511-3(c)]
11. T[511-7(b)]
12. F (514-2)
13. T (514-2)
14. F [516-2(a)]
15. F [517-100(a)(1)]
16. F(518-4)
17. T [555-9(a)(3)]

II. (Multiple Choice)

1. (b)(500-2)
2. (c)[500-7(b)]
3. (b) [501-4(b)]
4. (a) [501-5(a)(1)]
5. (c) [501-5(a)(2)]
6. (c) [501-6(b)(4)]
7. (b) [501-9(a)(3)]

8. (a) (502-5)
9. (b) (502-8)
10. (c) (502-11)
11. (c) [503-9(c)]
12. (b)(511-7)
13. (c) (514-2)
14. (a) (514-2)
15. (b) (514-5)

III. (Fill in the Blanks)

1. Class I
2. Class, division, group (500-3)
3. 5 [501-4(a)]
4. 18 (501-5)
5. 12 (501-5(a), Exception)
6. 80 [501-8(a)]
7. Class I, Division 1 (501-14)
8. Approved [502-10(b)(1)]
9. MC (503-3)
10. II, 2 (503-11)
11. Hard [511-6(b)]
12. Class I, Division 1 [511-3(b)]
13. 18 (514-2)
14. First [514-6(a)]
15. 2 (514-8)

CHAPTER 12 TEST

I. Basic DC Theory

1. 1.67 amperes
2. 51.43 volts, 176.4 watts
3. 10 amperes, 6 ohms
4. 13.33 ohms
5. 50.6 amperes

II. Conductors

1. 10383 cmils
2. .121 ohm
3. 955 MCM
4. .772 ohm
5. .050 ohm
6. .7854
7. 125 mils or .125 inch

III. AC Theory

1. .016667 second
2. .00278 second
3. 50 ohms, .133 henry
4. 9.23 amperes

5. .3846
6. 5.32 amperes, 106.5 volts, 566 watts
7. 28.8 kvars
8. 62.5 kilovolt-amperes, 260 amperes
9. 13. 510 watts
10. 240 kilowatthours
11. 33.33 kilovolt-ampere reactive, 14.5 kilovolt-ampere reactive

IV. Equipment in AC Circuits

1. Currents = 208.33 amperes, 260.42 amperes
   Overall values = .94 power factor, 106.8 kilovolt-amperes, 445 amperes
2. 48.11 amperes
3. 23.9 volts, 11.9%; 10.5 volts, 4.57%; 2.91 volts, 0.6%
4. No. 2/0
5. 360.8 amperes, 832.7 amperes
6. 1800 revolutions per minute, 3600 revolutions per minute
7. 54.47 horsepower
8. 41%, 58%, 52%, 76%
9. 52.8 kilovolt-ampere reactive (I = 92.86 amperes)

FINAL EXAMINATIONS

Final Examination No. 1

  I. (True or False)

1. T [110-14(a)]
2. F (210-21)
3. F (348-12)
4. T (356-7)
5. T [370-18(a)(1)]
6. T [422-17(a)]
7. F (514-2)
8. F (430-4)
9. T [700-12(a)]
10. F (384-15)
11. F (384-14)
12. (a) T (645-3)
    (b) T
    (c) F
    (d) F
13. T [250-53 (b)]
14. F (210-6)
15. T (346-11)
16. T [250-76 (b)]

    (Multiple Choice)

1. (c) (374-2)
2. (a) [680-25(c)]
3. (b)
4. (a) [Table 300-19(a)]
5. (b)
6. (c) (Table 310-13)
7. (d)
8. (a)

  II. (Problems)

Problem 1. *See Detailed Calculations in Appendix B*

Problem 2.

(a) The current per unit is $\dfrac{4 \times 40\,\text{W}}{277\,\text{V} \times (.8)} = .72\,\text{A}$

The load should be considered continuous; hence,

load per unit $= .72 \times 1.25 = .9\,\text{A}$

For 15-ampere circuits,

number/circuit $= \dfrac{15\,\text{A}}{.9\,\text{A}} = 16.7$, or 16 per circuit

number of circuits $= 400$ units/16 circuits $= 25$ circuits, or in like manner, eighteen 20-ampere circuits

(b) The feeder is a 480Y/277-volt feeder and is required to carry

total current per phase $= \dfrac{400\ \text{units}\ \times\ 4\ \text{lamps}\ \times\ 40\,\text{W}}{\sqrt{3}\ \times\ 480\,\text{V}\ \times\ .8} = 96.22\,\text{A per phase}$

(Fill in the Blanks)

1. 3 (Table 430-37)
2. 6 ohms
3. Orange color (215-8)
4. 150 amperes
5. Resistance
6. 250 kilowatthours

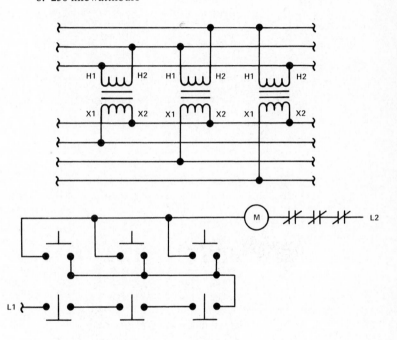

This requires No. 3 THW copper wire

Note: It is common to add 20% to the power of the lighting units to account for equipment losses. In this case, each unit would require $160 \times 1.2 = 192$ W and the current per unit would be 1.08 A/unit as a continuous load.

Problem 3.

Code Ref.	Motor Load or Other Load (480 V)	Current	Total Line Current
(Table 430-150)	Two 60 hp	2 × 77 A	154 A (Table 430-150)
	Two 30 hp	2 × 40 A	80 A
	Four 10 hp	4 × 14 A	56 A
	Four 3 hp	4 × 4.8 A	19.2 A
(430-24)	25% of largest motor	.25 × 77	19.25 A (430-24)
	10-kW heater	$I = \dfrac{10\ 000\ W}{\sqrt{3} \times 480} = 12.0\ A$	

$$\text{Required feeder ampacity} = 340.5\ A\ (\text{minimum})$$

The standard safety switch would be a 400-ampere switch in a NEMA 1 enclosure.

Problem 4.

Code Ref.	Motor Load (230 V)	Current	Total Line Current
[Table 430-148 (motors)]	Two 10 hp	2 × 50 A	100 A
	Two 5 hp	2 × 28 A	56 A
	Four 3 hp	4 × 17 A	68 A
	Three 1 hp	3 × 8 A	24 A
	Three ¾ hp	3 × 6.9 A	20.7 A
	Three ⅓ hp	3 × 3.6 A	10.8 A
(430-24)	25% of largest motor	.25 × 50 A	12.5 A
	Feeder current =		292 A

(Table 430-152) The largest branch-circuit protective device is obviously that of one of the 10-horsepower motors since all the motors have the same type of protection (nontime-delay fuse). For the 10-horsepower motor,

$$\text{fuse size} = 3 \times 50\ A = 150\ A$$

(430-62) Feeder protective device fuse size = 150 A + 50 A + 56 A + 68 A + 24 A + 20.7 A + 10.8 A = 379.5 A maximum

(240-6) Standard size = 350 A

Problem 5.

Code Ref.	Area of Wire	Total Area (in.²)
Chapter 9 Table 5	No. 4/0 THW 4 × .3904 in.²	1.562
	No. 2 THW 4 × .1473 in.²	0.5892
	No. 6 THW 4 × .0819 in.²	0.3276
	Total area =	2.479 in.²
Chapter 9 Table 4	Use 3-inch conduit	

Problem 6. *See Detailed Calculations in Appendix B*

Problem 7.

Code Reference	
(Chapter 9, Table 3B)	I. A 1-inch conduit will enclose the eight No. 8 THHN conductors
(Table 310-16)	II. The conductor must be derated for both conduit fill and ambient temperature. The conductors have an ampacity of 55 amperes normally
(Table 310-16, Note 8)	(a) Derating for 8 conductors: 70% for more than 7
(Table 310-16)	(b) Derating for 60°C ambient: The THHN conductors have a temperature rating of 90°C; if the ambient is 60°C, the correction factor is .71.

The ampacity (55 amperes) must be reduced by both factors

Final ampacity = 55 A × .7(fill) × .71 (temperature) = 27.34 A

FINAL EXAMINATION NO. 2

1. See Article 100, Definitions
2. 10 feet, 12 feet, 18 feet [230-24(b)]
3. Orange (215-8)
4. (a) Outlets in habitable rooms, bathrooms, hallways, attached garages, and at outdoor entrances (210-70)
   (b) Others required in storage or equipment areas
5. Bathrooms, kitchens, basements, garages, and outdoors [210-8(a)]
6. True [240-50(a)]
7. 18 (250-92)
8. 10 [250-81(a)]
9. No. 2, No. 1/0, No. 2/0, No. 4/0 (Table 310-16, Note 3)
10. False [300-3(b)]
11. 80% (Table 310-16, Note 8)
12. 12 (Table 346-12)
13. 20%, 75% (362-5, 362-6)
14. 6 [Table 370-6(b)]
15. 50, 1/3 (364-11)
16. 30, 30 (374-2, 374-5)
17. No. 18 (410-23)
18. 40 amperes, 250 volts, 6 feet (440-62, 64)
19. 135%, as small as practicable (460-8)
20. False [500-6(a)]
21. (b) [500-7(a)]
22. True (330-3)
23. Seal, union (514-6)
24. 18, 20 [514-2(b)]
25. 3 ohms, 22 ohms, 88 volts, 66 volts, 5 ohms

26. 69 horsepower
27. 1800 revolutions per minute
28. Diagram for Question 28.

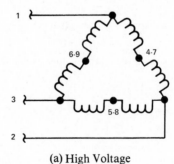

(a) High Voltage

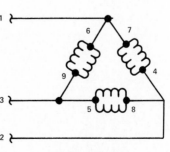

(b) Low Voltage

29. Diagram for Question 29.

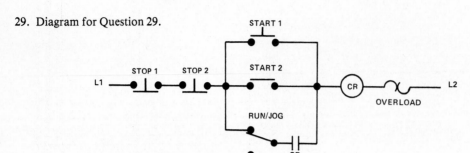

Note: Relay CR controls the motor control contacts (not shown) and the control circuit holding contacts CR.

II.   Problems

Problem 1. *See Detailed Calculations in Appendix B*

Problem 2. *See Detailed Calculations in Appendix B*

Problem 3.

25-horsepower motor

   FLA = 34 amperes (Table 430-150)
   Conductor ampacity = 1.25 × 34 A = 42.5 A (430-22)
   Branch-circuit protection = 3 × 34 A = 102 A; use 110-ampere fuse (Table 430-152) (240-6)
   Running overcurrent = 1.25 × 34 A = 42.5 A [430-32(a)(1)]
   Controller and disconnect: 25-horsepower rating (430-83)

30-horsepower Motor

   FLA = 40 amperes (Table 430-150)
   Conductor ampacity = 1.25 × 40 A = 50 A (430-22)
   Branch-circuit protection = 1.5 × 40 A = 60 A (Table 430-152)
   Running overcurrent = 1.15 × 40 A = 46 A [430-32(a)(1)]
   Controller and disconnect: 30-horsepower rating (430-83)

Feeder Circuit

   Ampacity = 1.25 × 40 A + 40 A + 34 A = 124 A (430-24)
   Feeder protection = 110 A + 40 A + 40 A = 190 A; use 175-ampere fuse [430-62(a)]

Problem 4. 9 feet [Table 110-34(e)]

Problem 5.

From Table 9, Chapter 9

R = 0.045 ohms/1000 ft
X = 0.048 ohms/1000 ft
$Z = \sqrt{R^2 + X^2}$ = 0.0658 ohms/1000 ft
Z = 0.0658 ohms/1000 ft × 800 ft/1000 ft
  = 0.0526 ohms

Problem 6.

(a) *40-hp, 230-volt, 3-phase motor*

   FLA = 104 amperes without capacitor
   kVA = $\sqrt{3}$ × 230 V × 104 A/1000 = 41.43 kVA
   kW = 41.43 kVA × .7 = 29 kW
   kvar = $\sqrt{(41.43)^2 - (29)^2}$ = 29.59 kvar

(b) Current = 29 kW/$\sqrt{3}$ × 230 V = 72.8 A (after correction)
   Running overcurrent device = 1.25 × 72.8 A = 91 A [430-32(a)], (460-9)

(c) Motor conductors = 1.25 × 104 A = 130 A (minimum)
   Capacitor current = 29.59 kvar/$\sqrt{3}$ × 230 V = 74.3 A
   Capacitor conductors = 1.35 × 74.3 A = 100.3 A minimum) (460-8)

DETAILED CALCULATIONS

4-1.2 Quiz Problem 1

Code Rule	Service Load Calculation		
		Line	Neutral
220-3(b)	General Lighting Load = 3 VA/ft² × 1500 ft²	4500	4500
220-16	Small Appliance Ckts = 2 × 1500 VA	3000	3000
	Laundry Circuit	1500	1500
	Total Ltg, Small Appl, and Laundry	9000	9000
220-11	Application of Demand Factors		
	First 3000W @ 100%	3000	3000
	Remainder (9000-3000) @ 35%	2100	2100
	General Lighting Demand Load	5100	5100
Table 220-19,	12-kW Range = 8000W; Neutral = .7 × 8000W	8000	5600
220-22	Service Load	13,100VA	10,700VA

Code Rule	Selection of Service Equipment
230-41	1. Service Ampacity = 13,100W ÷ 240V = 54.5A; Neutral = 10,700 ÷ 240V = 44.6A Required to use 100-ampere service
Table 310-16, Note 3	2. Conductors: Use two No. 4 type THW, one No. 8 type THW
230-90, 230-79	3. Overcurrent Protective Device and Disconnect: Use 100-ampere rating
Chapter 9	4. Conduit Size: Area of two No. 4 = .2174 in.² Area of one No. 6 = .0819 in.² Requires 1-inch size = .2993 in.²
Table 250-94	5. Grounding Electrode Conductor: Use No. 8 Copper

4-1.2 Quiz Problem 2

Code Rule	Service Load Calculation		
		Line	Neutral
220-3	General Lighting Load = 3VA/ft² × 2800 ft²	8400	8400
220-16	Small Appliance Circuits = 2 × 1500VA	3000	3000
	Laundry Circuit	1500	1500
	Total Ltg, Small Appl & Laundry	12,900	12,900
220-11	Application of Demand Factors		
	First 3000VA @ 100%	3000	3000
	Remainder (12,900-3000) @ 35%	3465	3465
	General Lighting Demand Load	6465	6465
Table 220-19	Range Load = 5 kW + 4 kW + 4 kW = 13 kW		
	Demand = 8 kW + .05 × 8 kW (13-12) =	8400	5880
220-22	(Neutral = .7 × 8400)		
220-18	4500-Watt Dryer	5000	3500
Table 430-148,	¼-hp disposal = 1.25 × 120V × 5.8A	870	870
430-25	Dishwasher	1200	1200
220-15, 220-21	Space heaters = 3 × 9 kW (Neglect A/C)	27000	0
	Service Load	48,935VA	17,915VA

Code Rule	Selection of Service Equipment
	1. Service Ampacity = 48,935VA ÷ 240V = 204A; Neutral = 17,915VA ÷ 240 = 75A
Table 310-16, 250-23(b)	2. Conductors: Use two No. 4/0 type THW, one No. 2 THW for neutral (based on size of grounding electrode conductor)
240-6	3. Overcurrent Protection and Disconnect: Use 225-ampere standard
Chapter 9	4. Conduit Size: Total area of conductors = .9281 in.² Requires 2-inch size conduit
Table 250-94	5. Grounding Electrode Conductor: Use No. 2 copper.

Code Rule	Feeder Load Calculation		
		Line	Neutral
220-30	General Lighting = 3VA/ft² × 2000 ft² Small Appliance = 2 × 1500VA Laundry	6000 3000 1500	6000 3000 1500 10,500 (Total Ltg.)
			3000 (First 3000 @ 100%) 2625 (Remainder @ 35%) 5625 (Ltg. Demand)
Table 220-19, Note 4, Note 1 220-22 Table 220-30	Cooking Equip. = 5 kW + 4.5 kW + 4.5 kW = Neutral Rating = 8 kW + .05 × 8 kW × 2 　　　　= 8800W 　　Total "Other Load" Application of Demand Factors First 10kVA @ 100% Remainder (24,500-10,000) @ 40% Heating or Air Conditioning Heating = .65 × 12 kW = 7.8 kW (Neglect) Air Conditioning @ 100% 　　Service Load	14,000 24,500 10,000 5,800 10,000 25,800VA	6160 (.7 × 8800W) 0 11,785VA (Neutral Load)
Code Rule	Selection of Feeder Equipment		
Table 310-16 240-6 Chapter 9 250-95	1. Feeder Ampacity = 25,800VA ÷ 240V = 108A; Neutral = 11,785VA ÷ 240V = 49A 2. Conductors: Use two No. 2 type THW, one No. 8 type THW (Minimum) 3. Overcurrent Protective Device: Use 125-Ampere Standard Size 4. Conduit: Requires 1¼-inch conduit 5. Equipment Grounding Conductor: Use conduit or No. 6 copper		

4-1.5 Quiz Problem 1

Code Rule	Branch Circuits Required	
220-4	General Lighting: $\dfrac{3VA/ft^2 \times 1500\ ft^2}{120V} = 37.5A$	Two 20-ampere, 2-wire circuits
	Small Appliance Circuits:	Two 20-ampere, 2-wire circuits
	Laundry Circuit:	One 20-ampere, 2-wire circuit
220-3	Dryer: 4500W ÷ 240V = 18.8A	One 20-ampere, 120/240-volt circuit
424-3	Space Heating: $1.25 \times \dfrac{14\ 000W}{240V} = 72.9A$	
	72.9A ÷ 5 = 14.6A/unit	Five 20-ampere, 240-volt circuits
Table 220-19	12-kW Range: 8000W ÷ 240V = 33.3A	One 40-ampere, 120/240-volt circuit
422-14 (b)	Hot Water Heater: $1.25 \times \dfrac{3000W}{240V} = 15.6A$	One 20-ampere, 240-volt circuit
440-32 440-22(a)	Air Conditioner: 1.25 × 13A = 16.25A Protect at 1.75 × 13A = 22.75A maximum	One 20-ampere, 240-volt circuit

Code Rule	Service Load Calculation		
		Line	Neutral
220-30	General Lighting = 3VA/ft² × 1500 ft²	4,500	4,500
	Small Appliance = 2 × 1500VA	3,000	3,000
	Laundry	1,500	1,500
			9,000 (Total Lighting, etc.)
			3,000 (First 3000 @ 100%)
			2,100 (Remainder @ 35%)
			5,100 (General Lighting Demand)
220-22,	12-kW Range	12,000	5,600 (.7 × 8000W)
Table 220-19	Dryer	4,500	3150 (.7 × 4500)
220-30(3)	Water Heater	3,000	-0-
	Total "Other Load"	28,500	
Table 220-30	Application of Demand Factors		
	First 10kVA @ 100%	10,000	
	Remainder @ 40%	7,400	
		17,400VA	
	40% of space heating	5,600	
	Service Load	23,000VA	13,850VA (neutral load)

Code Rule	Selection of Service Equipment
Table 310-16 Chapter 9 Table 250-94	1. Conductors: Use two No. 3 THW copper, one No. 6 THW copper. 2. Overcurrent Device and Disconnect: Use 100-ampere service. 3. Conduct: Total area of conductors = .3345 in.^{2}, use 1-inch conduit. 4. Grounding Electrode Conductor: Use No. 8 copper.

4-1.5 Quiz Problem 2

Code Rule	Branch Circuits Required	
220-4	General Lighting: $\dfrac{3VA/ft^2 \times 2800\ ft^2}{120V \times 20A} = 3.5$	Four 20-ampere, 2-wire circuits
	Small Appliance and Laundry:	Three 20-ampere, 2-wire circuits
Table 220-19, Notes 1 and 4	Cooking Equipment = 5 kW + 4 kW + 4 kW = 13 kW Demand Load = 8 kW + .05 × 8 kW × (13-12) = 8.4-kW 8400W ÷ 240V = 35A	One 40-ampere, 120/240-volt circuit
	Dryer: 4500W ÷ 240V = 18.7A	One 20-ampere, 240-volt circuit
Table 430-148	Disposal: 1.25 × 5.8A = 7.25A	One 15-ampere, 2-wire circuit
	Dishwasher: 1200VA ÷ 120V = 10A	One 15-ampere, 2-wire circuit
424-3	Space Heating: $1.25 \times \dfrac{9000W}{240V} = 46.9A$ each (3 units)	Three 50-ampere, 240-volt circuits
440-32 440-22(a)	Air Conditioning: 1.25 × 15A = 18.75A Maximum Overcurrent Protection = 1.75 × 15A = 26.25A	One 20-ampere, 2-wire circuit

Code Rule	Service Load Calculations			

Code Rule		Optional Calculation Line	Standard Calculation Line	Neutral
	General Lighting = 3VA/ft² × 2800 ft² =	8400	8400	8400
	Small Appliance = 2 × 1500VA	3000	3000	3000
	Laundry	1500	1500	1500
			12900	12900 Total
			3000	3000 (100%)
			3465	3465 (9900 @ 35%)
			6465	6465 (Demand)
220-19, 220-22	Cooking Equipment	13,000	8400	5880 (.7 × 8400)
220-18	Dryer	4,500	5000	3500
430-25	Disposal = 120V × 5.8A	696	870	870 (125%)
	Dishwasher	1,200	1200	1200
	Total "Other Load"	32,296		
Table 220-30	Application of Demand Factors			
	First 10-kVA @ 100%	10,000		
	Remainder @ 40%	8918		
220-15 220-21	Space Heating = 3 × 9kW @ 65% (Neglect A/C)	17,550	27,000	0
	Service Load	36,468VA	48,935VA	17,915VA

	Service Rating

1. Minimum Service Rating (Optional Calculation) = 36,486VA ÷ 240V = 152A
2. Minimum Service Rating (Standard Calculation) = 48,935VA ÷ 240V = 203.9A,
   Neutral = 17,915 ÷ 240V = 74.6A

4-2.1 Quiz Problem 1

Code Rule	Service Load Calculation		
		Line	Neutral
	General Lighting = 3VA/ft² × 1800 ft² × 7 =	37,800	38,700
	Small Appliance = 2 × 1500VA × 7	21,000	21,000
	Laundry = 1500VA × 7	10,500	10,500
	Total Ltg, Small Appl. & Laundry	69,300	69,300
	Application of Demand Factors		
	First 3000VA @ 100%	3,000	3,000
	Remainder (69,300−3000) @ 35%	23,205	23,205
	General Lighting Demand	26,205	26,205
Table 220-19	12-kW Range (Neutral @ 70%)	22,000	15,400
Table 430-148	Disposal = 6A × 120V × 7	5,040	5,040
	Dishwasher = 12A × 120V × 7	10,080	10,080
Table 220-18	Dryer = .65 × 5 kW × 7	22,750	15,925
	Two Air Conditioners = 2 × 12A × 240V × 7	40,320	0
Table 430-148	Air Handler Motor = 8A × 240V × 7	13,440	0
430-25	25% of Largest Motor = .25 × 12A × 240V	720	0
	Service Load	140,555VA	72,650VA

Code Rule	Service Rating
	1. Service Rating = 140,555VA ÷ 240V = 585.6A, Use 600-ampere standard.
	Neutral = 72,650VA ÷ 240V = 302.7A
220-22	**Further Demand Factor For Neutral**
	First 200 amperes @ 100% = 200
	Balance (303-200) @ 70% = 72
	272A
Table 310-16	2. Use Two 1250 MCM, One 300 MCM (or equivalents)

4-2.2 Quiz Problem 1

Code Rule	Service Load Calculation		
220-32		Line	Neutral (Standard)
	General Lighting = 3VA/ft² × 1000 ft² × 50 =	150,000	150,000
	Small Appliance = 2 × 1500VA × 50	150,000	150,000
	Laundry = 1500VA × 50	75,000	75,000
			375,000 Total
		3,000	3,000 @ 100%
		40,950	117,000 @ 35%
		63,750	255,000 @ 25%
		107,700	Demand Load
	Ranges = 10 kW × 50 (Neutral @ 70%)	500,000	43,750 .7 × [25 kW + ¾ (50)]
	Dryer = 5 kW × 50 (Neutral @ 70%)	250,000	43,750
220-17	Disposal = 6A × 120V × 50	36,000	27,000 (.75)
	Dishwasher = 12A × 120V × 50	72,000	54,000 (.75)
	Space Heating = 9.6-kW × 50	480,000	0
	Total Connected Load	1,713,000	
	Application of Demand Factors		
Table 220-32	1,708,500VA @ 26%	445,380	
	House Load = 1.25 × 10 × 200W	2,500	2,500
	Service Load	447,880VA	278,700VA

Code Rule	
	Load (amperes) = 447,800VA ÷ 240V = 1885, Neutral = 278,700VA ÷ 240V = 1161A
220-22	**Further Demand Factor For Neutral**
	First 200 amperes, @ 100% = 200
	Balance (1161-200) @ 70% = 672
	872 A

Code Rule	Branch Circuits Required	
220-3,4	General Lighting = $\dfrac{3\text{VA/ft}^2 \times 200\text{ ft}^2}{120\text{V}}$ = 5A	One 15- or 20-ampere, 2-wire circuit
220-4	Small Appliance Circuits (Since this motel is a dwelling)	Two 20-ampere, 2-wire circuit
	Laundry—Not required	
Table 430-148	1/3-hp Disposal = 1.25 × 7.2A = 9.0A	One 15- or 20-ampere, 2-wire circuit
	10-Ampere, 240-Volt Air Conditioner = 1.25 × 10A = 12.5A	
	Maximum rating of overcurrent device = 1.75 × 10A = 17.5A	One 15-ampere, 240-volt circuit

Code Rule	Feeder Load Calculation		
		Line	Neutral
220-10	General Lighting = $3\text{VA/ft}^2 \times 200\text{ ft}^2$	600	600
220-16	Small Appliance Circuits	3000	3000
	Total Ltg and Small Appl	3600	3600
Table 220-11	Application of Demand Factors		
	First 3000VA @ 100%	3000	3000
	Remainder (3600-3000) @ 35%	210	210
	General Lighting Demand Load	3210	3210
	Disposal = 120V × 7.2A	864	864
	Air Conditioner = 240V × 10A	2400	0
430-25	25% of largest motor = .25 × 240V × 10A	600	0
	Feeder Load	7074VA	4074VA

Code Rule	Selection of Feeder Equipment
215-2(a)	1. Feeder Ampacity = 7074VA ÷ 240V = 29.5A; Neutral = 4074VA ÷ 240V = 16.9A
	2. Conductors: Use Three No. 10 Type THW
430-63	3. Overcurrent Device = 15A (Air Cond.) + 7.2A (Disposal) + 13A (General Lighting) = 35.2A (maximum)

Code Rule	Service Load Calculation		
		Line	Neutral
Table 220-11	Lighting = 3VA/ft² × 200 ft² × 150	90,000	90,000
	Small Appliance Circuits = 2 × 1500VA × 150 =	450,000	450,000
	Total Ltg and Small Appl.	540,000	540,000
	Application of Demand Factors		
	First 3000 @ 100%	3,000	3,000
	117,000 @ 35%	40,950	40,950
	420,000 @ 25%	105,000	105,000
	General Lighting Demand Load	148,950	148,950
220-17	Disposals = 120V × 7.2A × 150 × .75	97,200	97,200
	Air Conditioners = 240V × 10A × 150	360,000	0
	25% of Largest Motor = .25 × 240V × 10A	600	0
	Service Load	606,750VA	246,150VA

Code Rule	Selection of Service Equipment
220-22	1. Service Ampacity = 606 750VA ÷ 240V = 2528A; Neutral = 246 150VA ÷ 240V = 1025A
	Further Demand Factor For Neutral
	200 + .7 (1025 − 200) = 778 Amperes
	2. Conductors: Use bus bars or parallel conductors
240-3, 240-6	3. Overcurrent device: Use standard size 2500-ampere device
Table 250-94	4. Grounding Electrode Conductor: No 3/0 copper

4-3 Quiz Problem 1

Code Rule	Motor Branch Circuit
Table 430-150, 430-22	1. Ampacity: 1.25 × 68A = 85A
	2. Conductors: Use No. 2 Type THW aluminum
Table 310-16	3. Overcurrent Device: 3.0 × 68A = 204A; use 200-ampere fuses
Table 430-152	4. Disconnecting Means: Use 25-hp device

Code Rule	Apartment Feeders	
		Line & Neutral
220-4	General Lighting = 3VA/ft² × 500 ft² × 10	15000
	Small Appliance Circuits = 2 × 1500VA × 10	30000
	Total Ltg and Small Appliance	45000
Table 220-11	**Application of Demand Factors**	
	First 3000VA @ 100%	3000
	Remainder (45000-3000) @ 35%	14700
	General Lighting Demand Load	17700VA

Code Rule	Service Load Calculation			
		Line A, C	Neutral	Line B
430-25	General Lighting Demand	76.9	76.9	0
	Motor Load	68.0	0	68.0
	25% of Motor	17.0	0	17.0
	Total Service Load	161.9A	76.9A	85.0A

Code Rule	Selection of Service Equipment
Table 310-16	1. Conductors: Use 4/0 aluminum for Lines A and C, Use No. 2 aluminum for Line B and Neutral
	2. Main Overcurrent Protection: Lines A and C = 200A (motor) + 76.9A = 276.9A (maximum)
430-63	Line B = 200 Amperes
	3. Apartment Feeder Panel: Use 80-ampere standard size

Code Rule	Branch Circuits Required	
	General Lighting: $\dfrac{3VA/ft^2 \times 1800 \ ft^2}{120V \times 20A} = 2.2$, or 3	Three 20-ampere, 2-wire circuits
	Small Appliance Circuits	Two 20-ampere, 2-wire circuits
	Laundry	One 20-ampere 2-wire circuit
Table 220-19 Note 4	Oven: 6000W ÷ 240V = 25A	One 30-ampere, 120/240-volt circuit
	Table-top Range: 6000W ÷ 240V = 25A	One 30-ampere, 120/240-volt circuit
	Clothes Dryer: 4800W ÷ 240V = 20A	One 25- or 30-ampere, 120/240-volt circuit
422-14(b)	Hot Water Heater: $\dfrac{1.25 \times 4500W}{240V} = 23.4A$ (continuous load)	One 25- or 30-ampere, 240-volt circuit
424-3	Central Heating: $\dfrac{1.25 \times 9000W}{240V} = 46.9A$	One 50-ampere, 240-volt circuit
440-22	Air Conditioner: 1.25 × 17A = 21.25 Overcurrent Device = 1.75 × 17A = 29.75A	One 30-ampere, 240-volt circuit
Table 430-148	Two ¼-hp motors: 1.25 × 5.8A = 7.25A	Two 15-ampere, 2-wire circuits

Code Rule	Service Load Calculation

		Line	Neutral	
	General Lighting = 3VA/ft² × 1800 ft²	5400	5400	
	Small Appliance Circuits = 2 × 1500VA	3000	3000	
	Laundry Circuit	1500	1500	
			9900 (Total)	
			3000	Demand Factors (First 3000)
			2415	(6900 @ 35%)
			5415	(Demand Load)
Table 220-19 Column C	Cooking Units = 2 × 6000W (Standard = .65 × 12-kW = 7800W)	12000	5460	(.7 × 7800W)
	Dryer	4800	3500	(.7 × 5000W)
	Hot Water Heater	4500	0	
	Motors = 2 × 5.8A × 120V	1392	1392	
			174	(25% of Largest motor)
	Total "Other Load"	32592		
220-30	Application of Demand Factors			
	First 10-kVA @ 100%	10000		
	Remainder (22592) @ 40%	9036		
	Space Heating = 9000W @ 65%	5850	0	
	Total Service Load	24886VA	15941VA	

Code Rule	Selection of Service Equipment
Table 310-16	1. Service Ampacity: 24886VA ÷ 240V = 104A amperes; Neutral = 15941VA ÷ 240V = 66.4A
	2. Conductors: Use two No. 2 type THW copper and one No. 6 type THW copper
	3. Overcurrent Protective Device: Use 110 ampere standard size (A 125-ampere service may be more practical.)

Test Chapter 4 Problem 2

Code Rule	Service Load Calculation		
		Line (VA)	Neutral (VA)
220-3,4 220-16	General Lighting Load = 3VA/ft² × 2000 ft² =	6000	6000
	Small Appliance Circuits = 2 × 1500VA =	3000	3000
	Laundry Circuit	1500	1500
	Total Ltg, Small Appl & Laundry Ckts	10500	10500
Table 220-11	Application of Demand Factors		

Code Rule	Service Load Calculation			
	First 3000 @ 100%	3000	3000	Neutral (Amperes)
	Remainder (10500-3000) @ 35%	2625	2625	
	General Lighting Demand Load	5625	5625	23.4
Table 220-19 Note 1	14-kW Range = 8 kW + .05 × 8-kW (14-12) (Neutral @ 70%)	8800	6160	25.7
	Dryer	5000	3500	14.5
220-15	Space Heating	16000	0	0
	Dishwasher	1200	1200	10.0
	Service Load	36625VA	16485VA	73.6A

Code Rule	Selection of Service Equipment

	1. Service Ampacity: 36625VA ÷ 240V = 152.6A
	     Neutral = 16485VA ÷ 240V = 68.7
	     (The computed neutral load is low by 4.9 amperes when the standard calculation is used. This difference is one-half the dishwasher load.)

Table 310-16	2. Conductors: use two No. 3/0 and one No. 3 type THW aluminum.
240-6	3. Overcurrent Protection: Use 175-ampere standard size.
Chapter 9, Tables 5, 4	4. Conduit: Area of conductors = 2 × .3288 in² + .1263 in² = .7839 in². Use 1½-inch conduit
250-94	5. Grounding: Use No. 6 copper grounding electrode conductor

Test Chapter 4 Problem 3

Code Rule	Service Load Calculation		
		Line	Neutral
220-3	General Lighting Load = 3VA/ft² × 900 ft² × 12	32400	32400
220-16	Small Appliance Circuits = 2 × 1500VA × 12	36000	36000
220-16	Laundry Circuits = 1500VA × 12	18000	18000
	Total Ltg, Small Appl & Laundry Ckts	86400	86400
Table 220-11	Application of Demand Factors		
	First 3000 @ 100%	3000	3000
	Remainder (86400−3000) @ 35%	29190	29190
	General Lighting Demand Load	32190	32190
220-19	Twelve 8.4-kW Ranges on 3-Phase Ckt = 23kW × 3/2 = 34.5kW (Neutral @ 70%)	34500	24150
	Bathroom Heater = 1.5kVA200 × 12	18000	18000
220-15	Central Heat	19000	0
	Service Load	103690VA	74340VA

Code Rule	Selection of Service Equipment

	1. Service Ampacity = 103690VA ÷ $\sqrt{3}$ × 208V = 287.8 amperes;
	     Neutral = 74340VA ÷ $\sqrt{3}$ × 208V = 206 amperes
	
	     Further Demand Factor For Neutral
	
	     200A + .7 × (206 − 200)A = 204 amperes

Table 310-16	2. Conductors: Use three 350 MCM and one No. 4/0 THW copper
240-6	3. Overcurrent Device: Use 300-ampere standard size
Chapter 9, Table 5, 4	4. Conduit: area of conductors = 3 × .6291 in.² + .3904 in.² = 2.2777 in.² 1.7842 in²; Use 3-inch conduit
Table 250-94	5. Grounding: Use No. 1/0 copper grounding electrode conductor

Code Rule	Feeder Load Calculation		
		Line	Neutral (Standard)
220-30	General Lighting Load = 3VA/ft² × 1000 ft² =	3000	3000
	Small Appliance Circuits = 2 × 1500VA =	3000	3000
	(No Laundry Facilities in units)		6000 (Total)
			Apply Demand Fac.
			3000 First 3000
			1050 Remainder @ 35%
			4050 Demand Load
Table 220-19	12-kW Range (Neutral @ 70% of 8 kW) =	12000	5600
	Dishwasher	1500	1500
	Total "Other Load"	19500	
Table 220-30	Application of Demand Factors		
	First 10 kW @ 100%	10000	
	Remainder (19500-10000) @ 40%	3800	
	Air Conditioner = 30A × 230V @ 100%	6900	0
	(Neglect Heating)		
	Service Load	20700VA	11150VA

Code Rule	Selection of Feeder Equipment
	1. Feeder Ampacity: 20700VA ÷ 240V = 86.25 amperes;
	Neutral = 11150VA ÷ 240V = 46.4 amperes
Table 310-16	2. Conductors: Use Two No. 3 and one No. 8 type THW copper
	3. Overcurrent Protection: Use 100-ampere feeder panels
Table 250-95	4. Equipment Grounding Conductor: Use No. 8 copper

Code Rule	Service Load Calculation		
		Line kVA	Neutral kVA (Standard)
220-32	General Lighting Load = VA/ft² × 1000 ft² × 25 =	75	75
	Small Appliance Circuits = 2 × 1500VA × 25 =	75	75
			150 (Total)
			(Apply Demand Fac.)
			3 (First 3@ 100%)
			40.95 (Next 117 @ 35%)
			7.50 (Remainder @ 25%)
			51.45 (Demand Load)
(Table 220-19 Neutral)	Ranges = 12 kW × 25 =	300	28 (.7 × 40-kW)
220-17	Dishwashers = 1.5kVA × 25 =	37.5	28.1 (.75 × 37.5)
	Air Conditioners (Neglect Heating) = 30A × 240V × 25 =	180	
	Total Connected Load	667.5	
Table 220-32	Application of Demand Factors		
	667.5kVA @ 35%	234	
	"House Loads"		
	Corridors = .5VA/ft² × 1000 ft²	0.5	0.5
	Washing Machines = 4 kW × 4	16	16
	Dryers = 6 kW × 4 =	24	16.8 (.7 × 24)
	Floodlights = 300W × 4 =	1.2	1.2
	Total Service Load	275.7kVA	142kVA

(Note: If any house load is continuous, an additional 25% of its load must be added to service load.)

Code Rule	Selection of Service Equipment
220-22	1. Service Ampacity: 275.7kVA ÷ 240V = 1149A Neutral = 142kVA ÷ 240V = 591.7A Further Demand Factor For Neutral 200A + .7 × (591.7 − 200) = 474.2A
300-20, 310-4	2. Conductors (500 MCM THW copper): 1149A ÷ 380A = 3.02, or 4 circuits in four separate conduits, Neutral = 591.7 ÷ 4 = 148A, use No. 1/0 type THW in parallel

5-1 Quiz Problem 1

Code Rule	Lighting and Receptacle Branch Circuits Required	
	A. Lighting and Receptacle Circuits for each floor— 3-floors required (balanced load on 3-phase circuits)	Summary
Table 200-3(b)	1. 265-Volt Lighting: 1.25×3.5VA/ft$^2 \times 22000$ ft$^2 = 96250$VA. or, $\frac{96250VA}{3} = 32083$VA per phase Number of 30-ampere circuits per phase = $\frac{32083VA}{265V \times 30A} = 4.03$ or 5 circuits	Fifteen 30-ampere, 265-volt circuits (5 per phase)
220-3(c)	2. 120-Volt Receptacles: 22000 ft$^2 \times 1$VA/ft$^2 = 22000$VA or, $\frac{22000VA}{3} = 7333$VA per phase Number of 20-ampere circuits per phase = $\frac{7333VA}{120V \times 20A} = 3.06$, or 4 circuits	Twelve 20-ampere, 120-volt circuits (4 per phase)
220-13	Transformer load is 10000 VA + .5 (22000 − 10000 VA) = 16000 VA 3. A standard 25-kVA transformer is required to step down 460V to 208Y/120V.	
Table 220-3(b)	B. Basement Lighting 1. Storage Area: 1.25×0.25VA/ft$^2 \times 5000$ ft$^2 = 1562.5$VA or, $\frac{1562.5VA}{3} = 520$VA per phase Number of 20-ampere circuits per phase = $\frac{520VA}{120V \times 20A} = .2$, or 1 per phase	Three 20-ampere, 120-volt circuits (1 per phase)
	2. Machinery Room Lighting 1.25×10000VA $= 12500$VA or, $\frac{12500VA}{3} = 4166$VA per phase Number of 20-ampere circuits per phase = $\frac{4166VA}{120V \times 20A} = 1.7$, or 2 per phase	Six 20-ampere, 120-volt circuits (2 per phase)
	3. Transformer load = 1250VA + 10000VA = 11250VA A standard 15-kVA transformer is required to step down 460V to 208Y/120V.	

Code Rule	Motor Branch Circuits Required	

430-22, Table 430-150, Table 430-152, 430-52, 240-6	Note: Use Time-delay fuses.
	
	Three 50-hp motors:
	    Conductor Ampacity = 1.25 x 52A = 65A
	    Overcurrent Protection = 1.75 X 65A = 91A
	              Use 100-ampere fuse
	
	Two 15-hp Motors:
	    Conductor Ampacity = 1.25 X 21A = 26.25A
	    Overcurrent Protection = 1.75 X 21A = 36.75A
	              Use 40-ampere fuse
	
	Five 10-hp Motors;
	    Conductor Ampacity = 1.25 X 14A = 17.5A
	    Overcurrent Protection = 1.75 X 14A = 24.5A
	              Use 25-ampere fuse
	
	Note: Each motor circuit requires a controller disconnect, overload protection, and an equipment grounding conductor.
		Summary
		
		Three 100-ampere, 460-volt, three-phase circuits. No. 6 THW conductors
		
		Two 40-ampere, 460-volt, three-phase circuits. No. 10 THW conductors
		
		Five 25-ampere, 460-volt, three-phase circuits. No. 12. THW conductors

Code Rule	Motor Feeder Load Calculation	
430-24, Table 430-150	Three 50-hp motors = 3 x 52A Two 15-hp motors = 2 X 21A Five 10-hp motors = 5 X 14A 25% of largest motor = .25 x 52A	Line 156 42 70 13 ‾‾‾‾‾‾‾‾ 281.0 Amperes

Code Rule	Selection of Feeder Equipment

Table 310-16 430-62, 240-6
 	
430-112
 	
250-95
	1. Conductors: Use 300 MCM THW copper.
	2. Feeder Over Current Protection: 100A (largest branch-circuit device) + 130A (two 50-hp motors) + 42A (15-hp motors) + 70A (five 10-hp motors) = 316A maximum; use 300-ampere standard size.
	3. Disconnecting Means: Use single disconnecting means (motors all in same room);
	    Rating = (3 X 50 hp) + (2 x 15hp) = (5 x 10hp)
	         = 230 hp minimum
	4. Equipment Grounding Conductor: Use No. 4 copper.

5-1 Quiz Problem 1 (concluded)

Code Rule	460Y/265-Volt Service Load Calculation		

Note: All loads assumed to be balanced.

	Line (VA)	Neutral (VA)
General Lighting = 1.25 × 3.5VA/ft² × 22000ft² × 3 =	288750	231000
Receptacles = 10kVA + .5(56000) = 38000VA =	38000	38000
Basement Lighting = 1.25 × 0.25VA/ft² × 5000 ft²	1562.5	1250
Machinery Room Lighting = 1.25 × 10000VA	12500	10000
Total Lighting and Receptacle Load	340812.5	280250

	Line (Amperes)	Neutral (Amperes)
Lighting and Receptacles = 340812.5VA ÷ $\sqrt{3}$ ÷ 460V	427.8	351.7
Three 50-hp motors = 3 × 52A =	156	0
Two 15-hp motors = 2 × 21A =	42	0
Five 10-hp motors = 5 × 14A =	70	0
25% of largest motor = .25 × 52 =	13	0
	708.8	351.7

Code Rule	Selection of Service Equipment
430-63, 240-6	1. Conductors: Use parallel conductors or bus bars. 2. Overcurrent Protection: 100A (largest motor device) + 130A (two 50 hp) + 42A (15 hp) + 70A (10 hp) + 427.8 = 769.8A; use 800-ampere standard size.
230-79	3. Main Disconnecting Means: 708.8 amperes minimum.
250-94	4. Grounding Electrode Conductor: Use No. 3/0 copper.

5-2 Quiz Problem 1

Code Rule	Service Load Calculation	Phase			
		A (amp.)	C (amp.)	B (amp.)	Neutral (amp.)
Table 220-3(b) 220-3(c)	A. Lighting and Receptacles				
	Lighting = 1.25 × 2VA/ft² × 1500 ft² ÷ 240V =	15.6	15.6	0	12.5
	Receptacles = 8 × 180VA ÷ 240V =	6.0	6.0	0	6.0
	Single-phase load	21.6	21.6	0	18.5
Table 220-20	B. Cooking Equipment				
	4-kW Warmer = 4000W ÷ 240V = 16.7A	16.7	0	16.7	0
	5-kW Dishwasher = 5000 ÷ 240V = 20.8A	0	20.8	20.8	0
	2.4kW Toaster = 2400W ÷ 240V = 10A	10	10	0	10
	10-kW Waffle Iron = 10000W ÷ 240V = 41.7A	41.7	41.7	0	41.7
Table 430-148	Vegetable Peeler	5.8	0	0	5.8
	20-kW Range = 20000W ÷ $\sqrt{3}$ × 240V	48.1	48.1	48.1	0
	10-kW Fryer = 10000W ÷ $\sqrt{3}$ × 240V	24	24	24	0
Table 430-150	Food Chopper	5.2	5.2	5.2	0
	Cooking Equipment Load	151.5	149.8	114.8	57.5
	C. Service Load				
	Lighting and Receptacles	21.6	21.6	0	18.5
	Cooking Load @ 65%	98.5	97.4	74.6	37.4
	25% of Largest Motor	1.3	1.3	1.3	0
	Total Service Load	121.4	120.3	75.9	55.9

Code Rule	Selection of Service Equipment
Table 310-16	1. Conductors: Use two No. 1 (phase A, C), one No. 4 (phase B), and one No. 6 (Neutral).

5-2 Quiz Problem 2

Code Rule	Service Load Calculation	
220-41	Application of Demand Factors	Line
	Largest Load (Blg. No. 1) @ 100% =	80
	Second Largest Load (Blg. No. 3) @ 75% =	
	.75 × 70A =	52.5
	Third Largest Load (Blg. No. 2) @ 65% =	
	.65 × 40A =	26
	Farm Load (less dwelling)	158.5
	Farm Dwelling Load	60
	Total Farm Load	218.5 Amperes

5-2 Quiz Problem 3

Code Rule	Motor Branch-Circuit Calculation
440-32 440-22, 240-6 440-52 440-41 440-12 Table 430-150 Table 430-151	130-Ampere, 208-Volt, 3-Phase Motor-Compressor Conductor Ampacity = 1.25 × 130A = 162.5A minimum, use No. 2/0 THW copper. Branch-circuit Protection = 1.75 × 130A = 227.5A, use 250-ampere standard. Overload Protection = 1.4 × 130A = 182A maximum. Controller Rating: Based on compressor nameplate values. Disconnecting Means: Ampere rating = 1.15 × 130A = 149.5A minimum. Horsepower rating = Larger of- horsepower equivalent of rated-load current = 50 hp, or horsepower equivalent of locked-rotor current = 40 hp Use 50-hp rating.

Code Rule	Motel Unit Feeder Load Calculation		
		Line	Neutral
Table 220-11 430-25	Lighting = 2VA/ft² × 240 ft² = Air Handler Motor = 1.25 × 2A × 208V =	480 520 1000VA	480 0 480VA
215-2	Ampacity = 1000VA ÷ 208V = 4.8 amperes One 15- or 20-ampere, 3-wire circuit is required for each unit		

Code Rule	Service Load Calculation			
	Note: Assume balanced loads.	Line (VA)	Line (Amperes)	Neutral
Table 220-3(b)	Lighting = 2VA/ft² × 240 ft² × 100 =	48000	–	48000
Table 220-11	Application of Demand Factors			
	First 20000 @ 50%	10000		10000
	Remainder (48000-20000) @ 40%	11200		11200
		21200	58.8	21200
220-3	Air Handler Motors = 2A × 208V × 100 =	41600	115.4	0
	Outside Receptacles = 10 × 180VA =	1800	5	1800
430-25	Motor-compressor = $\sqrt{3}$ × 130A × 208V	46835	130	0
	25% of largest motor = .25 × 46835 =	11708	32.5	0
		123143	341.7	23000

5-2 Quiz Problem 3 (continued)

Code Rule	Selection of Service Equipment

Table 310-16
430-63
	1. Ampacity (208V) = 123143VA ÷ $\sqrt{3}$ × 208V = 342 amperes;
	    Neutral = 23000 ÷ $\sqrt{3}$ × 208V = 64 amperes
	    Ampacity (480V) = 123143VA ÷ $\sqrt{3}$ × 480V = 148 amperes
	
	2. Conductors: Use three No. 1/0 type THW copper conductors for primary
	
	3. Main Overcurrent Protection = 250A (motor-compressor device) + 58.8 (lighting) + 115.4 (air handler) + 5A (receptacles) = 429.2A @ 208V or, 429.2 × 208V ÷ 480V = 186A @ 480V; use 175-ampere standard size
	
	4. Transformer kVA rating = 21.2 kVA (lighting) + 41.6 kVA (air handler) + 1.8 kVA (receptacles) + 46.8 kVA (motor-compressor) = 111.4 kVA; use 112.5 kVA commercially available unit.

215-2	5. 208/120-Volt Feeder: Use 400-ampere size panel and No. 500 MCM THW copper ungrounded conductors and a neutral no smaller than No. 1/0. Use No. 3 copper equipment grounding conductor.

5-3 Quiz Problem 1

Code Rule	Feeder Load Calculation		

550-13(a)	A. Lighting and Appliance Load
	
	Lighting load = 3VA/ft² × 50ft × 10 ft = 1500
	Small Appliance = 2 × 1500VA = 3000
	Laundry = 1500
	    Total ltg. small appl. & laundry 6000
	
	Application of Demand Factors
	
	First 3000 @ 100% 3000
	Remainder (6000-3000) @ 35% 1050
	    Lighting and Appliance Demand 4050 volt-amperes		

	B. Total Load	Line A (Amperes)	Line B (Amperes)
550-13(b)	Lighting and Appliance = 4050W ÷ 240V	16.9	16.9
	5-kW Range = .8 × 5000W ÷ 240V	16.7	16.7
	Water Heater = 1000W ÷ 120V =	8.3	0
	Dishwater = 500W ÷ 120V =	0	4.2
	Disposal = 500W ÷ 120V =	0	4.2
	Air conditioner (neglect heating)	7	7
	25% of largest motor = .25 × 7A	1.75	1.75
	Total	50.7	50.8

Test Chapter 5 Problem 1

Code Rule	Lighting Load Calculation	
Table 220-3(b)	Lighting load =1.25 × 0.25VA/ft² × 100000 ft² = 31250VA Application of Demand Factors	
Table 220-11	First 12,500 @ 100%	12500
	Remainder (31250-12500) @ 50%	9375
	Total Demand Load	21875 VA

Test Chapter 5 Problem 2

Code Rule	120/240-Volt Branch Circuits Required	
Table 220-3(b)	Lighting = $\dfrac{1.25 \times 3.5\text{VA/ft}^2 \times 9000 \text{ ft}^2}{120\text{V} \times 20\text{A}} \doteq 16.4$, or 17	**Summary** Seventeen 20-ampere 2-wire circuits
220-3(c)	Receptacles = $\dfrac{1.25 \times 80 \times 180\text{VA}}{120\text{V} \times 20\text{A}} = 7.5$, or 8	Eight 20-ampere, 2-wire circuits
220-3(c) Exception 3	Show Window = $\dfrac{200\text{VA/ft} \times 60 \text{ ft}}{120\text{V} \times 20\text{A}} = 5$	Five 20-ampere, 2-wire circuits
	6-kW Sign = $\dfrac{1.25 \times 6000\text{VA}}{120\text{V} \times 20\text{A}} = 3.1$, or 4	Four 20-ampere, 2-wire circuits
		Total: 41 circuits

Code Rule	120/240-Volt Feeder Load Calculations		
Table 220-3(b)			**Load**
		Transformer	Feeder
220-13	Lighting = 1.25 × 3.5VA/ft² × 9000 ft² = 39375VA	31500	39375
	Receptacles = 80 × 180VA = 14 400 VA	12200	12200
	Demand = 10000 VA + .5 (4400) VA		
	Show Window = 12000VA	12000	12000
	Sign = 6000VA × 1.25 = 7500	6000	7500
	Total Load	61700VA	71075VA

Code Rule	Selection of Feeder Equipment
	1. Feeder Protection = 71075VA ÷ 240V = 296 amperes (secondary).
	2. Overcurrent Protection (Conductors): 300 ampere standard
	3. Conductors: 61700 VA ÷ 240V = 257A; use 300 MCM Neutral = 200 A + .7 (257-200) A = 240 A; use 250 MCM
250-95, 250-26, Table 250-94	4. Grounding: Use No. 4 equipment grounding conductor; No. 2 copper bonding jumper (equipment grounding conductor to neutral); No. 2 copper grounding electrode conductor
	5. Transformer Rating = 61700 VA minimum; use 75 kVA
240-6	6. Primary Feeder Protection (480V) = 71075 W ÷ 480V = 148A (150A standard)
	7. Rated Transformer Current: Primary = 75 kVA ÷ 480V = 156.3A; Secondary = 75 kVA ÷ 240V =312.5V
450-3(b)	8. Individual Overcurrent Device For Transformer Primary Not Required

Code Rule	Motor Branch-Circuit Calculations
430-22, Table 430-150	**7½-hp, 480-Volt, Three-Phase Motor** Conductor Ampacity = 1.25 × 11A = 13.75A; Use No. 14 THW copper conductors
430-52	Overcurrent Protection: 2.5 × 11A = 27.5A; use 30-ampere standard size breaker
430-32	Overload Protection: 1.15 × 11A = 12.65A maximum Controller and Disconnecting Means Ratings: 7½-hp minimum
	21-Ampere, 480-Volt, Three-Phase Motor Compressor Conductor Ampacity = 1.25 × 21A = 26.25A; Use No. 10 THW copper conductors
	Circuit Protection: 1.75 × 21A = 36.75; Use 35-ampere standard size
440-52(a)	Overload Protection = 1.4 × 21A = 29.4A
440-41(a) 430-150 430-151	Controller Rating: 15 hp or 21-ampere full-load and 120-ampere locked-rotor current (Locked rotor current based on 15-hp equivalent)
440-12	Disconnecting Means: Ampere rating = 1.15 × 21A = 24 amperes Horsepower Rating = 15 hp

Code Rule	Service Load Calculation			
			Phase	
		A	B	C
430-25	120/240-Volt Loads = 71075VA ÷ 480V =	148	148	0
	7½-hp Motor	11	11	11
	21-Ampere Motor Compressor	21	21	21
	25% of largest motor = .25 × 21A =	5.3	5.3	5.3
	Total Service Load	185.3	185.3	37.3A

Code Rule	Selection of Service Equipment
250-23(b), Table 250-94	1. Conductors: Use two 3/0 and one No. 8 type THW copper. 2. Main Overcurrent Protection = 35A (compressor device rating) + 11A (7½-hp motor) + 148A (120/240-volt loads) = 194A; use 200-ampere standard size for Phases A and B, in same manner, use 60 ampere for Phase C. 3. Grounding: Grounded conductor brought to service and grounding electrode conductor should be No. 4 copper conductors. See 250-26 for grounding of separately derived systems.

Test Chapter 5 Problem 3

Code Rule	Cafeteria Loads
	A. 277-Volt Lighting Load: (Assume load is balanced and continuous) $1.25 \times 150 \times 100VA = 18750VA$ Feeder load (amperes) $= 18750VA \div \sqrt{3} \times 480V = 22.5A$ Overcurrent protection: Use 25- or 30-ampere, 3-pole breaker Conductors: Use Three No. 10 THW copper conductors; neutral $= 18A$, use No. 12

B. 120/208-Volt Loads: (Assume loads are balanced and continuous)

Ranges $= 6 \times 12$ kW $\times 1.25 =$ 90
Deep Fryers $= 3 \times 4$ kW $\times 1.25 =$ <u>15</u>
<u>105 kW</u>

Application of Demand Factors

105 kVA @ 65% $= 68.25$ kVA

	Phase A	Phase B	Phase C	Neutral
Demand Load (amperes) $= \dfrac{68.25 \text{ kVA}}{\sqrt{3} \times 208V}$	189.4	189.4	189.4	189.4
Dishwasher $= 7.5$kW $\div 120V =$	<u>62.5</u>	<u>0</u>	<u>0</u>	<u>62.5</u>
Total Load (Secondary)	251.9A	189.4A	189.4A	251.9A
Secondary Conductors (THW)Copper	250 MCM	No.3/0	No. 3/0	250 MCM
Fuses (Secondary)	300A	200A	200A	—
Primary Load (Secondary $\times$ 208V $\div$ 480V $=$	109.1A	82A	82A	—
Primary Conductors (THW Copper)	No. 2	No. 4	No. 4	—
Fuses (Primary)	110A	90A	90A	—

(Table 220-20)

Transformer Rating $= \sqrt{3} \times 251.9A \times 208V \div 1.25 = 72.6$-kVA
Continuous Rating; use 75-kVA unit
Grounding Electrode Conductor: Use No. 2 copper

Code Rule	Equipment System

		Line
	100-kVA Hot Water Generator $= \dfrac{100000 \text{ VA}}{\sqrt{3} \times 480V} =$ (Assume non continuous)	120.3
Table 430-150 Table 430-152	10-hp Pump $=$ Circuit Protection: $2.5 \times 14A = 35A$	14

	5-hp Pump $=$
	  Circuit Protection $= 2.5 \times 7.6 = 19A$;
	  Use 20-ampere standard
		7.6

440-22	Compressors $= 2 \times 77A$
	  Circuit protection $= 1.75 \times 77 = 134.75A$
	  Use 125-ampere standard size
		154

	25% of largest motor $= .25 \times 77A =$	<u>19.3</u> 315.2 Amperes

Code Rule	Selection of Feeder Equipment
430-62 430-63	1. Feeder Conductors: Use three 400 MCM THW copper 2. Overcurrent Protection $= 125A$ (largest device) $+ 77A$ (other compressor) $+ 14A$ (10-hp pump) $+ 7.6A$ (5-hp pump) $+ 120.3A$ (hot water generator) $= 343.6A$; Use 350-ampere standard size based on conductor size

Code Rule	Ward Area Loads		
		Line	Neutral
	Lighting = 1000 × 75W	75000	75000
	Receptacles = 150 × 180 VA	27000	27000
	Total	102000	102000
Table 220-11, 220-13	Application of Demand Factors		
	First 50,000 @ 40% =	20000	20000
	Remainder (102000−50000) @ 20% =	10400	10400
	Feeder Demand Load	30400VA	30400VA
	Note: The lights and receptacles in the ward area are treated as noncontinuous loads subject to the application of demand factors specified in the code.		

Code Rule	Service Load Calculation		
		Phases		
		A	B	C
	277-Volt Cafeteria Lighting	22.5	22.5	22.5
	120/208-Volt Cafeteria Loads	109.1	82	82
	Equipment System	315.2	315.2	315.2
	Ward Area = 30,400VA ÷ $\sqrt{3}$ × 480V =	36.6	36.6	36.6
	Critical Branch = 100 kVA ÷ $\sqrt{3}$ × 480V =	120.3	120.3	120.3
	Life Safety Branch = 90 kVA ÷ $\sqrt{3}$ × 480V =	108.3	108.3	108.3
	Total Service Load	712 A	684.9A	684.9A

Code Rule	Selection of Service Equipment
430-63 430-62(b)	1. Conductors: Use parallel conductors 2. Main Overcurrent Protection = 343.6A (Equipment System Rating) + 22.5A + 109.1A + 36.6A + 120.3A + 108.3A = 740.4A; use 700 ampere standard size for phases or base rating on conductor size.

Final Examination No. 1 Problem 1

Code Rule	Branch Circuits Required	
		Summary
Table 220-3(b) 220-4	1. Apartment Type (a)	
	Lighting = 3VA/ft² × 1800 ft² ÷ 120V = 45A	(3) 20A, 120V
	Two small appliance circuits	(2) 20A, 120V
	Laundry	(1) 20A, 120V
Table 220-19 422-14	12-kW Range = 8000W ÷ 240V = 33.3A	(1) 40A, 120/240V
	Hot water heater = 1.25 × 5000W ÷ 240V = 26A	(1) 30A, 120/240V
	Dryer = 4500W/240V = 18.8A	(1) 20A, 240V
424-3	Six Space Heaters = 1.25 × 2000W ÷ 240V = 10.4A	
	Use two heaters per 30A circuit	(3) 30A, 240V
440-32	Four Air Conditioners = 1.25 × 4.9A = 6.1A ea.	(4) 15A, 240V
	2. Apartment Type (b)	
	Lighting = 3VA/ft² × 1200 ft² ÷ 120V = 30A	(2) 20A, 120V
	Two small appliance	(2) 20A, 120V
	Laundry	(1) 20A, 120V
	10-kW Range = 8000W ÷ 240V = 33.3A	(1) 40A, 120/240V
	Hot water heater = 1.25 × 5000W ÷ 240V = 26A	(1) 30A, 240V
	Dryer = 4500W/240V = 18.8A	(1) 20A, 120/240V
	Six Space Heaters = 1.25 × 2000W ÷ 240V = 10A ea.	(3) 30A, 240V
	Four Air Conditioners = 1.25 × 4.9A = 6.1A ea.	(4) 15A, 240V
	3. Apartment Type (c)	
	Lighting = 3VA/ft² × 900 ft² ÷ 120V = 22.5A	(2) 20A, 120V
	Two small appliance circuits (no laundry)	(2) 20A, 120V
Table 220-19, Column C	Range = .8 × 8.75 kW ÷ 240V = 29.1A	(1) 40A, 120/240V
	Two Air Conditioners = 1.25 × 4.9A = 6.1A ea.	(2) 15A, 240V
	Four Space Heaters = 1.25 × 8.6A = 10.9A ea.	(2) 30A, 240V

Code Rule	Apartment Feeders					
		Type (a)		Type (b)		Type (c)
		Line	N	Line	N	Line	N

Code Rule		Type (a) Line	Type (a) N	Type (b) Line	Type (b) N	Type (c) Line	Type (c) N
	Lighting @ 3VA/ft²	5400	3000	3600	3600	2700	2700
	Small Appliance	3000	3000	3000	3000	3000	3000
	Laundry	1500	1500	1500	1500	0	0
	Total	9900	9900	8100	8100	5700	5700
	Application of Demand Factors						
	First 3000 @ 100%	3000	3000	3000	3000	3000	3000
	Remainder @ 35%	2415	2415	1785	1785	945	945
	Demand Load	5415	5415	4785	4785	3945	3945
	Range (Neutral @ 70%)	8000	5600	8000	5600	7000	4900
	Dryer	5000	3500	5000	3500	0	0
220-21, 220-15	Heating (Neglect A/C)	12000	0	12000	0	8000	0
422-14(b),	Water Heater	5000	0	5000	0	0	0
220-10	Service Load	35415VA	14515VA	34785VA	13885VA	18945VA	8845VA
	Selection of Feeder Equipment						
	1. Ampacity @ 240V =	148A	60.5A	144.9A	57.9A	78.9A	36.9A
	2. Conductors (THW Aluminum)	No. 3/0	No. 4	No. 3/0	No. 4	No. 3	No. 8
	3. Overcurrent Protection	150A	–	150A	–	90A	–
Chap. 9	4. Conduit Size	1 ¼ inch		1 ¼ in		1 inch	

Final Examination No. 1 Problem 1 (continued)

Code Rule	Service Load Calculation (Standard)		
		Line	Neutral
	Lighting: Area = (6 × 1800 ft²) + (4 × 1200 ft²) + (2 × 900 ft²) = 17400 ft² Load = 3VA/ft² × 17400 ft² =	52200	52200
	Small Appliance Circuits = 12 × 3000VA	36000	36000
	Laundry Circuits = 10 × 1500 VA	15000	15000
	Total Ltg. Small Appl. & Laundry	103200	103200
220-11	Application of Demand Factors		
	First 3000VA @ 100%	3000	3000
	Remainder (103200-3000) @ 35%	35070	35070
	General Lighting Demand Load	38070	38070
Table 220-19, Column A 220-18 220-21	12 Ranges (Neutral @ 70%)	27000	18900
	Water Heaters = 10 × 5 kVA (Ignore 220-17)	50000	0
	Dryers = 10 × 5 kVA × .5 (Neutral @ 70%)	25000	17500
	Heating = 68 units @ 2 kVA ea.	136000	0
		276070VA	74470VA

Code Rule	Selection of Service Equipment
220-22	1. Ampacity: 276070VA ÷ 240V – 1150 amperes; Neutral = 74470VA ÷ 240V = 310.3A Further Demand Factor for Neutral 200 + .7 (310.3-200) = 277.2 amperes 2. Conductors: Use parallel conductors 3. Overcurrent Protection: Not to exceed rating of conductors.
Table 250-94	4. Grounding Electrode Conductor: Use No. 3/0 copper.

Code Rule	Service Load Calculation (Optional)		
		Line	Neutral
220-32	Lighting = 3VA/ft² × 17400 ft²	52200	52200
	Small Appliance Circuits = 12 × 3000VA	36000	36000
	Laundry Circuits = 10 × 1500VA	15000	15000
	Ranges = (6 × 12 kVA + (4 × 10 kVA) + (2 × 8.75 kVA) =	129500	129500
	Water Heaters = 10 × 5 kVA =	50000	0
	Dryers = 10 × 4.5 kVA =	45000	45000
	Heating (Neglect A/C) = 68 × 2 kVA =	136000	0
	Total Connected Load	463700	277700
Table 220-32	Application of Demand Factors		
	12 units @ 41%	190117VA	113857VA

Code Rule	
220-32(a)(3)	1. Ampacity = 190117VA ÷ 240V = 792.2A Neutral = 113857VA ÷ 240V = 474.4A Use results of optional calculation for ungrounded conductors and results of standard calculation for neutral.

Code Rule	Branch Circuits Required	
		Summary
Table 220-3(b)	General Lighting: (The area is not given, it must be assumed that the actual load is used because it is greater than the minimum required by the Code.	
	Floor Lamps: 21 × 200W × 1.25 = 5250W	
	number of 20A circuits = $\dfrac{5250W}{20A \times 120V}$ = 2.18	Three 20A 2-wire
	Other Lamps: 21 × 150W × 1.25 = 3938W	
	number of 20A circuits = $\dfrac{3938W}{20A \times 120V}$ = 1.64	Two 20A 2-wire
220-3(c)	Receptacles: 21 × 180VA × 1.25 = 4725VA	
	number of 20A circuits = $\dfrac{4725VA}{20A \times 120V}$ = 1.97	Two 20A 2-wire
330-2(c) Exception No. 3	Show Window: 40 ft × 2000VA/ft = 8000VA	
	number of 20A circuits = $\dfrac{8000VA}{20A \times 120V}$ = 3.33	Four 20A 2-wire

Code Rule	Service Load Calculation		
		Line	Neutral
	Floor Lamps = 21 × 200W × 1.25	5250	4200
	Other Lamps = 21 × 150W × 1.25	3938	3150
	Receptacles = 21 × 180VA × 1.25	4725	3780
	Show Window = 40 ft × 200VA/ft =	8000	8000
	Service Load	21913 VA	19130VA

Code Rule	Selection of Service Equipment
600-6(b)	1. Ampacity = 21913VA ÷ 240V = 91.3 amperes; Neutral = 79.7A
	2. Conductors: A minimum of 91.3 amperes is required
	3. Use 100-ampere service
	Note: The load for the required outside branch-circuit has not been included since it was not mentioned in the problem. A first floor store would require it.

382

Final Examination No. 2 Problem 1

Code Rule	Feeder Load		
		Standard	
		Line	Neutral
220-3.(b)	General Lighting Load = 3VA/ft² × 2000 ft²	6000	6000
220-16(a)	Small Appliance Circuits = 2 × 1500VA	3000	3000
220-16(b)	Laundry Circuit	1500	1500
		10500	10500
	Application of Demand Factors		
Table 220-11	First 3000 @ 100%	3000	3000
	Remainder (10500-3000) @ 35%	2625	2625
	General Lighting Demand Load	5625	5625
220-19	Range Load 7.2 kW + 7.2 kW = 14.4 kW		
Notes 1, 4	8000W + 8000W × .05 [14-12] =	8800	6160
220-22	Neutral @ 70%		
220-18	Dryer (5 kW, 240V)	5000	3500
	Disposal = 120V × 6A	720	720
	Dishwasher = 120V × 12A	1440	1440
220-21,	Heating (Neglect A/C)	9600	0
220-15			
220-14	25% of largest motor (disposal)	180	180
	Feeder Load	31365 VA	17625 VA

	Feeder Ampacity 31365VA ÷ 240V = 130.7A
	Neutral 17625VA ÷ 240V = 73.4VA

Code Rule	Service Load Calculation
220-201	Note: The secondary are considered to be the service conductors.

	120/240-Volt Loads	Load
220-32(c)	General Lighting Load = 3VA/ft² × 2000 ft² × 100 units	600
	Small Appliance Circuits = 2 × 1500VA × 100 units	300
	Laundry Circuits 1500VA × 100 units	150
	Ranges 14.4 kW × 100 units	1440
	Dryers (5 kW) 5 kW × 100 units	500
	Disposals 120V × 6A × 100 units	72
	Dishwasher 120V × 12A × 100 units	144
220-32(c) (5)	Heating (9.6 kW) 9.6 kW × 100 units	960
	Total Connected Load	4166 kVA @ 120/240V

Application of Demand Factors

4166 kVA @ 23% =	958.18 kVA

220-32(a) (Note: Neutral Load can be calculated by standard method)

Main fuse for 120/240 volt single phase loads:

$$\frac{958180VA}{240V} = 3992.42 \text{ amperes}$$

	480-volt load (house load)
Table 430-150,	30A compressor = 1.25 × 30A = 37.5A
430-22	
440-22(a)	Fuse = 1.75 × 30A = 52.5A (maximum)

Final Examination No. 2 Problem 2

Code Rule	Service Load	Line A	Line C	Line B	Neutral
Table 220-3(b) 220-10	Lighting = $\dfrac{1.25 \times 3\text{VA/ft}^2 \times 140000 \text{ ft}^2}{240\text{V}}$ = 2187.5A (Neglect Fixtures)	2187.5	2187.5	0	1750
220-3(c) 220-10	Receptacles = $\dfrac{1.25 \times 180\text{VA} \times 40}{240\text{V}}$ = 37.5A	37.5	37.5	0	30
220-12	Show Window = $\dfrac{200\text{VA/ft} \times 80 \text{ ft}}{240\text{V}}$ = 66.7A	66.7	66.7	0	66.7
600-6(b)	Sign Circuit (Assume connection between A and N)	16.0	0	0	16.0
Table 430-150	80 hp, 230V motor, 3 phase Interpolate Table: $I_L = 192\text{A} + \dfrac{(80\text{-}75)\text{ hp}}{(100\text{-}75)\text{ hp}} \times (248\text{-}192)\text{A} = 203.2\text{A})$	203.2	203.2	203.2	0
	25% of largest motor = .25 × 203.2A = Service Capacity	50.8	50.8	50.8	0
		2561.7A	2545.7A	254A	1862.7A

Transformer Rating (Rated Continuous)

Required Phase A-C kVA = $\dfrac{240\text{V} \times .8 \times 2291.7 + 120\text{V} \times 16\text{A (sign)}}{1000}$ = 441.93 kVA (less motor)

Required Motor Circuit kVA = $\dfrac{\sqrt{3} \times 240\text{V} \times 203.2\text{A}}{1000}$ = 84.47 kVA Three-phases

Total Phase A-C Rating = 441.93 kVA + (84.47 ÷ 3) kVA = 470.09 kVA

Bibliography

1. *National Electrical Code®*. National Fire Protection Association (470 Atlantic Ave.,
   Boston, Mass. 02210).
   BRANDON, MERWIN M., *The National Electrical Code and Free Enterprise.* Boston, Mass.:
   National Fire Protection Association, 1971.

Publications Covering the National Electrical Code®

SUMMERS, WILFORD I., *National Electrical Code® Handbook.* Boston, Mass.: National Fire
   Protection Association.

GARLAND, J. D., *National Electrical Code® Reference Book,* Fifth Edition, Based on the
   1987 Code. Englewood Cliffs, N.J.: Prentice-Hall, Inc., 1987.

GEBERT, KENNETH L., *National Electrical Code Blueprint Reading.* Chicago, Ill.: American
   Technical Society.

MCPARTLAND, J. F., *Making Electrical Calculations.* New York: McGraw-Hill Book
   Company.

SEGALL, B. Z., *Electrical Code Diagrams.* New Orleans: Peerless Publishing Company.

General References

SUMMERS, WILFORD I. (editor), *American Electricians Handbook.* New York: McGraw-Hill
   Book Company.

JOHNSON, ROBERT C., *Electrical Wiring.* Englewood Cliffs, N.J.: Prentice-Hall Inc., 1971.

LENK, JOHN D., *Handbook of Simplified Electrical Wiring Design.* Englewood Cliffs, N.J.:
   Prentice-Hall Inc., 1975.

Periodicals

Electrical Construction and Maintenance. Published monthly by McGraw Hill, Inc., New
   York, N.Y. (This excellent magazine covers topics of current interest in construction and emphasizes the *National Electrical Code* rules.)

IAEI News. Published by International Association of Electrical Inspectors, Park Ridge, IL.

Index